Physics of Laser Crystals

NATO Science Series

A Series presenting the results of scientific meetings supported under the NATO Science Programme.

The Series is published by IOS Press, Amsterdam, and Kluwer Academic Publishers in conjunction with the NATO Scientific Affairs Division

Sub-Series

I. Life and Behavioural Sciences	IOS Press
II. Mathematics, Physics and Chemistry	Kluwer Academic Publishers
III. Computer and Systems Science	IOS Press
IV. Earth and Environmental Sciences	Kluwer Academic Publishers
V. Science and Technology Policy	IOS Press

The NATO Science Series continues the series of books published formerly as the NATO ASI Series.

The NATO Science Programme offers support for collaboration in civil science between scientists of countries of the Euro-Atlantic Partnership Council. The types of scientific meeting generally supported are "Advanced Study Institutes" and "Advanced Research Workshops", although other types of meeting are supported from time to time. The NATO Science Series collects together the results of these meetings. The meetings are co-organized bij scientists from NATO countries and scientists from NATO's Partner countries – countries of the CIS and Central and Eastern Europe.

Advanced Study Institutes are high-level tutorial courses offering in-depth study of latest advances in a field.
Advanced Research Workshops are expert meetings aimed at critical assessment of a field, and identification of directions for future action.

As a consequence of the restructuring of the NATO Science Programme in 1999, the NATO Science Series has been re-organised and there are currently Five Sub-series as noted above. Please consult the following web sites for information on previous volumes published in the Series, as well as details of earlier Sub-series.

http://www.nato.int/science
http://www.wkap.nl
http://www.iospress.nl
http://www.wtv-books.de/nato-pco.htm

Series II: Mathematics, Physics and Chemistry – Vol. 126

Physics of Laser Crystals

edited by

Jean-Claude Krupa

CNRS-IN2P3, Institut de Physique Nucléaire,
Orsay, Paris, France

and

Nicolay A. Kulagin

Kharkiv National University for Radioelectronics,
Kharkiv, Ukraine

Kluwer Academic Publishers

Dordrecht / Boston / London

Published in cooperation with NATO Scientific Affairs Division

Proceedings of the NATO Advanced Research Workshop on
Physics of Laser Crystals
Kharkiv-Stary Saltov, Ukraine
26 August–2 September 2002

A C.I.P. Catalogue record for this book is available from the Library of Congress.

ISBN 1-4020-1675-1 (HB)
ISBN 1-4020-1676-X (PB)

Published by Kluwer Academic Publishers,
P.O. Box 17, 3300 AA Dordrecht, The Netherlands.

Sold and distributed in North, Central and South America
by Kluwer Academic Publishers,
101 Philip Drive, Norwell, MA 02061, U.S.A.

In all other countries, sold and distributed
by Kluwer Academic Publishers,
P.O. Box 322, 3300 AH Dordrecht, The Netherlands.

Printed on acid-free paper

Printed in the Netherlands.

PHYSICS OF LASER CRYSTALS

Contents

Preface

Physics of laser crystals has been constantly developing since the invention of the laser in 1960. Nowadays, more than 1500 wide-band-gap and semiconductors crystals are suitable for the production of the laser effect. Different laser devices are widely used in science, medicine and communication systems according to the progress achieved in the development of laser crystal physics. Scintillators for radiation detection also gained benefit from these developments.

Most of the optically active materials offer laser radiations within the 500 to 3000 nm region with various quantum efficiency which fit the usual applications. However, new crystals for laser emissions are needed either in the blue, UV and VUV - region or far IR- region, especially for medicine, computer microchip production and for undiscovered practical uses.

Scientific problems of the growth and properties of laser crystals are discussed in numerous books and scientific journals by many scientists working in the field. Therefore, we thought that joint discussions of the scientific and technical problems in laser physics will be useful for further developments in this area. We have proposed to held a Workshop on Physics of Laser Crystals for attempting to induce additional advances especially in solid state spectroscopy. This NATO Advanced Research Workshop (ARW) was hold in Kharkiv - Stary Saltov (Ukraine) on august 26th - September 2nd, 2002, and was mainly devoted to the consideration o f modern approaches and l ast results in p hysics of laser crystals. Ukraine is one of the countries that have powerful facilities for production of the laser crystals and different laser systems in operation.

52 participants (students included) from 10 countries delivered 65 communications within 10 scientific sessions: Crystals doped with RE ions, Crystals doped with Me ions, Growth and properties of laser crystals and scintillators, Charge and energy transfer, Theoretical methods and calculations, Layers and thin films, Laser generation, Clusters and nano-clusters, ESR and spin-relaxation data.

More generally, fundamental and applied problems concerning optically active materials were discussed during ARW.

We are very pleased to thank the NATO Science Committee and particularly Dr. F. Pedrazzini, STP Progamme Director at the Scientific and Environmental Affairs Division in Brussels for his incommensurate patience and very efficient help in holding this workshop in Ukraine.

J.-C. Krupa and N. Kulagin
Co-chairmen of PST ARW 978990

Addresses of Participants

Elisabeth Antic-Fidancev
Laboratoire de Chimie Appliquée de l'Etat Solide
ENSCP
11, Rue Pierre et Marie Curie
F-75231 Paris Cedex 05, France

Jean-Claude Krupa
Institut de Physique Nucléaire
91406 Orsay France

Jiri A. Mares
Institute of Physics ASCR
Cukrovarnicka 10
16253 Praha, Czech Republic

Fabienne Pellé
CNRS
1, Place Aristide Briand
F-92195 Meudon, France

Jorma Hölsä
University of Turku
Department of Chemistry
FIN-20014 Turku, Finland

Nina Mironova
Institute of Solid State Physics
University of Latvia
LV-1063 Riga, Latvia

Uldis Ulmanis
Institute of Solid State Physics
University of Latvia
LV-1063 Riga, Latvia

Oscar Malta
Depertamento de Quimica Fundamental
CCEN, UFPE
50590-470 Recife, Brazil

Taiju Tsuboi
Faculty of Engineering
Kyoto Sangyo University
Kamigamo, Kita-ku
Kyoto 603-8555, Japan

Aleksander Bagmut
NTU "Kharhov Polytechnical Institute"
21, Frunze str.
Kharkiv 61002, Ukraine

Michail Bondarenko
Kharkiv National University for Radioelectronics
Lenin av., 14
Kharkiv 61166, Ukraine

Svetlana El'khaninova
B. Berkin Institute for Low Temperature NANU
47, Lenin av.
Kharkiv 61103, Ukraine

Ivan Maksymov
Kharkiv National University for Radioelectronics
14, Lenin av.
Kharkiv 61166, Ukraine.

Vladimir Koshkin
NTU "Kharhov Polytechnical Institute"
21, Frunze str.
Kharkiv 61002, Ukraine.

Nicolai Kulagin
Kharkiv National University for Radioelectronics
av. Shakespeare, 6-48
Kharkiv 61045, Ukraine.

Andrey Marchenko
NTU "Kharhov Polytechnical Institute"
21, Frunze str.
Kharkiv 61002, Ukraine

Vladimir Miloslavskyy
Kharkiv National University
Kharkiv 61077, Ukraine

Leonid Litvinov
Institute of Single Crystals
Lenin av., 60
Kharkiv 61079, Ukraine.

Sergey Nikolaev
Usikov Institute for Radiophysics and Electronics NANU
12, Proskura str.
Kharkiv 61085, Ukraine

Jury Pedash
Research Institute for Chemistry
V.N. Karazin National University
4, Svoboda sq.
Kharkiv 61077, Ukraine

Valeryy Shevchenko
Usikov Institute for Radiophysics and Electronics NANU
12, Proskura str.
Kharkiv 61085, Ukraine

Andrey Marchenko
NTU "Kharhov Polytechnical Institute"
21, Frunze str.
Kharkiv 61002, Ukraine.

Andrey Semenov
Research Institute for Chemistry
V.N. Karazin National University
Svoboda sq., 4
Kharkiv 61077, Ukraine.

Vjacheslav Solov'ev
Dovrova Center NANU
60, Shevchenko av.
Kyiv 01032, Ukraine

Yaroslav Zhydachevskii
L'viv Politechnical National University
12, Bandera str.
L'viv, Ukraine

Sergei Basun
Ioffe Physics-Technical Institute RAS
Politechnicheskaja, 26
St.-Petersburg 194021, Russia

Vladimir Chernyshev
Department of Physics
Ural State University
51, Lenin av.
Ekaterinburg 610083, Russia

Georges Denisenko
Institute of Crystallography RAS
Leninsky av., 59
Moscow 117333, Russia

Svetlana Ivanova
S.I. Vavilov State Optical Institute
St.-Petersburg 199034, Russia

Ludmila Ivleva
Laser Materials and Technology Research Center
General Physics Institute RAS
38, Vavilov st.
Moscow 119991,GSP-1, Russia

Alisa Kosnatantinova
Institute of Crystallography RAS
Leninsky av., 59
Moscow 117333, Russia

Barot Namozov
Ioffe Physics-Technical Institute RAS
Politechnicheskaja 26,
St.-Petersburg 194021, Russia

Vijacheslav Osiko
Laser Materials and Technology Research Center
General Physics Institute RAS
38, Vavilov st.
Moscow 119991,GSP-1, Russia

Valentin Simonov
Institute of Crystallography RAS
Leninsky av., 59
Moscow 117333, Russia

Valery Tarasov
Kazan Physics-Technical Institute
Sibirsky trackt, 10/7
Kazan, Russia

Alexandra Tkachuk
S.I. Vavilov State Optical Institute
St.-Petersburg 199034, Russia

Airat Zhiganshin
Kazan State University
18, Kremlevskaja
Kazan 420008, Russia

Evgenii Zhiteitsev
Kazan Physics-Technical Institute
Sibirsky trackt, 10/7
Kazan, Russia

Ion Geru
Superconductivity and Magnetism Laboratory
State University of Moldova
60, Mateevici str.
MD2009 Chisinau, Moldova

Olga Kulikova
Institute of Applied Physics
5, Academiei,
Kishinev MD-028, Moldova

Ryzhevich Anatol
Institute of Molecular and atomic Physics NASB
70, F. Skorina av.
Minsk 220072, Belarus

Galina Semkova
Institute of Molecular and Atomic Physics NASB
70, F.Skorina av.
Minsk 220072, Belarus

Olga Bazyuk
Ukraninan State University for Water Management
11, Soborna str.
Rivne 33018, Ukraine

Aleksander Bukhanko
Donetsk Physico-Technical Institute NANU
72, Luxemburg
·Donetsk 83114, Ukraine

Olga Bystrova
Physics Department
Kharkiv National University for Radioelectronics
Lenin av., 14
Kharkiv 61166, Ukraine

Petr Grytsjuk
Research Institute for Chemistry
V.N. Karazin National University
4, Svoboda sq.
Kharkiv 61077, Ukraine

Oksana Khoroshun
Physics Department
Kharkiv National University for Radioelectronics
Lenin av., 14
Kharkiv 61166, Ukraine

Yakov Krivosheev
Donetsk National University
24, University str.
Donetsk 83055, Ukraine

Elena Kupko
Physics Department
Kharkiv National University for Radioelectronics
Lenin av., 14
Kharkiv 61166, Ukraine

Ivan Lysyi
Kamenets-Podolsky State Pedagogical University
61, Ogienko str.,
Kamenets-Podolsk 32300, Ukraine

Zhana Suprun
Physics Department
Kharkiv National University for Radioelectronics
Lenin av., 14
Kharkiv 61166, Ukraine

Nataly Timchenko
Physics Department
Kharkiv National University for Radioelectronics
Lenin av., 14
Kharkiv 61166, Ukraine

STRUCTURE AND FUNCTIONAL PROPERTIES OF CRYSTALLINE MATERIALS.

V. I. SIMONOV

Shubnikov Institute of Crystallography of RAS,
59 Leninsky prosp., Moscow, Russia

Abstract. Regular connections between atomic structure and physical properties of series of single crystals obtained with isomorphous substitutions have been established on the basis of precise structural studies. Fluorites doped with rare earths, crystals with sillenite structure, lithium niobate doped with zinc, $KTiOPO_4$ and $(Sr,Ba)Nb_2O_6$ solid solution structures have been considered. Understanding of composition – structure – properties relationships of crystals allows changing purposefully the physical properties of solid solutions via isomorphous substitutions. The domain of properties changing is always restricted. However, the interval of possible changes of properties such as non-linear optical characteristics, superionic conductivity and some others were measured.

Keywords: Crystal structures, solid solutions, structure – physical properties correlations.

1. Introduction

Material engineers are now facing increasing difficulties since the use of lasers in science, technology, medicine and other areas is progressively widening. Not only new laser materials are needed but also materials for electrooptics and photodetectors, non-linear optics for light modulators and optical frequency multipliers, photorecording materials for the direct optical recording and hologram registration. This list steadily increases because each field of application of lasers makes demands to new materials. Structural crystallography, which studies the atomic structure of crystal materials using X-ray, neutron and electron diffraction, opens the ways for establishing correlations between chemical composition, atomic structure and physical properties of the crystals.

Understanding of composition – structure – property relations is important for the solid-state physics when transiting from phenomenology to the microscopic theory for properties and effects occurring in the crystals under external action. Such information helps for the material science to escape the laborious try - and - error method and to turn to the purposeful synthesis of the required crystalline

1

J.-C. Krupa and N.A. Kulagin (eds.), Physics of Laser Crystals, 1–21.

compounds. Modification of the known and used methods in crystal production by means of isomorphic replacements, which is based on the knowledge of composition – structure – property connections, is important for material science also. The field of possible modification of the crystalline material properties is always limited, but for some of the characteristics it can reach orders of magnitude.

Precise X-ray structural studies, which not only allow to establish the coordinates of the atoms in the unit cell of the crystal with high precision, but also to register surely the parameters of thermal vibrations of the atoms taking into account the anisotropy and anharmonicity of these vibrations, are very laborious. Sometimes these studies allow judging not only atomic, but also electronic structure of single crystal, using X-ray diffraction data. The works at such a level can be performed with single crystals only. Working partnership between specialists in synthesis and crystal growing, experts in atomic structure of crystals and those who study the properties of these samples, is needed for success. Establishing regular relations between composition, structure and properties of crystals requires studies of series of crystals with isomorphic replacements. Only series allow building up trends and common regularities in connection with atomic structure and physical properties of the crystals.

There is another interesting approach to the structure – properties problem. Under variation of temperature or pressure, phase transitions may occur in the crystal and open a real possibility to lay down correlations between structure and properties. At a first-order phase transition, structure and properties of the crystal are changing. Therefore, studying of the structure and properties before and after the first-order phase transition allows seeing the structural changes that determine the corresponding modification of properties. The task becomes more difficult if a second-order phase transition occurs in the crystal, when the electronic structure changes at the minimum atom removal. In this case, more high-precise structural studies near the phase transition are needed.

The electronic structure changing affects the chemical bonds of the atoms in the crystal, and leads to abnormality in the interatomic distances and parameters of thermal vibrations of atoms especially. These parameters are determined by the geometry and the character of the chemical bonds of each atom in the structure. Thus structural studies of transition in high-temperature superconductors into superconducting state allowed establishing the ways of transfer of charge carriers from normal layers to the superconducting layers in the corresponding crystals. In the present work on the basis of structural studies of crystalline materials, which are of interest of laser technique, the possibilities of adjustment of the certain physical properties of such crystals by means of isomorphic replacements have been analyzed. The establishment of structure – properties connections allows to use the information, which is accumulated in the structural data banks, much more efficiently. Oxford bank of organic and metal-organic compounds structures and bank of inorganic crystalline structures in Karlsruhe give the possibility of structural characteristics search of the advanced materials with the properties that are of interest for the researcher.

2. Fluorites

As early as the beginning of the last century it was found that in CaF_2 minerals from different deposits there are isomorphic admixtures of rare earths and additional $Ca_{1-x}^{+2}R_x^{+3}F_{2+x}$ fluorine ions, which compensate the valence difference . By now dozens of compounds with fluorite structure $M_{1-x}R_xF_{2+x}$ (M = Ca, Sr, Ba, Cd,P b; R = Sc, Y and any of rare earths) were synthesized, studied and have found an application in different fields of technique. Single crystals of solid solutions, which have been grown in MF_2-RF_3 systems, find an application for lasers, materials for the nuclear radiation detectors, infrared optics and superionic conductors with fluorine anions conductivity. The detailed information about the state diagrams of these systems and voluminous bibliography can be found in the recently published two volumes of B.P. Sobolev`s monograph [1,2].

In the part about fluorites we are interested in atomic structure of not simple fluoride compounds, but of nonstoichiometric single crystals of solid solutions, which are used for the sensitized laser arrays. The structure of $M_{1-x}^{+2}R_x^{+3}F_{2+x}$ single crystals was studied using X-ray structural and neutron structural analysis. As was found, rare earth elements R are not evenly statistically distributed in the sites of the basic cations M, but form clusters with a certain structure. More than ten different atomic models were suggested for these crystals in the literature. What surroundings the generate laser radiation ion will have in the corresponding solid solution depends on the cluster structure.

The structure of CaF_2 fluorite without isomorphic replacements is very simple. Each Ca^{2+} ion is located in the center of the regular cube made of 8 F atoms. The cubes, which are occupied by calcium, unify through common edges into a three-dimensional unit laying with the alternation of occupied and empty cubes. The structure is characterized by the cubic space group symmetry Fm3m with a = 5.462 A unit cell parameter. At every replacement of bivalent calcium cation by the trivalent rare earth, one more additional fluorine atom incorporates into the structure for the compensation of the extra charge.

The idea of the model where additional fluorine atoms simultaneously but with different probability occupy two independent crystallographic positions on two- and threefold symmetry axes appeared in the literature starting since 1964 year. It is very difficult to refine the structure using X-ray techniques when the occupancy of fluorine sites is small in the presence of rare earth elements, especially if not calcium but more heavy barium is the basis in the structure. The fact is that the intensity of X-ray dispersion of the atom is proportional to the square of electrons number in this atom. For such crystals, which contain the atoms with a very different atomic numbers in the periodic system, the structural problem can be solved more surely using neutron diffraction.

Let's take $Ba_{0.73}Pr_{0.27}F_{2.27}$ compound as an example. Amplitudes of elastic coherent scattering of neutrons by atomic nucleus are close for this compound and they are $b_{Ba}= 0.525\,10^{-12}$ cm, $b_{Pr} = 0.445\,10^{-12}$ cm, $b_F = 0.565\,10^{-12}$ cm. The neutron scattering by fluorine is maximal in this case. In order to localize the atomic s with small occupancy, the difference maps of distribution of electron (X-rays) or nucleus (neutrons) density in the crystal are used in structural analysis. Surely localized basic atoms of the structure are moved off from these maps. Diagonal section y-z = 0 of nucleus density distribution in the $Ba_{0.73}Pr_{0.27}F_{2.27}$ crystal (without isomorphic mixture of atoms (Ba, Pr) in the point of origin and F atom in its basic position with (1/4 1/4 1/4) coordinates) is shown in Fig. 1a. in the section one can see neatly the heavy maximum, which corresponds to the position of additional fluorine on the twofold

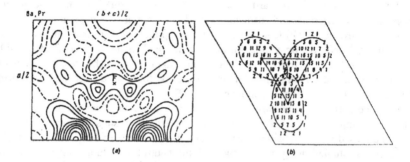

Figure 1. $Ba_{0.73}Pr_{0.27}F_{2.27}$ structure. Difference nuclear density distribution: y-z = 0 plane diagonal section (a), x+y+z = 1.236 plane perpendicular to threefold symmetry axis section (b).

symmetry axis and partially flowing together with it more low peak on the threefold symmetry axis. Maps of this type are the evidence of the presence of additional fluorine atoms in the two mentioned sites with different occupancy probability. However, if we will plot for the same structure x+y+z = 1.236 section normal to symmetry axis 3 and passing through the additional fluorine atom on this threefold axis, then we will see the picture that is shown on Fig. 1b. This section shows unambiguously, that additional fluorine atoms are localized statistically only on twofold axes, while on the threefold axis the fringe regions of the correspondent peaks are superimposed on each other. This statement was tested by studying the crystal structure under low temperature. The decrease of the thermal vibrations amplitude resulted in the more sharp peaks and the whole disappearance of the density on the threefold symmetry axis. Thus, the additional fluorine atoms of $Ba_{0.73}Pr_{0.27}F_{2.27}$ structure localize statistically in the only crystallographically independent position on the twofold symmetry axes.

Let's consider another example of solid solution with the fluorite structure $Sr_{0.69}La_{0.31}F_{2.31}$. Refinement of this structure was done using neutron diffraction method similarly. Diagonal section of the difference atomic density of this structure is shown on Fig. 2. It demonstrates clearly the statistical allocation of the additional fluorine atoms in the only crystallographic position similarly, but on the threefold symmetry axes. Positive and negative extremums near the basic fluorine position with (1/4 1/4 1/4) coordinates attract attention on the section (Fig. 2). There may be two possible reasons of such maximums origin. It may denote unharmonicity of thermal vibrations of fluorine atoms or the splitting of the position of these atoms into two with small reciprocal dispersion along threefold symmetry axis. This dispersion may be the result of the presence of rare earth cations and of relaxation of the part of fluorine atoms from the basic fluorine position, which correct this way their distance to the R^{+3} cation.

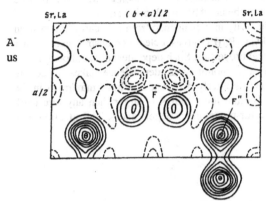

A⁻
us

Figure 2. $Sr_{0.69}La_{0.31}F_{2.31}$ structure. Difference nuclear density distribution on the diagonal section y-z=0

The limited experimental data array ($sin\vartheta/\lambda \leq 0.78$ [1]) that we had, didn't allow to solve this problem uniquely. Account of unharmonicity and insertion of fluorine positions cleaned the difference synthesis from these extremums equally good.

The detailed analysis of the large amount of solid solutions with fluorite structure leads to the conclusion that allocation of additional fluorine atoms takes place in the only independent crystallographic position. In one type of the structures, the twofold symmetry axis is that very position, and in others it is the threefold symmetry axis. Realization of one or another type of allocation of additional fluorine atoms is determined by the ionic radii of M^{+2} and R^{+3} cation ratio. If the radii ratio is r(M)/r(R) > 0.95, then additional fluorine atoms allocate on the twofold symmetry axes. When radii ratio is r(M)/r(R) < 0.95 then, additional atoms localize on the threefold axes only. Establishment of the fact that in $M_{1-x}^{+2}R_x^{+3}F_{2+x}^{-1}$ solid solutions the additional fluorine atoms allocate on the two- and threefold symmetry axes allowed to construct two types of clusters using the crystal chemistry methods. The result of this construction is shown on Fig. 3.

The left branch of the picture demonstrates constructing of the cluster in r(M)/r(R) > 0.95 case and allocating of additional fluorine atoms on the twofold symmetry axes. At the same time 8 atoms that form the regular cube remove from the basic fluorine position. 12 fluorine atoms (8 of them compensate F atoms that leaved the basic position, and 4 of them compensate the extra charge of R^{+3} cations) build in the formed cavity. This local reconstruction in the allocation of fluorine atoms leads to the transformation of 6 cubes (which form the three-dimensional cross with the empty cube in the center) occupied by cations into Thomson cubes, with one face rotated round the normal onto 45°. The distance from Thomson cube to the vertex is less than the corresponding distances in previous regular cubes and Thomson cubes prove to be preferential for occupying by rare earth cations, which have ionic radii less than radii of M^{+2} cations. Four R^{+3} ions and two M^{+2} ions are allocated statistically in six Thomson cubes. This R_6F_{36} cluster build in the fluorite matrix easily because all the exterior edges F-F stay without changes practically.

The right branch in Fig. 3 demonstrates constructing of the cluster in the case when cations radii have inverse relation. In this case 1 F atom removes from the basic fluorine position and 4 fluorine atoms build in the in the formed cavity. At the same time 4 cubes reconstruct into the deca-apexes figure where three more large cations R^{+3} and one M^{+2} allocate statistically. Cluster R_4F_{26}, which build in the fluorite matrix, forms with the supplement of the neighboring fluorine atoms.

Analysis of superionic conductivity in dependence from rare earth in two solid solution systems $Ba_{0.9}R_{0.1}F_{2.1}$ and $Ca_{0.9}R_{0.1}F_{2.1}$ gives a clear demonstration of cluster structure influence over the physical properties. Such dependences are shown in Fig. 4a. In the $Ba_{0.9}R_{0.1}F_{2.1}$ system the monotonous dependence takes place. In $Ca_{0.9}R_{0.1}F_{2.1}$ system the dependence is utterly different and have maximum in the middle of the lanthanide series. Barium ionic radius is larger then any rare earth ionic radius and that is the reason for monotonous dependences. The situation with calcium is another.

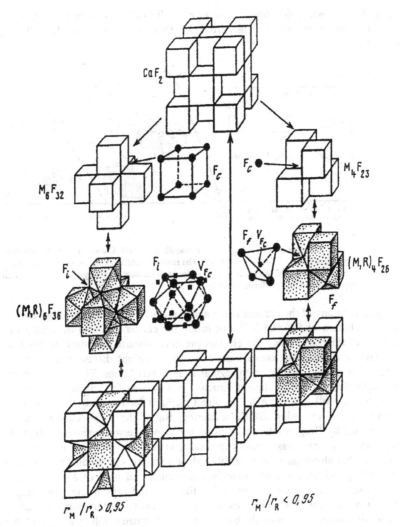

Figure 3. $M_{1-x}^{+2}R_2^{+3}F_{2+x}^{-1}$ fluorites with rare earths. Formation of two types of clusters in dependence from ionic radii ratio $r_m:r_R$.

Because of the lanthanide compression rare earths of the first part of series have ionic radius larger than calcium, and for the R^{+3} ions of the second part of the series it is smaller. Therefore, in the middle of the lanthanide series in $Ca_{0.9}R_{0.1}F_{2.1}$ system the change of cluster structure occurs (Fig. 4b). It is the reason for the difference of ionic conductivity dependences from rare earths in Ba- and Ca- matrixes.

The structures of nonstoichiometric fluorite solid solutions with further complication of clusters structure are known. Under $Ba_{1-x}Ho_xF_{2+x}$ solid solutions investigation it was found that besides R_6F_{36} clusters in the single crystals there are

R_6F_{37} clusters with allocation of one more additional F atom in the center of the cluster in the empty cubic octahedron [3]. There is information about the R_6F_{38} clusters

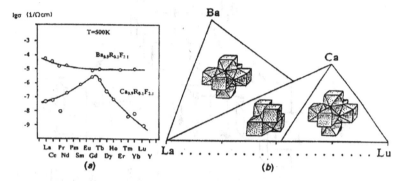

Figure 4. Superionic conductivity in $Ba_{0.9}R_{0.1}F_{2.1}$ and $Ca_{0.9}R_{0.1}F_{2.1}$ solid solutions in dependence   from rare earth (a). The reason for different characters of dependences: in Ba-system one type of cluster with any rare earth element, in Ca-system changing of cluster type occurs in the middle of the series.

existence. In any of the clusters listed above there are defects caused by statistics of the distribution of R^{+3} and M^{+2} ions inside the cluster and also by localization of additional fluorine atoms. The clusters presence disturbs the ideal structure of the crystal and it leads to appearance of elastic coherent diffuse scattering on the diffraction pattern together with regular Bragg's reflections. This diffuse scattering is totally determined by clusters structure and is used for refining of their atomic structure. Up till now there are no direct methods for receiving the structural data about the clusters using diffuse scattering. But if the rough cluster structure is determined by Bragg's intensities, using try – and – error method or Monte Carlo method, then using least-squares method it is possible to refine not only cluster structure but also relaxation removal of fluorine atoms surrounding the cluster from its basic positions [4]. Unfortunately, the accuracy of modern X-ray and neutron diffraction studies does not allow localizing in the crystal the generating laser radiation ions, which amount is estimated in percent parts. Structural characteristics of sensitized laser arrays on the basis of nonstoichiometric solid solutions of fluorite phases are stated above. Defects variety in these matrixes allows evaluating the multivariant possible allocation of generating laser radiation ions in it.

3. Sillenites

Sillenites were named after Swedish chemist Sillen L.G. The accepted chemical formula for sillenites was $Bi_{12}MO_{20}$ (here M=Si, Ge, Ti, (Fe,P), (Bi,Ga), (Bi,Zn), (Bi,V) and others). These crystalline materials are characterized by cubic symmetry. They have I23 space group without center of inversion, the unit cell parameter for

$Bi_{12}GeO_{20}$ compound is a = 10.148(3) A. By now about 50 crystalline materials of different chemical composition with sillenites type structure were synthesized. These compounds are interesting because of their physical properties, which make them very promising for using in piezotechnology, acousto- and optoelectronics. Combination of the low speed of ultrasound propagation with a large piezoelectric coefficient ($BiGeO_{20}$ has d_{111} ninefold than quartz) allows using them as delay lines, resonators and a surface wave amplifiers. High photosensitivity in combination with electrooptic effect open up possibilities of using of the sillenites as spatio-temporal light modulators and also for recording and reproduction of phase holograms. Index of refraction of the material modulates because of the photoelasticity under ultrasound wave influence. Sillenites single crystal becomes optical phase modulator, which period can be regulated by the frequency of the supplied ultrasound wave.

Precision X-ray and mostly neutron diffraction studies of the atomic structure of sillenites with different chemical composition allow asserting that these materials have a great future as broad-spectrum materials. Extraordinary multiplicity of the atomic structure details of sillenites with different chemical composition, which modify the ideal initial model of the structure, is the reason for such a forecast. The first X-ray study of $Bi_{12}GeO_{20}$ single crystals [5] had shown that the distorted $[BiO_5E]$ octahedrons make the basis of the structure. The lone electron pair $6s^2$ of the Bi^{+3} cation occupies one of the vertexes of these octahedrons. Every pair of such groups is united through the common edge O-O into dimer. Dimers and $[GeO_4]$ tetrahedrons are united through the common oxygen vertexes into the three-dimensional framework with cubic symmetry. The structure of $[BiO_5E]$ groups and the fragment of sillenites structure are shown in Fig. 5. X-ray structural studies of sillenites $Bi_{12}MO_{20}$ with M = (Bi,Zn) and (Bi,Fe), were done later. Far-reaching conclusions were done that all the sillenites have the same structure. Exactly 20 oxygen atoms, which occupy their sites in the structure without any defects, correspond to the $Bi_{12}MO_{20}$ chemical formula. At the same time, "average" valency of the isomorphic cations admixture in M site is exactly equal to four. In order to fulfill this condition the authors had to consider that bismuth in M site has oxidation state equal to five Bi^{+5}, but not three as in the basic bismuth site. This conclusion was also argued by the fact that Bi^{+3} ion with its lone electron pair is too large for allocation in M-tetrahedra. From these considerations $Bi_{12}^{+3}[Bi_{2/3}^{+5}Zn_{1/3}^{+2}]O_{20}$ and $Bi_{12}^{+3}[Bi_{1/2}^{+5}Fe_{1/2}^{+3}]O_{20}$ formulas were given to the compounds.

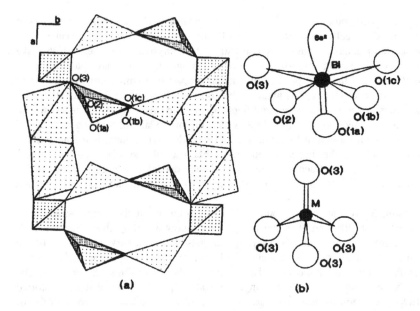

Figure 5. Projection of the sillenites $Bi_{12}MO_2$ structure fragment and building of [BiO$_5$E] and [MO$_4$] groups making the structure.

Before our structural studies of sillenites the structure defects in titanium and one of three basic oxygens (which belongs to the tetrahedral surrounding of titanium) were fixed in $Bi_{12}TiO_{20}$ compound using neutron diffraction structural data. At the same time cubic symmetry of the structure have been saved statistically. We have done the neutron diffraction structural studies of the series of sillenites with different chemical composition. First of all $Bi_{12}GeO_{20}$ single crystals were studied. This study completely confirmed the structure of the ideal sillenites with stoichiometric composition which was found before.

Our results of Ti-sillenites refinement have confirmed the structure deficiency. The fragment of $Bi_{12}Ti_{0.90}O_{19.8}$ structure with defects in titanium and oxygen are shown on Fig. 6a. Sillenites $Bi_{12}(P,Fe)O_{20}$ single crystals were the next subject for study. Isomorphic entry of P^{+5} (ionic radius 0.35 A) and Fe^{+3} (ionic radius 0.64 A) cations into the same site is impossible in respect to classic crystal chemistry. Usually isomorphism occurs if ionic radii differ not more than for 15%.

It turned out, that isomorphism of phosphorus and iron is possible in the sillenites [6]. The fact that Bi-O distances very widely in the Bi-polyhedrons is the reason for it. Thus, in $Bi_{12}GeO_{20}$ compound the distances from bismuth to the 5 nearest oxygens are (in A): 2.072(1), 2.215(1), 2.221(1), 2.622(1) and 2.624(1). Bi-O distances to the oxygens common with P- and Fe-tetrahedrons in Bi-polyhedrons of $Bi_{12}(P,Fe)O_{20}$ structure are different.

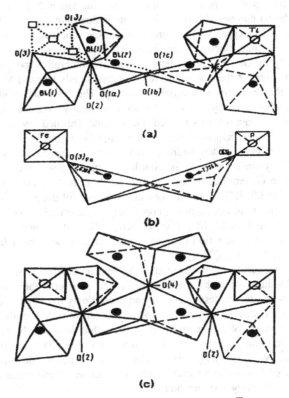

Figure 6. Fragments of $Bi_{12}Ti_{0.90}O_{19.8}$ (a), $Bi_{12}(P_{0.59}Fe_{0.35}\square_{0.06})O_{20}$ (b) and $Bi_{12}(Bi_{0.03}V_{0.89}\square_{0.08})O_{20.27}$ (c) sillenites structures.

They are equal to 2.756(1) and 2.639(1) Å correspondingly. The rest four distances Bi-O in different Bi-polyhedrons are the same (in Å): 2.075(1), 2.208(1), 2.210(1) and 2.612(1). Such a displacement of one of the oxygens (allowable for Bi-polyhedron) is enough to the M-O distances in regular P- and Fe- polyhedrons be equal to 1.534(1) and 1.814(1) Å correspondingly. The difference of these distances is practically equal to the difference between ionic radii of P^{+5} and Fe^{+3} cations. During the refining of $Bi_{12}(P,Fe)O_{20}$ structure it turned out, that all the basic oxygens 100% occupy their sites. The situation with cations content in $[MO_4]$ tetrahedrons is more difficult. It tuned out, that this site is occupied by phosphorus 59%, iron 35% and 6% of sites in M tetrahedrons proved to be vacant. Therefore the chemical formula of these sillenites is $Bi_{12}(P_{0.59}Fe_{0.35}\square_{0.06})O_{20}$. The fragment of this structure with P- and Fe-tetrahedrons is shown on Fig. 6b.

The results of precise structural solving of sillenites with tetrahedral M-sites partially occupied by bismuth proved to be of especial interest. We have studied the compounds with $Bi_{12}(Bi,Ga)O_{20}$, $Bi_{12}(Bi,Fe)O_{20}$ и $Bi_{12}(Bi,Zn)O_{20}$ initial formulas. We will disregard the process of structure solving and will discuss the results only. The oxidation number of bismuth in M site is three but not five. [MO_4] tetrahedron loses one of its oxygen vertexes and it solves the problem of allocation of large ion Bi^{+3} and its lone electron pair. Thus, M sites in these structures are occupied statistically by Ga-, Fe- or Zn-tetrahedrons and umbrella-type groups [BiO_3]. The cubic symmetry of these sillenites remains statistically because M-tetrahedrons lose any of 4 their vertexes with equal probability. The deficiency of these structures in oxygens, which form [MO_4] tetrahedrons, fixes neatly over neutron diffraction data. Chemical formula of the studied Ga-sillenite (refined by diffraction data) is $Bi_{12}(Bi_{0.50}^{+3}Ga_{0.50}^{+3})O_{19.50}$. Positions of Bi and Ga in the M site are split. Gallium stays in the center of tetrahedron, but bismuth is displaced from this center over three-fold symmetry axis in the fourth missing oxygen direction. All four Ga-O distances in Ga-tetrahedron are equal to 1.879(1) Å. Bismuth is distant from three oxygens for 2.018(1) Å, and its lone electron pair fill the place of missing oxygen. Atomic structure of Fe-sillenite is completely isostructural to Ga-sillenite. Its refined chemical formula is $Bi_{12}(Bi_{0.50}^{+3}Fe_{0.50}^{+3})O_{19.5}$ [7]. In the structure with bivalent zinc the bismuth and zinc ratio in M-site changes. That is why the deficiency in oxygen changes too $Bi_{12}(Bi_{0.67}^{+3}Zn_{0.33}^{+2})O_{19.33}$.

Another result was received after the neutron diffraction study of sillenite with V^{+5} in M-site. In the three above mentioned structures the effective valency of cations in M-site was less than four and accordingly the number of oxygens in the formula unit was less than 20. And in sillenites with vanadium structures the effective valency of cations in M-site is more than four and the number of oxygens in the formula exceed 20. The improved formula of this compound is $Bi_{12}(Bi_{0.03}^{+3}V_{0.89}^{+5}\square_{0.08})O_{20.27}$. Here, as usual, vacancies are denoted as squares. The fragment of V-sillenite structure with the position of additional oxygen occupied with 9% probability is shown on Fig.6c.

The above-stated results of structural studies of sillenites show that stoichiometric composition $Bi_{12}MO_{20}$ and the ideal structure without defects belong to completely isostructural $Bi_{12}GeO_{20}$ and $Bi_{12}SiO_{20}$ compounds. These two materials have already found their place in the solid solution physics and material science. The accessibility of regulating of the M cations effective valency and oxygen amount in sillenites open up unimproved opportunities for the purposeful changing of physical properties of these materials. It is clear that stated-above mechanisms of wide isomorphism in the sillenites do not become exhausted with listed. The sillenites family is rich with representatives. And even already synthesized compounds are waiting for the physical properties researchers. The joint efforts of physicists and crystallographers are needed in order to establish the relationships between chemical composition, atomic structure and physical properties of these crystalline materials. Knowing of these correlations will open the ways for purposeful synthesis of new sillenites and also modification of well-known sillenites by means of isomorphic replacements.

4. Lithium Niobate

Numerous studies and practical application of lithium niobate $LiNbO_3$ make needless the supplementary presentation of this crystalline material. X-ray study of $LiNbO_3$ single crystals was done as early as 1966 year by S. Abrahams. Two pictures with the ideal lithium niobate structure depicted as balls and polyhedrons are presented in Fig.7 7.

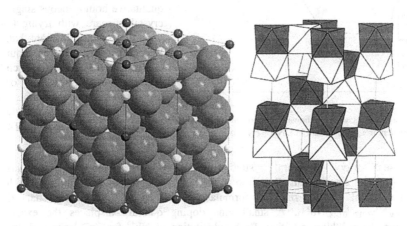

Figure 7. $LiNbO_3$ structure. Two ideas of atomic structure presentation in balls and polyhedrons.

Closest hexagonal packing of O atoms makes the basis of the structure. Nd and Li cations regularly allocate in the octahedral cavities of this packing. One third of the octahedrons is occupied by niobium, one third by lithium and one third stays empty. The acentric space group of symmetry of the structure is R3c, the unit cell parameters are a = 5.149(1) Å, c = 13.862(2) Å. Six formula units with $LiNbO_3$ composition dispose in it. Fundamentally important result of the structural study is that lithium niobate single crystals grow with deviation from stoichiometry. Excess quantity of niobium, which isomorphically replaces lithium in its positions, is always fixed in these crystals. Formation of vacancies in Li-sites compensates the difference between Nb^{+5} and Li^{+1} oxidation states. Four lithium vacancies form with every replacement act and the chemical formula of lithium niobate must be written as $(Li_{1-5x}Nb_x)NbO_3$.

Lithium niobate with stoichiometric composition can be obtained by crystal annealing for 800 hours in Li_2O atmosphere. Lithium niobate crystals with different content of excess niobium were grown and studied. Single crystals of congruous composition with an excess of niobium content of about 1.5 at. % proved to be the best to be used in optics. Two publications [8, 9] prompt us to perform X-ray structural studies of lithium niobate. Dependences of correspondingly electrooptical coefficients and photorefraction from zinc doping lithium niobate quantity are established in these publications. The fact is that $LiNbO_3$ doping with magnesium, zinc, indium, scandium or some other elements suppresses photorefraction for orders of magnitude and enhances stability of the crystals to the intensive laser exposure. The dependence of linear electrooptical effect coefficient from zinc content LiNbO$_3$:Zn crystals is presented in Fig. 8. At about 1.5 at. % Zn the

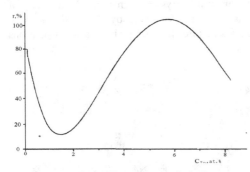

dependence passes the minimum, at about 6 at. % it riches maximum. Difficulties with production of the qualitative homogeneous single crystal appeare with trying to dope the crystal with more than 8 at. % Zn. The question about the structural reasons for such considerable deviations of properties from the monotonous dependence under the steady enhance of the doping Zn quantity appears.

Figure 8. Dependence of linear electrooptical effect coefficient from zinc content in LiNbO$_3$ crystals.

Precise X - ray structural studied of LiNbO$_3$:Zn single crystals with different zinc content allowed to answer this question [10]. Studies started with single crystal without zinc. Its formula was refined using diffraction data is $(Li_{0.940}Nb_{0.012}\square_{0.048})NbO_3$. A small zinc doping quantity displaces the excess niobium from lithium position. Each substitution of Nb^{+5} for Zn^{+2} goes with the occupancy of three vacancies by Li^{+1} cations. Thus on the first stage (up to the minimum on the curve in Fig. 8) the excess of niobium decreases, zinc and lithium quantum increases and vacancies in lithium positions disappear with a coefficient 3 at doping of lithium niobate with zinc. The process of such isomorphic zinc entry into the crystal stops when all the excess niobium is exhausted. In our case the crystals with $(Li_{0.976}Zn_{0.012}\square_{0.012})NbO_3$ composition is formed. Minimum is achieved in Fig.8 with this composition. At the further increase of zinc content, the mechanism of its entry into the crystal changes. Zn^{+2} begins to replace Li^{+1} in its position when all the excess niobium is replaced. Each replacement of lithium for zinc is accompanied by the appearance of one new vacancy. This process continue till 6 at.% Zn entry into the crystal (maximum in Fig. 8). The composition of the crystal in this case is $(Li_{0.88}Zn_{0.06}\square_{0.06})NbO_3$. Subsequent increase of zinc content in the crystal again lead to the replacement of the mechanism of its entry into the crystal. Zn^{+2} cations begin to displace Nb^{+5} from the basic niobium position. The process goes with a quick decrease of vacancy amount as on the stage of replacement of excess niobium by zinc. This mechanism of zinc entry into lithium niobate stops working at amount of 8 at.% Zn. The crystal composition is $(Li_{0.94}Zn_{0.06})(Nb_{0.98}Zn_{0.02})O_3$ in this case. The limit for the zinc entry into the crystal occurs because there is no more vacancies in the crystal. The existence of the crystals with self-compensation (when the ratio of zinc amount in the lithium and niobium positions is 3:1) was predicted in [11]. As was mantioned above, even if the attempts to entry more zinc into lithium niobate single crystals are a success, the homogeneity and optical properties of these crystals leave much to be desired. All above-stated limits in the zinc content may be removed somewhat at the changing of the mechanisms of its entry into the crystal in relation with the growing conditions. The example of lithium niobate crystals doped with zinc shows that precise structural data are very important for understanding the phenomena, which take place in the crystals under their doping.

5. Potassium Titanil Phosphate and its Solid Solutions

Initially $KTiOPO_4$ crystals attracted researcher attention by nonlinear optical characteristics and the availability of their using for laser radiation frequency doubling. Superionic properties and ferroelectric phase transition were revealed in these crystals later. Unique combination of properties and the opportunity to regulate the characteristics of these properties within certain limits by means of isomorphic replacements (potassium for Na, Tl, Rb or Cs; titanium for Ge, Sn or Nb; phosphorus for As) attract researcher attention to the solid solutions crystals with $KTiOPO_4$ structure till now. The publication [12] is the most complete review dedicated to these materials. $KTiOPO_4$ crystals are characterized by acentric orthorhombic space group $Pna2_1$. Their unit cell parameters are a = 12.817(1), b = 10.589(1), c=6.403(1) (in Å). The model of the structure is show in Fig. 9. $[TiO_6]$ octahedrons and $[PO_4]$ tetrahedrons conjoined through common oxygen vertices make the three-dimensional framework of the structure. Zigzag chains made of Ti-octahedrons extend parallel to [011] and [0 $\overline{1}$1] directions. Octahedrons in the chains and from different chains are connected by P-tetrahedrons. The framework is quite an openwork and there are screw-like channels, which general directions coincide with the directions of the 2_1 screw symmetry axes. Potassium ions are located in the channels of the structure and they are responsible for the superionic properties of these compounds. Ionic conductivity in the solid solutions with $KTiOPO_4$ structure may change within wide limits by means of isomorphic replacements. Splitting of Tl cations positions take place at the replacement of K atoms for Tl in the channels of the structure. Study of $TlTiOPO_4$ structure at the room temperature did not allow to make the unique choice between the model with splitting of thallium position and the model with strong anisotropy and anharmonicity of thallium thermal vibrations without splitting of positions. The question was solved using X-ray structural data, performed at 11K temperature. Helium temperature decreased thermal vibrations of the atoms and clearly revealed splitting of the thallium positions [13]. However, a sharp increase of ionic conductivity up to three orders is not the consequence of potassium replacement for other univalent cations, but the result of heterovalent replacement of Ti^{+4} for Nb^{+5}. At the same time a corresponding deficiency of univalent potassium is formed in the crystal for the valency balance. The graph of the resistance against temperature with the discontinuity for sesqui-orders at the transition of $K_{0.977}(Ti_{0.977}Nb_{0.023})OPO_4$ compound into superionic state is given in publication [14].

The question about establishing structural parameters that determine nonlinear optical properties of this type of solid solutions, is proved to be more difficult.

Figure 9. Model of KTiOPO$_4$ structure. Chains made of alternate Ti(1)-
and Ti(2)-octahedrons and [PO$_4$]-tetrahedrons fastening them are visible.
Potassium cations are located in the channels of the structure.

The analysis of all available data about structures and properties of the
corresponding solid solutions was necessary for its solving. This analysis revealed
clear correlation between nonlinearity and asymmetry in the location of titanium or
other replaced atoms in the oxygen octahedrons. In KTiOPO$_4$ two
crystallographically independent titanium atoms alternate in the mentioned above

Ti-octahedrons chains. Non-linear characteristics of solid solutions only decrease at any known replacements of titanium by other elements. Maximal intensity of the second harmonic for the laser made of yttrium-aluminium garnet doped with neodymium was obtained under the irradiation of $KTiOPO_4$ crystals without doping. This intensity is 400 times higher than for SiO_2 second harmonic at other equal status. In $[TiO_6]$ octahedrons one of the Ti-O distances is shorter than others (strong titanyl bond), in trans-position the attenuated bond with enlarged Ti-O distance is situated. Shortened and enlarged Ti-O bonds alternate in Ti-chains of studied structures. Difference in these bonds length is the very structural parameter, which directly correlates with optical non-linearity of the crystals. Distances in -O(9)=Ti(1)-O(10)=Ti(2)-O(9)= chains for some compounds with $KTiOPO_4$ structure from [15-17] are listed in the table.

Table. Interatomic Ti-O distances in the Ti-octahedrons chains in series of crystals with $KTiOPO_4$ structure.

Compounds	Interatomic distances, A				$\Delta_1(Ti\text{-}O)$	$\Delta_2(Ti\text{-}O)$
	-O(9)=	Ti(1)-O(10)	= Ti(2)-O(9)=			
$KTiOPO_4$	1.716	1.985	1.736	2.099	0.269	0.363
$K_{0.96}Ti_{0.96}Nb_{0.04}OPO_4$	1.732	1.959	1.750	2.077	0.227	0.327
$K_{0.93}Ti_{0.93}Nb_{0.07}OPO_4$	1.750	1.940	1.770	2.057	0.190	0.287
$K_{0.89}Ti_{0.89}Nb_{0.11}OPO_4$	1.771	1.908	1.801	2.190	0.137	0.218
$KTi_{0.94}Sn_{0.06}OPO_4$	1.744	1.948	1.792	2.046	0.204	0.254
$KTi_{0.47}Sn_{0.53}OPO_4$	1.857	1.915	1.911	1.989	0.058	0.078
$KTi_{0.25}Sn_{0.75}OPO_4$	1.955	1.940	1.940	1.952	0.015	0.012
$KSnOPO_4$	1.978	1.975	1.961	1.957	0.003	0.004

As is obvious the differences of $\Delta(Ti\text{-}O)$ distances decrease with Nb content increasing. $KSnOPO_4$ compound has practically symmetrical $[SnO_6]$ octahedrons that answer to the complete disappearance on the second harmonic after irradiation of these crystals. $KTiOPO_4$-$KSnOPO_4$ system makes the continuous series of solid solutions with $KTiOPO_4$ structure. Even small amount of tin, which isomorphically replaces titanium, essentially decreases non-linear properties of the crystals. The temperature of ferroelectric phase transition decreases from 900 C to 400 C simultaneously.

Establishment of structural conditionality of physical properties for the rich family of $KTiOPO_4$ solid solutions opens the ways for purposeful regulating of these properties by means of isomorphic isovalent and heterovalent replacements.

6. $(Sr,Ba)Nb_2O_6$ Solid Solutions

Crystals of solid solutions with $(Sr_xBa_{1-x})Nb_2O_6$ composition attract attention by the number of physical properties, which open up possibilities of their using as piroelectrics, piezoelectrics and materials with non-linear optical characteristics. These are relaxor ferroelectrics. It is possible to regulate within certain limits the ferroelectric phase transition temperature, the smearing of the transition and other physical properties of these crystals by means of changing the barium and strontium abundance. These tetragonal crystals do not have symmetry center and are characterized by P4bm space group. Unit cell parameters of $(Sr_{0.50}Ba_{0.50})Nb_2O_6$ crystals are a = 12.461(2) Å, c=3.9475(3) Å [18]. Atoms of five formula units with $(Sr_xBa_{1-x})Nb_2O_6$ composition are located in this cell. First X-ray study was done for the crystals with $(Sr_{0.73}Ba_{0.27})Nb_2O_6$ composition. Later precise structural studies of solid solutions single crystals with x = 0.33, 0.50, 0.61 and 0.75 were fulfilled [19]. Solid solutions with the structure of interest for us exist only in the 0.25 < x < 75 concentration range. Limits of the interval may be removed under variations of single crystals growing conditions. Structure of such compounds is shown in Fig. 10. Three-dimensional framework made of two types of crystallographically independent Nb-octahedrons is the basis of the structure. Niobium from $[Nb(1)O_6]$-octahedron localizes at the intersection of the planes of reflection symmetry mm. Two Nb(1)-octahedrons fall at the unit cell. Nb(2) atoms are in the general octuple position in accordance with 8 of such octahedrons in the unit cell. As can be seen from Fig. 10 three channel types pierce the framework in parallel with c axis. The channels of the triangled section are always empty. In the middle channels of the square section only strontium localizes at all the studied compositions of solid solutions. Point symmetry of its position is 4, cell multiplicity is 2. All barium is located in the widest channels of pentagonal section in the position with m symmetry and multiplicity 4. The part of strontium localizes in the same channels. Composition with minimal strontium amount $(Sr_{0.33}Ba_{0.67})Nb_2O_6$, when both strontium and barium are situated in their channels, is exceptional. Summary position multiplicity in these channels is 6, and number of (Sr_xBa_{1-x}) atoms per cell is only 5. In other words, this type structure exists only with vacancy defects. And as the structural studies have shown, positions both in square and pentagonal channels are located statistically. In the structures with x = 0.33; 0.50; 0.61 and 0.75 occupancy of square channels by strontium is 70.5; 72.0; 72.5 and 71.5 % correspondingly. To all appearances, the structure can not exist with smaller occupancy of these channels, and it determines the lower limit x > 0.25 of the discussed solid solutions. Wide channels occupancy for the same consequence of compositions is 84.0 %; 62.3 %; 48.7 % and 30.9 % in barium. According to our information, positions of Ba and Sr atoms are split in the wide channels. Barium always are located in the planes of reflection symmetry. Sr is displaced from these planes into the general position and statistically with equal probability localizes on the two sides from m planes. Taking into account this circumstance, the occupancy of wide channels by strontium is 0; 26.2 %; 40.4 % and 57.2 %. At the same time, the occupancy of wide channels by Ba and Sr is 84.0 %; 88.5 %; 89.1 % and 88.1 %. In compounds with x = 0.50; 0.61 and 0.75, where Sr atoms are present in the wide channels, they localize at 0.335 A; 0.305 Å and 0.262 Å distances respectively from Ba position. Thus, quantitative characteristics of disorder in $(Sr_xBa_{1-x})Nb_2O_6$ solid solutions monotonically depend on Sr:Ba ratio in the crystal. As regarding properties, increase of strontium quantity results in the same Nb(2)-octahedrons make independent chains of octahedrons joined by common oxygens, along c axis.

Figure 10. Model of $(Sr_xBa_{1-x})Nb_2O_6$ solid solutions structure. Three-dimensional framework made of $[NbO_6]$-octahedrons. The channels with Sr cations (dark spheres) and Ba (light spheres) allocated are visible.

The positions of these oxygens joined octahedrons are split and each of them occupy two places with 50 % probability. The distances between split oxygen positions increase monotonically along the strontium amount increase. Values of splitting of bridge oxygens in Nb(1)-octahedrons are (in Å): 0.508; 0.538; 0.567 and 0.636 for the series of solid solutions with x = 0.33; 0.50; 0.61 and 0.75. These values for Nb(2)-octahedrons are: 0.387; 0.401; 0.458 and 0.567 Å correspondingly. The difference of Nb-O distances to bridge oxygens characterizes numerically the acentricity of Nb atoms position in octahedrons. One of these distances is minimal and the largest Nb-O distance is in trans-position. In the compound with x = 0.33, these distances have the following values in Nb(1)-octahedrons chains –O = 1.82 Å,

= Nb(1)-2.17 Å --O= and in Nb(2)-octahedrons chains $-O = 1.87$ Å $= Nb(2)$-2.15 Å $-O =$. The differences in these bonds lengths are 0.35 Å and 0.28 Å correspondingly. These crystals with $(Sr_{0.33}Ba_{0.67})Nb_2O_6$ composition have maximal double refraction and maximal value of d_{33} component of the tensor of quadratic optical susceptibility, which determines the intensity of the second harmonic generation. The differences of these bonds lengths decrease with increasing of strontium amount in solid solutions: for x = 0.50 $\Delta[Nb(1)-O]$ and $\Delta[Nb(2)-O]$ are equal to 0.34 Å and 0.26 Å. In compounds with x = 0.61 and 0.75 these pairs of differences are 0.30 Å and 0.19 Å; 0.13 A and 0.10 Å. Components d_{33} of the tensor of the quadratic optical susceptibility for the second harmonic generation of the 1.06 μm radiation for $(Sr_xBa_{1-x})Nb_2O_6$ solid solutions with x = 0.33; 0.61 and 0.75 are given in [18]. These values are equal to $d_{33}= 38 \cdot 10^{-9}$, $22 \cdot 10^{-9}$ and $17 \cdot 10^{-9}$ CGSE. Clear correlation between them becomes apparent at their comparison with the above-stated structural characteristics of the corresponding solid solutions.

7. Conclusion

The above-stated results of precise structural studies of several classes of known crystalline materials allow fundamental understanding of the structural conditionality of some physical properties of materials. Practically the attempts to establish the relationships between atomic structure and properties of the important crystals for laser generation were made. All the set of rigid body properties are determined by electron, atomic and real structure of the corresponding crystals. Therefore, joint efforts from specialists in crystal formation and growth, professionals in the above mentioned three levels of crystal structure and, of course, specialists in each of properties under interest for this particular problem, are needed. A mutual understanding of the possibilities and operating methods of each branch is necessary for a creative collaboration. The purpose of the work stated above is to demonstrate that specialists in structural crystallography can bring useful information in solving general problems.

8. References.

1. Sobolev, B.P. (2000) *The Rare Earth Trifluorides. Part 1. The High Temperature Chemistry of the Rare Earth Trifluorions*, Institut d'Estudis Catalans, Barcelona, Spain.
2. Sobolev, B.P. (2001) *The Rere Earth Trifluorious. Part 2. Introduction to Materials Science of Multicomponent Metal Fluoride Crystals,* Institut d'Estudis Catalaus, Barcelona, Spain.
3. Golubev, A.M., Ivanov-Shits, A.K. and Simonov, V.I. et al. (1990) A Structural Model for Fluoride Ionic Transport in $Ba_{1-x}Ho_xF_{2+x}$ Solid Solutions (x > 0.1), *Solid State Ionics* 37, 115-121.
4. Hull, S., and Wilson, C.C. (1992) The Defect Structure of Anion-Excess $(Ca_{1-x}Y_x)F_{2+x}$ with x = 0.06, *J. Solid State Chem.* 100, 101-114.
5. Abrahams, S.C., Jameson, P.B., Bernstein, J.L. (1967) Crystal Structure of Piezoelectric Bismuth Germanium Oxide $Bi_{12}GeO_{20}$, *J. Chem. Phys.* 47, 4034-4042.
6. Radaev, S.F., Simonov, V.I., Kargin, Yu.F. and Skorikov, V.M. (1992) New Data on Structure and Crystal Chemistry of Sillenites $Bi_{12}M_xO_{20-\delta}$, *Eur. J. Solid State Inorg. Chem.* 29, 383-392.

7. Radaev, S.F., Muradyan, L.A., Simonov, V.I. (1991) Atomic Structure and Crystal Chemistry of Sillenites: $Bi_{12}(Bi_{0.50}^{3+}Fe_{0.50}^{3+})O_{19.50}$ and $Bi_{12}(Bi_{0.67}^{3+}Zn_{0.33}^{2+})O_{19.33}$, *Acta Cryst.* **B47**, 1-6.

8. Abdi, F., Aillerie, M., Fontana, M., Bourson, P., Volk, T., Maximov, B., Sulyanov, S., Rubinina, N., Wohlecke, M. (1999) Influence of Zn Doping on Electrooptical Properties and Structure Parameters of Lithium Niobate Crystals, *Appl. Phys.* **B68**, 795-799.

9. Volk, T., Maximov, B., Chernaya, T., Wohlecke, M., Simonov, V. (2001) Photorefractive Properties of $LiNbO_3$:Zn Crystals Related to the Defect Structure, *Appl. Phys.* **B72**, 647-652.

10. Chernaya, T.S., Maximov, B.A., Volk, T.R., Rubinina, N.M., Simonov, V.I. (2001) Zn Atons in Lithium Niobate and Mechanisms of Their Entrance into the Crystal, *JETP Letters* **73**, 110-113.

11. Donnerberg, H., Tomlinson, S.M., Catlow, C.R.A., Schirmer, O.F. (1991) Computer-Simulation Studies of Intrinsic Defects in $LiNbO_3$ Crystal, *Phys. Rev.* **B44**, 4877-4883.

12. Satyanarayan, M.N., Deepthy, A., Bhat, H.L. (1999) Potassium Titanil Phosphate and its Isomorphs: Growth, Properties and Applications, *Critical Reviews in Solid State and Materials Sciences*, **24**, 103-191.

13. Blomberg, M.K., Merisalo, M.J., Sorokina, N.I., Lee, D.Yu., Verin, I.A., Voronkova, V.I., Yanovskii, V.K., Simonov, V.I. (1998) Structural Study of TlTiOPO$_4$ Single Crystals at 11K, *Crystallography Reports.* **43**, 748-757.

14. Losevskaya, T.Yu., Kharitonova, E.P., Voronkova, V.I., Yanovskii, V.K., Stefanovich, S.Yu., Sorokina, N.I., Simonov, V.I. (1999) Superionic Transitions in $K_{1-x}Ti_{1-x}Nb_xOPO_4$ and $K_3Nb_3B_2O_{12}$ Crystals *Crystallography Reports* **44**, 90-92.

15. Andreev, B.V., D'yakov, V.A., Sorokina, N.I., (1991) n-Irradiated KTiOPO$_4$: Precise Structure Studies, *Solid State Communication*, **80**, 777-781.

16. Yanovskii, V.K., Voronkova, V.I., Losevskaya, T.Yu., Stefanovich, S.Yu., Ivanov, S.A., Simonov, V.I., Sorokina, N.I. (2002) Growth and Properties of Nb-or Sn-doped KTiOPO$_4$ Crystals, *Crystallography Reports* **47**, S99-S104.

17. Alekseeva, O.A., Blomberg, M.K., Molchanov, V.N., Verin, I.A., Sorokina, N.I., Losevskaya, T.Yu., Voronkova, V.I., Yanovskii, (2001) Refinement of the $K_{0.96}Ti_{0.96}Nb_{0.04}OPO_4$ Structure, *Crystallography Reports* **46**, 642-646.

18. Chernaya, T.S., Maksimov, B.A., Volk, T.R., Ivleva, L.I., Simonov, V.I. (2000) Atomic Structure of $Sr_{0.75}Ba_{0.25}Nb_2O_6$ Single Crystal and Composition–Structure–Property Relation in (Sr,Ba)Nb$_2$O$_6$ Solid Solutions, *Physics of the Solid State* 42, 1716-1721.

19. Chernaya, T.S., Volk, T.R., Verin, I.A., Ivleva, L.I., Simonov, V.I. (2002) Atomic Structure of $(Sr_{0.50}Ba_{0.50})Nb_2O_6$ Compounds, *Crystallography Reports*, **47**, 213-216.

20. Oliver, J.R., Neurgaonkar, R.R., Cross, L.E., (1988) A Thermodynamic Phenomenology for Ferroelectric Tungsten Bronz $Sr_{0.6}Ba_{0.4}Nb_2O_6$, *J. Appl. Phys.* **64**, 37-47.

UV-VUV LASERS AND FAST SCINTILLATORS

J.-C. KRUPA[1] and V. N. MAKHOV[2]

[1]Institut de Physique Nucléaire, CNRS-IN2P3,
91406 Orsay Cedex, France
[2]Lebedev Physical Institute,
53, Leninsky Prospect, Moscow 119991, Russia

Abstract: Important domains of application for luminescent materials are now emerging where intense Ultra-Violet (UV) and Vacuum Ultra-Violet (VUV) emissions are needed. Within this relatively high energy range, progress in the development of new photon emitting materials is directly related to our understanding of the physical processes of energy absorption and relaxation in solids. The most suitable materials are probably the large band gap inorganic lattices activated with rare earth ions. Optical excitation in these combined systems results either in a direct excitation of the luminescence center or excitation of the host lattice that partly transfers the stored energy to the luminescence activator. Every steps of the luminescence process enter in competition with non radiative losses or undesired luminescence that have to be minimized in order to get a high luminescence efficiency. Determination of the dominant energy transfer and energy loss mechanisms can be performed by time resolved luminescence spectroscopy using UV-VUV synchrotron radiation excitation.

Keywords: Rare earths, UV-VUV absorption-emission, synchrotron radiation

1. Introduction

Luminescent materials, also called phosphors, are solids which convert energy into electromagnetic radiations of various frequencies. A distinction should be made between phosphors emitting in the UV (Ultra-Violet) and VUV (Vacuum Ultra-Violet) range and phosphors emitting in the visible when excited by UV-VUV photons. The latter ones are required for two important technological developments: the Hg-free fluorescent lamps for lighting and the colour plasma display panels mainly for colour television flat screens. For both applications, VUV excited phosphors are applied in noble gas discharge devices which emission (at 172 nm for Xe discharge) is used to excite the luminescent materials. Three basic colours are required for both applications: blue, green and red. However, it is worthy noticing that compared to the well-known Hg emission, noble gas (Xe) discharge is less efficient and yields photons with higher energy. In order to obtain acceptable energy conversion efficiency, luminescent materials are needed that generate more than one visible photon per one VUV incident photon. Such a process can take place in the so-called quantum cutting materials in which the VUV photon is cut into two or more visible photons. In addition, the phosphor must have a very high absorption coefficient (>90%) at wavelengths

23

J.-C. Krupa and N.A. Kulagin (eds.), Physics of Laser Crystals, 23–33.
© 2003 *Kluwer Academic Publishers. Printed in the Netherlands.*

corresponding to the characteristic lines of noble gas discharge and the luminescence decay time must be shorter than the frame time of the device in display applications.

Another important application of luminescent materials emitting in the UV or VUV domains concerns mainly the nuclear medicine imaging, where a claim of new inorganic scintillators is constantly formulated. Scintillating materials convert X- or γ-rays into lower frequency UV radiations which can be easily detected by the most conventional photomultiplier tubes or by UV, VUV photosensitive vapours in multi-wire proportional chambers. Scintillator materials should meet the basic requirements of a high emission yield and a fast response to lower the radiation doses delivered to the patients. The improvement of the X-ray storage phosphors and by extension of the thermo-luminescence dosimeters for ionising radiation monitoring is also a great challenge opened in the same field of nuclear medicine.

At last, tuneable UV-VUV solid state lasers for different applications such as the photolithography for computer microchips production, micro-mechanics or nano-machining are hopefully expected in the near future. Concerning the computer chip capabilities, it is well known that the size and the power (number of diodes and transistors) of the chips are directly related to the engraving resolution that depends on the wavelength of the radiation used to shine the photosensitive resins deposited on the buffer. Accordingly, short wavelength radiations are in demand.

For all these applications, rare earth ions embedded in different crystalline hosts constitute one of the most promising research directions. The search for such new phosphors either excited by VUV and XUV radiations or UV-VUV photon emitters will certainly progress with a better understanding of the luminescence excitation mechanisms. With the full knowledge of the processes occurring in rare earth activated inorganic lattices within the energy excitation range considered here, the behaviour of new luminescent materials can be predicted and optimised for applications.

Therefore, progress in the development of these new luminescent materials is directly related to our knowledge of the physical processes associated with high energy photon absorption and energy relaxation in solids that will be described below. Synchrotron radiation as a fully tuneable intense light source is a powerful tool to explore the UV and VUV energy region at sub-nanosecond time scale. Its time structure allows lifetime measurements of the radiatively decaying levels that provide in particular, information on the kinetics of the energy transfer.

2. Optical Spectroscopy of Rare Earth doped Luminescent Materials

Rare earth ($RE^{3+, 2+}$) ions are widely applied for emission especially when they are doped into solid compounds. Fluorides with their large forbidden energy gap are exceptional materials where electronic levels of the doping rare earth ions involved in the luminescence processes are well displayed within the optical window of the host and therefore can be easily investigated.

Many wide band gap crystals doped with different rare earth ions show fast and intense emission [1] when properly excited, with a decay time of the order several tens of nanosecond due to the parity-allowed $5d \rightarrow 4f$ transitions. These transitions occur between Stark electronic levels belonging to the first excited configuration and the ground state one, in the isolated rare earth ion diluted in a convenient host matrix. The broad and intense band features characterizing these d-f transitions, mainly vibronic in character, suggest on the one hand, that the rare earth ion-solid system can be used as an active medium for tunable solid state UV-VUV lasers. On the other hand, because of the fast UV emission associated with d-f transitions in Ce^{3+}, Pr^{3+} and Nd^{3+} ions, such doped crystals can also be used as fast scintillators [2].

2.1. ABSORPTION AND EMISSION PROCESSES IN THE RARE EARTH CENTRE

The ground state configuration of trivalent lanthanide ions is $4f^n$ that means that only the 4f shell is under completion and contains the optically active electrons. The 4f-electrons belong to an inner shell well protected from the surrounding interactions by $5s^2$, $5p^6$ close outermost shells and because of a weak covalence when the rare earth is engaged in a compound, the 4f-electrons in compounds do not lose their atomic character. Then, optical excitations in these ions embedded in solids occur between quasi-atomic states resulting from the degeneracy removing of $^{2S+1}L_J$ free-ion levels by a non spherical crystal field perturbation provided by the ion environment [3]. The symmetry of this electrostatic interaction is deduced from the crystallographic arrangement of the charges belonging to the immediate surrounding of the cation substituted by the rare earth ion into the host lattice. The crystal field magnitude depends in first order, on the near neighbor distances to the cation (the so-called point charge model) as well as the degree of covalence in the bonding or the spatial overlapping of the charge distributions.

In the RE ground configuration, the intra-configuration electric dipole transitions are forbidden by the Laporte parity conservation rule stating that the parity of the final state must change after photon absorption (or emission) involving oscillations of an electric dipole. Only small admixtures of the wave functions of opposite parity into the $4f^n$ wave functions allow the parity rules to be partly relaxed. The rule relaxation is accounted for the odd terms of the crystal field development which operates within the degenerate free-ion manifolds and at a lower extent to the odd symmetry terms of the crystal vibrations. Then, intra-configuration transitions are mainly forced electric dipole radiations occurring between crystal field states of the rare earth ions in the condensed media and consequently, the corresponding lines in optical spectra are weak and very sharp. These characteristics seem not very attractive for a strong absorption spreads over a broad energy domain as it is required for efficient phosphors. But in turn, these narrow forced transitions within the open f-shell give rise to relatively long (μs, ms) luminescence lifetime whose duration is convenient for display taking into account the eye sensitivity. Emissions are energetically good enough for getting almost saturated colors as it is required for a natural-like color display. The red color is usually obtained through the hypersensitive $^5D_0 \rightarrow {}^7F_2$ electric dipole transitions in Eu^{3+} which are in general of the highest intensity for a non centro-symmetric crystallographic site and the blue one through the d-f transitions in Eu^{2+} stabilized in strontium or barium crystalline hosts [4]. The saturated green color is still difficult to obtain with rare earth ion activators. Tb^{3+} emission from $^5D_4$ level looks rather yellow in different matrices and needs to be filtered with a loss in efficiency.

In the high energy range, UV and VUV incoming photons can induce electronic transitions from the $4f^n$ ground state toward 5d-states belonging to the first excited configuration, $4f^{n-1}5d$, of the rare earth ion. These electric dipole transitions promoting one f-electron from f- to d-orbitals occur between two configurations of opposite parity and therefore, they are totally allowed at first order. The resulting absorption bands [5] are very intense and their intensity is directly related to the rare earth concentration. In addition, due to a greater radial extension of the d-orbitals, the energy of the f-d bands are very sensitive to the crystal field interaction on one hand and they are strongly coupled with the vibration modes of the matrix on the other hand. Consequently, they appear in the optical spectra as very wide because they are mainly vibronic in character.

The crystal field splitting of the five-fold orbitally degenerate 5d-levels and their coupling with the remaining $4f^{n-1}$ states gives rise to a great number of levels within the first excited configuration and their energy location is an important factor to consider in phosphor modeling. From the magnitude of the splitting depends the relative position of the lowest d-level toward either the closest 4f-level or the conduction band (CB) of the inorganic host, which can also be involved in the energy transfer process to the emitting center. In particular, it is worthwhile considering the variation of the energy of the lowest 4f-5d transition versus the number of f-electrons in the 4f-shell of the rare earth ions [6]. The energy of the lowest f-d transition follows the variation of 3+/4+ redox potential along the lanthanide series and is related to the ability of the trivalent rare earth ion to lose one f-electron. Accordingly to the stabilization energy of the 4+ state, the lowest f-d transition in Ce^{3+} and Tb^{3+} ions appears at rather low energy since their ground configurations show only one electron more than the empty and the half-filled shell respectively.

2.2. PHOTON ABSORPTION INVOLVING CHARGE DISPLACEMENTS

2.2.1. *The Ligand to Metal Electron Transfer*

The electron de-localization from ligand to metal can be described in the frame of the molecular orbital model for a cluster formed by the rare earth cation and its first coordination sphere. This charge transfer can be modeled as an electronic transition from states having essentially a ligand character (2p in oxides and fluorides) to states localized on the rare earth ion, i.e. transitions between the last occupied molecular orbital (HOMO) of the host to unoccupied quasi atomic 4f, 5d and 6s states of the rare earth ion .

The energy change between f^n and f^{n+1} configurations involved in this charge transfer (CT) can be estimated via the spin-pairing energy theory developed by Jorgensen [7]. The probability of the electronic transfer from a given ligand of the nearest neighborhood toward the rare earth ion is directly connected to the electro-negativity of the ligand and the electron affinity of the rare earth ion, the latter one depending on the degree of occupation of the open f-shell. The electron affinity follows the 2+/3+ redox potential variation along the lanthanide series in agreement with the stabilization energy variation of the 2+ state within this series. The tendency to reduction to 2+ state is readily understandable for Eu^{3+} with a $4f^6$ configuration, which will gain easily one electron in order to reach the $4f^7$ more stabilized configuration of the half-series. Consequently, the charge transfer involving Eu^{3+} and different ligands especially oxygen and sulfur, will appear at quite low energy (around

250 nm, 4000 cm^{-1} for oxygen). The same conclusion stands for Yb^{3+} showing a $4f^{13}$ configuration (one electron lesser than the full shell).

In addition to the electron-affinity of the metallic ion, the energy of the charge transfer process depends strongly on the electro-negativity of the ligand. Fluorides for example, place this transition at the highest energy, much more higher than oxides and sulfides and even higher than the 4f-5d transitions except for Eu^{3+} as was already mentioned. In the particular case of a fluorine environment, the electron transfer from the ligand to the central ion occurs in the VUV energy range [8] and the corresponding absorption band in the optical spectra can be hidden by a strong overlapping with f-d absorption bands.

The last consideration on the charge transfer band energy position rests on the ligand-metal ion distance. Short distances accounted for covalent bonding provide low energy charge transfer whereas more ionic bonding results in higher energy CT. Accordingly, for the same ligand, the CT energy depends on the coordination number that determines the ligand -metal ion distance.

In a rare earth doped luminescent material, the ligand to metal charge transfer is interpreted as an electron transfer from the valence band states of the host to excited states of the luminescence center (Fig. 1). Electron acceptors such as Sm^{3+}, Eu^{3+}, Tm^{3+} and Yb^{3+} with a relatively high electron-affinity will develop in the solid, a short range potential to attract electron from the valence band (VB). In other words, they are hole donors and the resulting excited charge transfer state $Ln^{2+} + h^{+}(VB)$ is generally energetically widely spread because of a strong coupling to the host lattice vibrations. ion, the f-d transitions in the rare earth ion can promote one electron from the $4f^n$ ground state to excited 5d-states located in the conduction band (CB) of the host and in case of a strong coupling between these 5d-states and the continuum of the solid, the promoted electron can be completely de-localized in the conduction band giving rise to $Ln^{4+} + e^-$ (free) auto-ionization state (Fig. 1). ion, the f-d transitions in the rare earth ion can promote one electron from the $4f^n$ ground state to excited 5d-states located in the conduction band (CB) of the host and in case of a strong coupling between these 5d-states and the continuum of the solid, the promoted electron can be completely de-localized in the conduction band giving rise to $Ln^{4+} + e^-$ (free) auto-ionization state (Fig. 1).

2.2.2. The Auto-Ionization Process

The auto-ionization process is as follows: in the combined system, host + rare earth The propensity of the rare earth ion to give up one electron should be regarded as its hole acceptor capability. It means that when these ions are embedded in solids, a more or less intense short range potential for hole attraction is developed, whose attraction strength depends upon the stabilization energy of the 4^+ state. As was already mentioned, Ce^{3+} and Tb^{3+} whose ground configuration contains one f-electron more than the empty and half-filled f-shell respectively will behave as strong electron donors or hole acceptors in phosphors based on these luminescence activator ions.

28

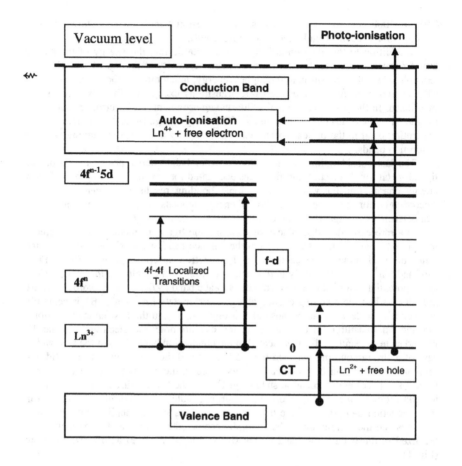

Figure 1. Diagram of different photon absorption mechanism in the Ln^{3+} + host system

2.2.3. *The Host Absorption*

When the incoming photon energy is higher than the solid band gap, E_g, the photon absorption can promote one electron from the valence band to the conduction band, leading to the so-called intrinsic absorption of the solid. During this interaction, free charge carriers are created: one hole in the valence band and one electron in the conduction band with a certain amount of kinetic energy. Both free charge carriers can either move independently through the lattice or can form excitons when the electron remains trapped by the coulomb attraction of the hole left behind him in the valence band. In the latter case, the correlated electron + hole neutral carrier is diffusing through the lattice as a quasi particle.

Mobile charge carriers and excitons can also be trapped by local potentials created by impurities or lattice defects. Many solids exhibit a phenomenon of self-trapping of charge and neutral carriers by creating around the carrier a potential well due a self-induced lattice deformation. The migration of such carriers trapped in deformations may only proceed through the hopping diffusion mode under thermal activation till they reach an impurity center for recombination [9]. The excess of recombination (e^- + h^+) energy can be transferred non radiatively or radiatively to the rare earth excited states and in the expected case, the energy relaxation will result in the desired emission. The self-trapping of the carriers is clearly seen in figure 2 showing the decay profile of Pr^{3+} luminescence in YF_3. At high excitation energy, the rise time in the luminescence decay indicates that the recombination is delayed by the trapping of one of the charge carriers. For example V_k centers (F_2 molecule) are formed after hole trapping in the fluoride crystal affecting the transfer kinetics (Fig. 2).

The free excitons are very mobile and they can travel through many lattice constants during their lifetime that is shorter than 10^{-9} s even at low temperature. In reverse, the slow migration of self-trapped excitons (STE) in the lattice proceeds, as was already mentioned via the hopping diffusion mode from one localized deformation state to another. But, due to their longer lifetime, STE migration range is approximately of the same order as for a free exciton. Then, the average distance covered by the exciton in the lattice is approximately 10 nm. It corresponds to the average distance between two rare earth centers for an homogeneous doping yield of 0.1% [6].

At higher photon energy, the kinetic energy of the fast (hot) primary photoelectron and hole is enough for the creation of one or more secondary electronic excitations as the result of inelastic scattering on the valence or activator center electrons. This step called "electronic excitation multiplication", is followed by the thermalization of the hot charge carriers by using lattice individual phonons or plasmons described as collective motions of atoms and electrons in the solid. At the end of the thermalization step, the solid is left with a thermal population of free electrons in the conduction band and a population of holes in the valence band. It is the way how the energy can be stored into the matrix.

3. Relaxation of Energy Stored in the Doped Solid

3.1. ENERGY RELAXATION IN THE RARE EARTH CENTRE

In phosphors, when a 5d-state of the rare earth ion or a CT state is populated, radiative and non-radiative pathways are open for energy relaxation toward the 4-f emitting level. The excitation energy can be transferred via intra-system energy crossing (5d/4f or CT/4f) generally phonon assisted. The energy transfer mechanisms are commonly explained on the basis of the simple configurational coordinate model involving phonon participation and showing Stokes shift [10]. When the solid is heavily doped, the excitation energy can migrate through the rare earth sub-lattice before to be trapped by the impurity. The efficiency of such a migration process depends upon the magnitude of the square overlap integrals between absorption and emission bands.

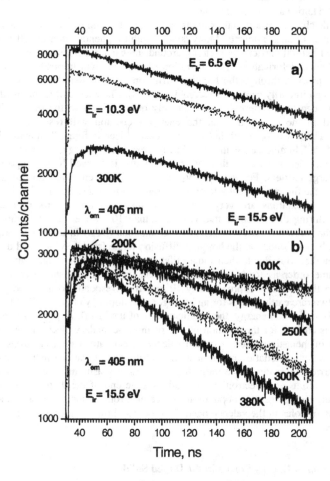

Figure 2. Decay curves of Pr^{3+}-emission in YF_3:Pr^{3+}(1%) under intra-center (6.5 eV), excitonic (10.5 eV) and band-to-band (15.5 eV) excitation at various temperatures

For scintillators, the direct 5d → 4f transitions in rare earth ions with a fast decay (few ns) are of great interest. These allowed emissions occur when exists an energy gap between the lowest 5d-level of the excited configuration and the nearest 4f-level belonging to the ground configuration. This situation always occurs for Ce^{3+} and often for Pr^{3+}, Nd^{3+}, Er^{3+} and Tm^{3+} in selected matrices. In the case of ions with more than seven electrons in the 4-f shell such as Er^{3+}, Tm^{3+}, a slow component (μs) due to spin forbidden 5d-4f transitions (Fig. 3) strongly restricts possible applications of Er^{3+} and Tm^{3+} doped materials as fast scintillators [11].

3.2. ENERGY STORED IN THE RARE EARTH + MATRIX COMBINED SYSTEM

As was already mentioned, the auto-ionization process forms "Ln^{4+} + e⁻(free)" state. The relaxation of this state can be interpreted in the model of the exciton trapped on the impurity center by the capture of a free-electron leading to Ln^{4+} + e⁻ (bounded) state, followed by the capture of the hole. The recombination of the hole and electron on the rare earth center provides energy that can be transferred to the 4f emitting level via for instance, dipole-dipole interaction in the case of allowed transitions or higher order multipolar interactions for the forbidden ones.

3.3. ENERGY STORED INTO THE MATRIX: HOST SENSITIZED LUMINESCENCE

The excitation spectrum of the rare earth luminescence in different host lattices shows at energies close to the host band gap, Eg, a broad asymmetric band whose intensity depends on the efficiency of the energy transfer from the lattice to the luminescence center. This « excitonic » band appears for photon energy lower than the band gap and its intensity decreases when the energy of the incoming photons increases (Fig. 4). This means that the rare earth emission enters in a strong competition with non-radiative luminescence quenching processes or undesirable luminescence generation. This competition turns to be more and more important when the energy of the incoming photons becomes higher than the threshold of the intrinsic absorption and in many cases

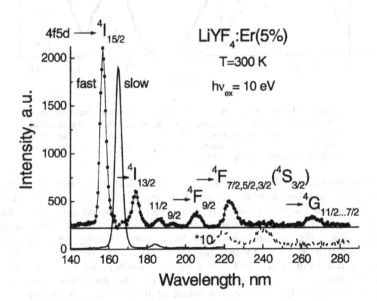

Figure 3. Emission spectrum of Er^{3+} in LiYF₄

the luminescence of the activator is totally quenched for high energy photons. It seems that the luminescence quenching phenomenon is due to two combined factors: i) the reduction of the exciton formation cross-section when the primary photon energy is increasing in favor of free electron and hole which are traveling deeper in the lattice increasing their probability to be trapped by luminescence killers and ii) the reduction of the on-center recombination probability of the charge carriers when their initial kinetic energy is increasing.

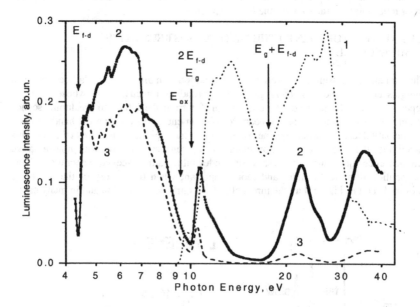

Figure 4. VUV spectroscopy in CeF₃: 1 - absorption spectrum calculated from the reflectivity spectrum, 2- luminescence excitation spectrum from a cleaved surface, 3- the same spectrum obtained with a polished surface. The peculiar energies are indicated on the figure: E_{f-d} : lowest 4f-5d transition in Ce^{3+}, E_{ex} : valence band exciton formation energy, Eg : band gap

It is worthwhile also considering that the surface quenching by recombination on surface killers increases with the reduction of the depth of the primary photon penetration when the incoming energy is in resonance with the intrinsic band to band transition characterized by a very high absorption coefficient. Then, it is obvious that the surface defects play an important role in the phosphor efficiency, especially in this energy region (Fig. 4). Generally speaking, the energy transfer mechanisms from the host to the rare earth centers and to the surface quenchers are identical. As was mentioned already, they are controlled by dipole-dipole interactions and energy migration processes. Then, for the host sensitized luminescence, the competition b between the two kinds of energy acceptor will determine the phosphor efficiency for high energy incoming photons. Experimentally, the surface quenching is revealed by dips in luminescence excitation spectra and accelerations of the luminescence decays.

Broad bands often occurring in excitation spectra at approximately 2 Eg and 3 Eg are due to electronic excitation multiplication effects. For high concentrations of rare earth ions, in particular in stoichiometric compounds (e.g. CeF_3), the so-called impact mechanism of electronic excitation multiplication can become the dominant process of the energy transfer from the host lattice to the emission centers (see Fig. 4 at energy $(E_g + E_{f-d})$).

4. Conclusion

The most direct source of information on the electronic energy levels in solids is provided by optical spectroscopy that have been developing for a long time. However, though energy levels and mechanisms are nowadays well identified, full and rigorous modeling of the VUV excited phosphor quantum efficiency is still unsuccessful because of a large number of parameters difficult to estimate. For example, the unknown concentration of defects within the first atomic layers of the crystal is a real problem when trying to evaluate the undesired premature carrier recombination. Therefore, a systematic study of different host lattice-activator combinations and the determination of their quantum efficiencies under VUV excitation is still necessary. For this purpose, fluorides are certainly, the most suitable materials to study. These compounds are excellent school materials since their wide band gap (or optical window) permits by incorporating luminescent impurities, to investigate both, the emitting center and the host electronic structures as well as to explore the excited state dynamics involved in the energy relaxation before and during the luminescence process.

5. References

1. Becker, J., Gesland, J. Y., Kirikova, N.Yu., Krupa, J.C., Makhov, V.N., Runne, M., Queffelec, M., Uvarova, T. and Zimmerer, G. (1998) Fast VUV emission of rare earth ions (Nd^{3+}, Er^{3+}, Tm^{3+}) in wide band gap crystals, *J. Alloys and Comp.* **275**, 205-208.
2. Makhov, V. N., Kirikova, N. Yu., Kirm, M., Krupa, J.C., Liblik, P., Lushchik, A., Lushchik, Ch., Negodin, E., Zimmerer, G. (2002) Luminescence properties of YPO_4:Nd^{3+}: a promising VUV scintillator material, *Nucl. Inst. Meth. in Phys. Res. A*, **486**, 437-442.
3. Hufner, S. (1978) *Optical spectra of transparent rare earth compounds*, Academic Press, New York.
4. Blasse, G. and Grabmaier, B.C. (1994) *Luminescent Materials*, Springer-Verlag, Berlin.
5. Krupa, J.C. and Queffelec, M. (1997) UV and VUV optical excitations in wide band gap materials doped with rare earth ions: 4f-5d transitions *J. of Alloys and Comp.* **250**, 287-292.
6. Belsky, A.N., Krupa, J.C. (1999) Luminescence excitation mechanisms of rare earth doped phosphors in the VUV range, *Displays* **19**, 185-196.
7. Jörgensen, C.K. (1962) *Orbitals in Atoms and Molecules*, Academic Press, New York.
8. Gerard, I., Krupa, J.C., Simoni, E. and Martin, P. (1994) Investigation of charge transfer O^{2-}- Ln^{3+} and F^--Ln^{3+} in LaF_3 and YF_3, *J. of Alloys and Comp.* **207**, 120-127.
9. Song, K.S. and Williams, R.T. (1992) *Self-Trapped Excitons*, Springer-Verlag, Berlin.
10. Struck, C.W. and Fonger, W.H. (1991) *Understanding luminescence spectra and efficiency using Wp and related functions*, Inorg. Chem.Concepts **13**, Springer-Verlag, Berlin.
11. Makhov, V. N., Kirikova, N. Yu., Kirm, M., Khaidukov, N. M., Negodin, E., Zimmerer, G., Lam, S. K., Lo, D., Krupa, J.C., Gesland, J. Y. (2002) VUV spectroscopy of crystalline emitters based on 5d-4f transitions in rare earth ions, *Surface Review and Lett.* **9**, 621-626.

PERSISTENT LUMINESCENCE MATERIALS

T. AITASALO[1,2], J. HÖLSÄ[1*], J.-C. KRUPA[3],
M. LASTUSAARI[a] and J. NIITTYKOSKI[a,b]
[1] *University of Turku, Department of Chemistry,*
FIN-20014 Turku, Finland.
[2] *Graduate School of Materials Research, Turku, Finland.*
[3] *Institut de Physique Nucléaire, CNRS-IN2P3, F-91406 Orsay Cedex,*
France.

Abstract. The luminescent efficiency of rare earth ions is usually drastically lowered when defects are present in the host lattice. Persistent luminescence is the most recent rare earth application based on lattice defects. Typical materials are the Eu^{2+} doped alkaline earth aluminates, $MAl_2O_4:Eu^{2+}$ (M = Ca and Sr). The trivalent R^{3+} ions as co-dopants enhance greatly the duration and intensity of persistent luminescence. As a result of very slow thermal bleaching of the excitation energy from the lattice defects acting as traps, the new persistent luminescent materials yield luminescence still visible to naked eye for more than ten hours. Despite the seemingly simple stoichiometry and structure of these materials, the determination of persistent luminescence mechanism(s) presents a very complicated problem. This report presents in detail some of the factors affecting the luminescence properties of the Eu^{2+},R^{3+} doped MAl_2O_4. The possible mechanisms involved with different defect centers and interactions between them and the emitting Eu^{2+} ion are discussed based on the results of systematic investigations carried out on the preparation, composition, structure and different luminescence properties.

Keywords: Persistent luminescence, aluminates, europium

1. Introduction

Rare earths (R) are used extensively in luminescent materials which employ practically any kind of energy as the excitation source. Non-doped rare earth compounds with well-defined stoichiometry can be employed, but most frequently the rare earths enter as dopants into different host matrices. In order to avoid any decrease in efficiency, a very good homogeneity is then required of the crystal lattice. In most cases, the efficiency of luminescence is drastically lowered when any kind of lattice defects are found in the host lattice. One of the consequences of the presence of lattice defects is the long afterglow which is exhibited by some potentially efficient luminescent materials. This feature is usually considered disadvantageous among the properties of a phosphor when

35

J.-C. Krupa and N.A. Kulagin (eds.), Physics of Laser Crystals, 35–50.
© 2003 *Kluwer Academic Publishers. Printed in the Netherlands.*

the practical applications are considered [1]. Accordingly, the luminescence applications based on afterglow are rare. Despite the usually undesired afterglow, the persistent luminescence materials have already been used since tens of years. The material used so far, copper and cobalt doped ZnS is, however, extremely sensitive to moisture [2]. Moreover, the duration of the afterglow is of the order of a few hours only and because of the potentially hazardous use of radioactive elements (*e.g.* Pm^{3+}) as an auxiliary excitation source these materials are environmentally hazardous. The use of the rare earth luminescence in persistent luminescence materials is one of the latest application of the rare earth elements (Table).

Table. Luminous Phosphors for Phosphorescent Paints [2].

Composition	Lumines-cence color	Lumines-cence wavelength at peak / nm	Afterglow brightness (after 10 min) / mcd m^{-2}	Afterglow persistence time / min
$CaSrS:Bi^{3+}$ (Sr, 10-20%)	Blue	450	5	Semi-long (about 90)
$CaAl_2O_4:Eu^{2+},Nd^{3+}$	Blue	440	35	Long (over 1000)
ZnS:Cu	Yellow-green	530	45	Semi-long (about 200)
ZnS:Cu,Co	Yellow-green	530	40	Long (over 500)
$SrAl_2O_4:Eu^{2+}$	Green	520	30	Long (over 2000)
$SrAl_2O_4:Eu^{2+},Dy^{3+}$	Green	520	400	Long (over 2000)
$CaS:Eu^{2+},Tm^{3+}$	Red	650	1.2	Short (about 45)

Light source: 1000 lx, 5 min.

Since the mid 1990's a completely new generation of persistent luminescent phosphors has been developed and partly entered into the commercial market [3]. These new phosphors originally included only the Eu^{2+} doped alkaline earth aluminates, $MAl_2O_4:Eu^{2+}$ (M = Ca and Sr) [4-6], but other, more complex aluminates, *e.g.* Eu^{2+} or Ce^{3+} doped melilite based aluminosilicates ($Ca_2Al_2SiO_7:Eu^{2+}$, $CaYAl_3O_7:Eu^{2+},Dy^{3+}$ *etc.* [7-10]) have been studied. Other ceramic materials of interest include calcium magnesium triple silicates ($Ca_3MgSi_2O_8:Eu^{2+},Dy^{3+}$ [11]) as well as Mn^{2+} doped zinc gallate ($ZnGa_2O_4:Mn^{2+}$ [12]). Eu^{2+} doped silicate or borate glasses have been investigated extensively, as well [13-23]. The Eu^{2+} doped alkaline earth aluminates remain, however, most important persistent phosphors, especially because the afterglow is greatly enhanced by co-doping with some R^{3+}-ions as Dy^{3+} and Nd^{3+} [24-29]. The new persistent luminescent materials much superior to ZnS:Cu,Co (Fig.1) yield luminescence still visible to naked eye for more than ten hours in the dark after exposure to radiation (sunlight, UV radiation, fluorescent lamp, *etc.*). These materials are used *e.g.* in traffic signs, emergency signage, watches and clocks as well as in textile printing.

The overall mechanism of the persistent luminescence is now quite well agreed on to involve the formation of traps followed by a subsequent thermal bleaching of the traps and emission from the Eu^{2+} sites. Despite the seemingly simple stoichiometry and structure of the alkaline earth aluminates, the determination of persistent luminescence

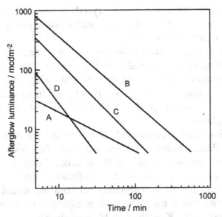

Figure 1. Comparison of the persistent luminescence lifetimes of A: $SrAl_2O_4:Eu^{2+}$, B: $SrAl_2O_4:Eu^{2+},Dy^{3+}$, C: $SrAl_2O_4:Eu^{2+},Nd^{3+}$ and D: $ZnS:Cu,Co$ [24].

mechanisms seems to present a very complicated problem. Accordingly, no general agreement has been achieved on the detailed mechanisms involved and several interesting and even exciting mechanisms have been proposed. Especially, the mechanisms resulting in the prolonged and enhanced afterglow when R^{3+} ions have been introduced into $MAl_2O_4:Eu^{2+}$ as co-dopants, are either ignored or are contradictory [30-34].

In this report, after presenting an initial review on the status on the field, the factors affecting the luminescence properties of the $MAl_2O_4:Eu^{2+}$ phosphors are described in detail. The possible mechanisms involved with different defect centers and interactions between them and the emitting Eu^{2+} ion are discussed based on the results of systematic investigations carried out on the preparation, composition, structure and different luminescence properties (UV-, laser-, persistent luminescence and thermoluminescence) of the Eu^{2+},R^{3+} doped alkaline earth aluminates. The understanding of the mechanisms of the persistent luminescence is crucial when new, even more efficient materials will be developed in a systematic way. In order to successfully treat with this task, multidisciplinary investigations including solid state chemistry and physics, materials science and different spectroscopic methods are required.

2. Experimental

The polycrystalline Eu^{2+} (and Na^+ and R^{3+}) doped alkaline earth aluminates were prepared by a solid state reaction between aluminium oxide, Al_2O_3, calcium (or strontium) carbonate, $CaCO_3$ (or $SrCO_3$), sodium carbonate, Na_2CO_3, and rare earth oxide, R_2O_3 (4N), powders. The mixtures were annealed in a reducing ($N_2 + 12\% H_2$) atmosphere at 1250 (or 1300) C for 6 (or 4) hours, respectively. B_2O_3 (1 mole per cent) was used as a flux. The amounts of Eu^{2+} and R^{3+} (Na^+) were 1 and 2 mole per cent of the host cation, respectively. Prior to heating the reagents were ground using a ball mill to form a homogeneous mixture. The phase and structural purity of all samples was checked by X-ray powder diffraction. Single crystals or fibres of single crystals seem to have homogeneity problems and contain easily additional impurities and/or defects and thus they were not considered in this work.

The sol-gel process is an efficient technique for the preparation of phosphors due to the good mixing of starting materials and relatively low reaction temperature. $CaAl_2O_4:Eu^{2+}$ was prepared with a sol-gel method using aluminum isopropoxide $(Al(OC_3H_7)_3)$, calcium nitrate $(Ca(NO_3)_2 \cdot 4H_2O)$ and europium oxide (Eu_2O_3) as starting materials. The gel was first dried for 10 h at 180 °C in air and then for 20 h at 850 °C in a $N_2 + 12\%$ H_2 atmosphere.

The low resolution luminescence and afterglow spectra were measured at room temperature using a Perkin Elmer LS-5 spectrometer. The wide band (20 nm) UV-excitation ($\lambda_{exc} = 350$ nm) was provided by a xenon lamp. Prior to the afterglow measurements the materials were exposed to radiation from a conventional (11 W) tricolor fluorescent lamp for 10 s. The delay between the initial irradiation and the afterglow measurements was 3 min. The high resolution luminescence and excitation spectra were measured at 77 K with a SLM Aminco SPF-500C spectrofluorometer equipped with a 300 W xenon lamp.

Thermo-luminescence (TL) is a useful method to estimate the energy levels for trapped carriers, *i.e.* the number and the depth of the traps. However, the TL measurements are difficult to carry out and, especially, the results are even more difficult to interpret because of *e.g.* overlapping peaks and re-trapping. The thermo-luminescence glow curves were measured with a Risø TL/OSL-DA-12 system at the temperature range between 25 and 400 C employing the linear heating rates of 2, 4, 5, 6, 8 and 10 C·s^{-1}. Global TL emission from UV to 600 nm was monitored. Prior to the TL measurements, the samples were exposed to irradiation from a 60 W incandescent lamp for 10 s. The delay between the exposure and TL measurements was 3 min. Although several methods can be used in the analysis of trap depths (peak shape method, heating rate method, curve fitting and preheat analysis) only the last one can seriously deal with many overlapping peaks, *i.e.* with traps of similar energy.

3. Results and Discussion

3.1. PERSISTENT LUMINESCENCE OF $MAl_2O_4:Eu^{2+}$

Together with the photo-stimulated luminescence [35], the persistent luminescence is based on the use of lattice defects for storing the excitation energy. However, in contrast to the phenomenon of photo-stimulated luminescence which has been studied exhaustively and the materials are well established and characterized, the corresponding features of the persistent luminescence are not at all well characterized. Both phenomena deal with very complicated mechanisms and involve lattice defects (vacancies, colour centers) as well as the possibility to have both correlated and uncorrelated centers. These two phenomena have some common features, too, as the use of the Eu^{2+} luminescence.

There remain, however, several problems in the persistent luminescence as trap identity, the energy balances, the nature of the charge carriers, in brief, the whole of the detailed mechanisms is left unknown. Some mechanisms for the persistent luminescence have been published but they are deficient and often contradictory.

However, these mechanisms gave the initial idea to the present studies: the mechanisms published proposed the formation of single valent Eu^+ and tetravalent Nd^{4+} and/or Dy^{4+} [30-34] - or, at least, the presence of their virtual counterparts as charge transfer states and excited 5d electron configurations. The presence of the Eu^+ or Nd^{4+} (and/or Dy^{4+}) species in ambient conditions and in oxide materials is, however, not acceptable and the suggested reduction of Eu^{2+} to Eu^+ (and oxidation of Nd^{3+}/Dy^{3+} to Nd^{4+}/Dy^{4+}) would require tremendous amounts of energy. The following examples illustrate the energies involved in the formation of the proposed species. For example, the energy of the lowest CTS band for the Eu^{3+} ion situates usually above 30,000 cm^{-1} and the corresponding CTS bands for the Eu^{2+} ion would be at a much higher energy. Although the presence of Eu^{2+} can be stabilized in the Ca^{2+} (or Sr^{2+}) site by the lattice energy there is no such energy available for the stabilization of the Eu^+ ion. Moreover, the 5d bands for Dy^{3+} are at 50,000 cm^{-1} [36]. The band gap E_g in oxide materials situates above or at 40,000 cm^{-1}. The mechanisms proposed so far involve thus highly dubious processes in solid state and clearly impossible under mild near UV and visible excitation available at ambient conditions. The detailed mechanisms of persistent luminescence remain thus as open questions.

3.2. STRUCTURE OF MAl$_2$O$_4$ AND PERSISTENT LUMINESCENCE

The present studies started from the identification of the luminous centre. All $MAl_2O_4:Eu^{2+}$ (M = Ca and Sr) phosphors show strong broad band emission characteristic to the Eu^{2+} ion in the blue/green visible range under UV-excitation. The Eu^{2+} luminescence which, in general, can vary from UV to red depending on the host lattice, are, in the cases of $CaAl_2O_4:Eu^{2+}$ and $SrAl_2O_4:Eu^{2+}$ centered at 440 and 520 nm, respectively (Fig.2 and 3).

It is thus pretty certain that the broad band emission, excitation and absorption of $MAl_2O_4:Eu^{2+}$ is due to transitions between the $^8S_{7/2}$ ($4f^7$) ground state and the crystal field components of the excited $4f^65d^1$ configuration of the Eu^{2+} ion [6]. The luminescence lifetime for the persistent luminescence is very long (hours) in contrast to the ordinary Eu^{2+} doped materials with lifetimes of 700 ns to 1.2 μs [37]. Despite the enormous difference in the time scale, both the UV-excited and persistent luminescence occur at a practically identical spectral range. The band shape and width of the UV-excited luminescence and persistent luminescence spectra were found identical, too (Fig.2 - 4).

The same luminescence properties point out that the luminous center in both cases is the same Eu^{2+} ion. The shape of the emission bands at room temperature suggests that there is only one luminous centre. However, a survey of the crystal structures of both $CaAl_2O_4$ and $SrAl_2O_4$ reveals that these monoclinic structures [38, 39] (space groups $P2_1/n$ (Z = 12) and $P2_1$ (Z = 4), respectively) offer three and two M^{2+} sites for the Eu^{2+} ion, respectively. In the monoclinic $CaAl_2O_4$ there are one M^{2+} site with a distorted tricapped trigonal antiprism coordination polyhedron (CN = 9) and two sites with distorted octahedron coordination polyhedra (CN = 6). The average Ca – O distances, 2.420 and 2.434 Å, for the six-coordinated Ca-sites are clearly shorter than the average distance for the nine-coordinated Ca-site (2.784 Å). With the help of the ionic radii [40] it is clear that the Eu^{2+} ions – at least at low concentrations - prefer the large nine-coordinated Ca^{2+} site in $CaAl_2O_4$ (Fig.5). This is in agreement with the observation of only one emission band.

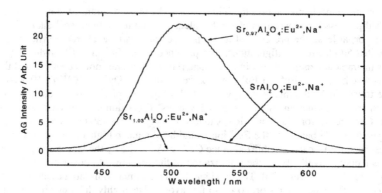

Figure 2. UV excited (350 nm) luminescence and persistent luminescence spectra of $CaAl_2O_4$: Eu^{2+} at room temperature.

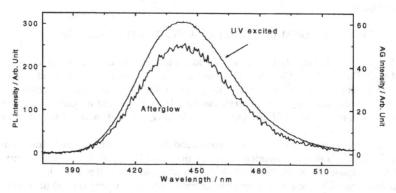

Figure 3. Effect of stoichiometry on the UV excited (350 nm) luminescence of $SrAl_2O_4$: Eu^{2+},Na^+ at room temperature.

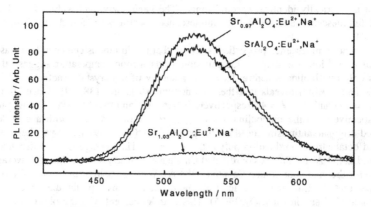

Figure 4. Effect of stoichiometry on the persistent luminescence of $SrAl_2O_4$: Eu^{2+},Na^+ at room temperature.

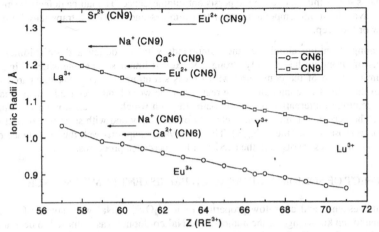

Figure 5. Comparison of the ionic radii of the cations in $MAl_2O_4:Eu^{2+},R^{3+},Na^+$ (M = Ca, Sr) [40].

The situation is different for the $SrAl_2O_4$ host: although the crystal structures of both $CaAl_2O_4$ and $SrAl_2O_4$ belong to the same stuffed tridymite type, there are only two nine-coordinated Sr-sites with distorted tricapped trigonal antiprism coordination in the $SrAl_2O_4$ structure. The average Sr – O distances for these two sites are close to each other (2.876 and 2.838 Å) and thus no preferential site occupancy for the Eu^{2+} ions should be caused by the crystal structure. At low temperature, two band emission from $SrAl_2O_4:Eu^{2+}$ has been observed, indeed, but at room temperature there is only one band due to energy transfer between the two sites [41].

It is well known that some rare earth R^{3+} ions as Dy^{3+} and Nd^{3+} prolong and enhance considerably the persistent luminescence of $MAl_2O_4:Eu^{2+}$ [24]. Even if this effect will be treated in detail later, the sites occupied by the RE^{3+} ions in the MAl_2O_4 lattices are of interest. Comparison of the ionic radii of the M^{2+} and R^{3+} ions (Fig.5) shows that the RE^{3+} ions fit rather well into the nine-coordinated sites in both host lattices. The smaller R^{3+} ions can, however, occupy the six-coordinated Ca-sites in $CaAl_2O_4$, too. These structural considerations show that already the starting point for the studies of the persistent luminescence in $MAl_2O_4:Eu^{2+}$ is rather complicated.

3.3. EFFECT OF STOICHIOMETRY ON PERSISTENT LUMINESCENCE

An excess of barium has been shown to quench the persistent luminescence of $BaAl_2O_4:Eu^{2+},Dy^{3+}$ [28] whereas the Eu^{2+} and Dy^{3+} doped strontium aluminates with an Al/Sr ratio 2 have been shown to have the strongest persistent luminescence when the ratio was varied from 1 to 12 [42]. The variation of the composition in such a range may, however, lead easily to the formation of impurity phases. In this work, Eu^{2+} doped alkaline earth aluminates with deficit, stoichiometric or excess amounts of alkaline earth were co-doped with Na^+ ions ($M_xAl_2O_4:Eu^{2+}$, Na^+;M = Ca or Sr, x = 0.97, 1.00 or 1.03).

Among calcium aluminates, the stoichiometric compound had the strongest UV-excited luminescence as well as persistent luminescence intensity whereas the non-stoichiometric compounds had lower intensities. The non-stoichiometry in the calcium aluminates seems thus to quench the persistent luminescence. The lattice defects formed do not favor the room temperature persistent luminescence since the traps may be too shallow or too deep.

Among strontium aluminates the compounds with the deficit and stoichiometric amounts of strontium had equally strong UV-excited luminescence intensities (Fig.3). The luminescence of the compound with excess strontium is very weak compared to those of the other two compounds. The reason for the weak luminescence intensity may be the creation of alternative relaxation paths through which the excitation energy may be lost. The persistent luminescence intensity of the compound with strontium deficit is superior to those of the others (Fig.4). This observation supports well the importance of the cation vacancies involved in the persistent luminescence processes.

3.4. EFFECT OF SODIUM CO-DOPING ON PERSISTENT LUMINESCENCE

The luminescence and afterglow properties of $MAl_2O_4:Eu^{2+},Na^+$ were studied for the first time to the knowledge of the authors. The Na^+ co-doping was supposed to decrease the number of cation vacancies. The persistent luminescence intensities of calcium aluminates with and without Na^+ co-doping are equally strong. All calcium aluminates have the typical persistent luminescence time dependence where two or even three processes may be involved. The stoichiometric compound had the longest and the non-stoichiometric compounds had equal to each other but somewhat shorter persistent luminescence decay time. The results suggest that the sodium ion is not entering into the cation vacancies which idea is supported by the ionic radii of Na^+ being larger than that of Ca^{2+} (Fig.5).

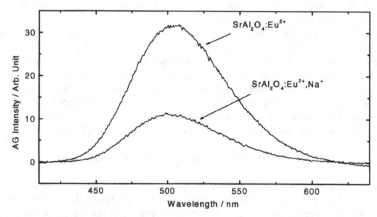

Figure 6. Effect of sodium co-doping on the persistent luminescence of $SrAl_2O_4$: Eu^{2+} at room temperature.

In the strontium aluminate host, the Na^+ co-doping quenched efficiently the persistent luminescence intensity (Fig.6). Only the compound with deficit strontium had the typical persistent luminescence decay whereas the persistent luminescence of the other two compounds is quenched logarithmically (Fig.7). In the compound with excess strontium the lack of cation vacancies may be dominantly responsible for the quenching of persistent luminescence. Also the Na^+ co-doping and possible alternative relaxation paths of excitation energy may quench the afterglow. In the compound with strontium deficit the quenching effect of Na^+ is not so significant but this may be due to the high amount of crystal defects which enhance the persistent luminescence.

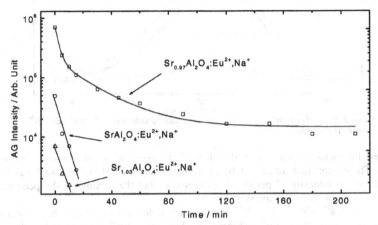

Figure 7. Effect of stoichiometry on the persistent luminescence lifetimes of $SrAl_2O_4$: Eu^{2+},Na^+ at room temperature.

3.5. EFFECT OF RARE EARTH CO-DOPING ON PERSISTENT LUMINESCENCE

The afterglow of the Eu^{2+} ion has been shown to be enhanced by co-activation with only some trivalent rare earth ions, *e.g.* Dy^{3+} and Nd^{3+} [30-32]. The Eu^{2+} ion acts as a luminescent centre with luminescence at the blue and green regions for $CaAl_2O_4$:Eu^{2+} and $SrAl_2O_4$:Eu^{2+}, respectively, whereas the R^{3+} ions are assumed to act as traps in these phosphors [24]. The intensity of the persistent luminescence has been found to depend strongly on the depth of these traps. The previous studies have, however, been concentrated only on the best persistent luminescence materials, *i.e.* $CaAl_2O_4$:Eu^{2+},Nd^{3+} and $SrAl_2O_4$:Eu^{2+},Dy^{3+}. A systematic and exhaustive study of the phosphors doped only with Eu^{2+} and, on the other hand, co-doped with the whole R^{3+} series is missing, even if this kind of study is expected to reveal new facts about the persistent luminescence mechanism.

The R^{3+} ion co-doping was not found to change the persistent luminescence band position nor the shape of the spectrum (Fig.2 and 8). The emission can thus be concluded to originate from the same Eu^{2+} centre as for the non-co-doped $CaAl_2O_4$:Eu^{2+}. The luminescence from the R^{3+} or Eu^{3+} ions has not been observed which indicates that neither direct excitation of R^{3+} nor energy transfer from Eu^{2+} to R^{3+} occurs The persistent luminescence intensity was found to depend strongly on the co-doping RE^{3+} ion: the Ce^{3+}, Pr^{3+}, Nd^{3+}, Tb^{3+}, Dy^{3+}, Ho^{3+} and Tm^{3+} ions enhance – more or less - the persistent luminescence of $CaAl_2O_4$:Eu^{2+} (Fig.8). The most intense

44

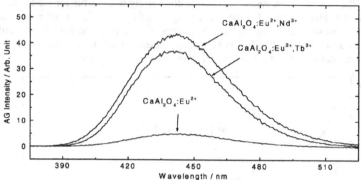

Figure 8. Effect of rare earth co-doping on the persistent luminescence of $CaAl_2O_4$: Eu^{2+}, R^{3+} at room temperature.

persistent luminescence was obtained with Nd^{3+} co-doping in agreement with previous studies. The enhancement of the persistent luminescence of $CaAl_2O_4$:Eu^{2+} was extended not only to the intensity of persistent luminescence but the lifetime of the persistent luminescence was increased, too (Fig. 9).

The Er^{3+} La^{3+}, Gd^{3+}, Lu^{3+}, Y^{3+} ions quenched to varying extent the persistent luminescence of $CaAl_2O_4$:Eu^{2+} whereas the Sm^{3+} and Yb^{3+} ions caused a drastic decrease in the intensity of the persistent luminescence of $CaAl_2O_4$:Eu^{2+}. The last result was interpreted in such a way that the Sm^{3+} and Yb^{3+} ions can be reduced to the divalent state and are thus able to fill the cation vacancies involved in the persistent luminescence processes. As a proof of the presence of divalent Sm^{2+} the Nd:YAG laser excitation produced the typical luminescence spectrum of the $4f^6$ configuration (Figure 10) with the easily recognizable $^5D_0 \rightarrow {}^7F_{0-2}$ transitions [43-45].

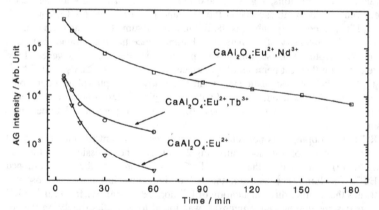

Figure 9. Effect of rare earth co-doping on the persistent luminescence lifetimes of $CaAl_2O_4$: Eu^{2+},R^{3+} at room temperature.

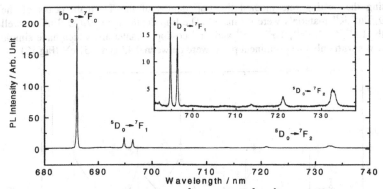

Figure 10. Photoluminescence of Sm^{2+} in $CaAl_2O_4:Eu^{2+},Sm^{3+}$ at 77 K (λ_{ex} = 532 nm).

The typical thermoluminescence glow curve of $CaAl_2O_4:Eu^{2+},(R^{3+})$ has one prominent, broad peak at *ca.* 80 °C followed by a tail up to 200 °C (Fig.11). The materials co-doped with Ce^{3+}, Pr^{3+}, Tb^{3+} and Ho^{3+} ions had the same TL main peak, but with increased intensity. The Nd^{3+}, Dy^{3+} and Tm^{3+} ions seemed to enhance strongly the high temperature tail of $CaAl_2O_4:Eu^{2+}$. The glow curve of $CaAl_2O_4:Eu^{2+},Nd^{3+}$ was very wide between 40 and 200 °C with a maximum at *ca.* 100 °C probably consisting of several maxima close to each other. The La^{3+}, Gd^{3+}, Er^{3+} and Lu^{3+} ions suppressed strongly the TL of $CaAl_2O_4:Eu^{2+}$ whereas the Sm^{3+}, Yb^{3+} and Y^{3+} ions created also new high temperature peaks. As a conclusion, it is quite clear that the easily reducible rare earths (Yb^{3+} and Sm^{3+}) suppressed TL intensity whereas the easily oxidable rare earths (Ce^{3+}, Tb^{3+} and Pr^{3+}) enhanced the main TL peak intensity. The "neutral" R^{3+} ions had clearly weaker effect on the TL peak intensity. The TL results support the observations made of the intensity of the persistent luminescence.

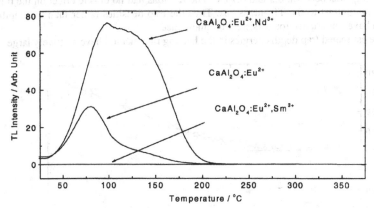

Figure 11. Effect of rare earth co-doping on the thermo-luminescence glow curves of $CaAl_2O_4$: Eu^{2+},R^{3+}.

Using the heating rate method the trap depths of the TL main peaks of the $CaAl_2O_4:Eu^{2+},R^{3+}$ materials were estimated. The trap depths for $CaAl_2O_4:Eu^{2+}$ as well as for the La^{3+}, Ce^{3+}, Pr^{3+}, Tb^{3+}, Ho^{3+} and Er^{3+} co-doped materials, which have simple glow curves with only one prominent peak, were between 0.42 and 0.51 eV (Fig.12).

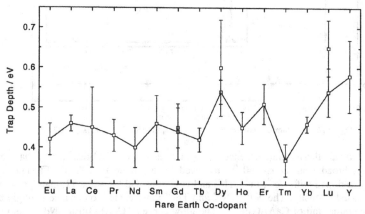

Figure 12. Trap depths calculated by the heating rate method for $CaAl_2O_4:Eu^{2+},R^{3+}$.

For the glow curves with two overlapping peaks (Gd^{3+}, Dy^{3+} and Lu^{3+} co-doped aluminates), which are, however, visually separable, the first trap energy was estimated between 0.44 and 0.54 eV and the second one between 0.45 and 0.65 eV. For the Nd^{3+} and Tm^{3+} co-doped materials with one very broad TL peak the trap energies corresponding to the main TL peak were estimated as *ca.* 0.4 eV. For the Sm^{3+}, Yb^{3+} and Y^{3+} co-doped materials, which had new high temperature peaks in the glow curve, the first trap was between 0.5 and 0.7 eV. The R^{3+} ions had no drastic effect on the trap depths, in contrast to the TL intensities. Thus the trap densities rather than the depths seem to have importance for persistent luminescence.

The estimated trap depth energies in the heating rate method have relatively large

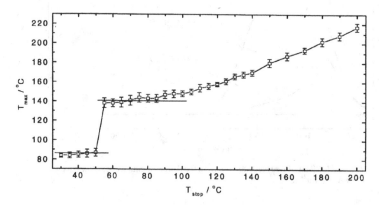

Figure 13. Preheating curve of $CaAl_2O_4:Eu^{2+},Dy^{3+}$.

errors (between 0.02 and 0.1 eV) and thus the preheating method was used to estimate preheating technique various trapping levels for the $CaAl_2O_4:Eu^{2+},R^{3+}$ materials were observed, indeed. As an example, $CaAl_2O_4:Eu^{2+},Dy^{3+}$ had at least one trap at *ca.* 80 °C. preheating technique various trapping levels for the $CaAl_2O_4:Eu^{2+},R^{3+}$ materials were observed, indeed. As an example, $CaAl_2O_4:Eu^{2+},Dy^{3+}$ had at least one trap at *ca.* 80 °C. (Fig.13), but it disappeared or was superimposed by stronger high temperature peaks already when using T_{stop} temperatures above 60 °C, followed by almost linear increase above 120 °C suggesting continuously distributed or closely overlapping trapping levels

3.6. MECHANISM OF PERSISTENT LUMINESCENCE

Although several studies of the persistent luminescence phenomenon have already been published, the mechanisms obtained from the $MAl_2O_4:Eu^{2+},R^{3+}$ materials have not been elucidated. Theories presented up to date are deficient or even contradictory. One reason for the uncertainty can be that the original afterglow of $MAl_2O_4:Eu^{2+}$ has almost totally been forgotten although the properties of this material are very similar to the more efficient $MAl_2O_4:Eu^{2+},R^{3+}$ ones. The rather schematic mechanisms presented involve either a direct or valence band assisted transfer of a hole from Eu^{2+} to a trap and its eventual recombination with the emitting Eu^{2+} center after thermal excitation. However, these mechanisms involve the creation of Eu^+ (and R^{4+}) ions which are highly dubious processes in solid state.

In this work, an alternative mechanism for the persistent luminescence in $MAl_2O_4:Eu^{2+}$ was searched for and is presented (Fig.14). The excitation energy to the traps is provided either directly or *via* migration in the conduction band. The electron traps are then bleached thermally at room temperature and feed electrons to the electron-hole recombination process which causes the excitation of the Eu^{2+} luminescence center by non-radiative energy transfer. The emission results from the normal de-excitation of the Eu^{2+} luminescence center.

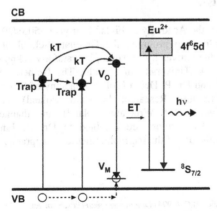

Figure 14. Proposed mechanism of the persistent luminescence of $CaAl_2O_4:Eu^{2+}$.

The whole process may require close contact between different defect centers (oxygen and calcium vacancies as well as the emitting Eu^{2+} ion) since the conduction band may be at too high energy to be used for electron migration. The mechanism proposed here needs no presence of monovalent Eu^+ ions. Moreover, the presence of cation (as well as oxygen) vacancies in calcium aluminates has been known already since the early 1970s [6]. The feeding of the Eu^{2+} persistent luminescence center is similar to the storage phenomena in wide band gap semiconducting materials, *e.g.* Si doped by Er^{3+} [46, 47] and in photosimulated phosphors [35]. However, further studies utilizing different experimental spectroscopic and other techniques are needed to prove and reveal the details of the proposed mechanism.

4. Conclusions

The systematic study of $CaAl_2O_4:Eu^{2+}$ and $SrAl_2O_4:Eu^{2+}$ with R^{3+} co-dopants other than Nd^{3+} and Dy^{3+} revealed important information about the mechanism of persistent luminescence so far ignored in the literature. It was concluded that neither reduction of Eu^{2+} to Eu^+ nor oxidation of Nd^{3+}/Dy^{3+} to Nd^{4+}/Dy^{4+} occurs in the persistent luminescence processes. However, due to reducing preparation conditions, the reduction of Sm^{3+} (and Yb^{3+}) produces Sm^{2+} (and Yb^{2+}) ions which remove the cation vacancies which act as hole traps. The R^{3+} co-doping in M^{2+} sites creates more lattice defects which enhance both the intensity and lengthen the life times of persistent luminescence. It is concluded that the role of lattice defects in $MAl_2O_4:Eu^{2+}$ as traps is very important for the persistent luminescence. The $4f^N$ energy level scheme of the R^{3+} ion may also be of importance but the details of this effect are not yet clear. The energy transfer or/and cross-relaxation which feeds the emitting Eu^{2+} centre underline the complicated nature of the final mechanism and emphasize the need to further studies.

5. Acknowledgements

Financial support from the Academy of Finland (project #5066/2000), European Union and Marie Curie Fellowship program, the Graduate School of Materials Research (Turku, Finland) and the University of Turku, is gratefully acknowledged. Discussions with Prof. A. Meijerink (University of Utrecht, The Netherlands) were found very useful. Prof. W Stręk and Dr. P. Dereń (Polish Academy of Sciences, Institute of Low Temperature and Structure Research, Wrocław Poland) as well as Prof. J. Legendziewicz (University of Wrocław, Poland) are thanked for the use of luminescence setups. The authors are indebted to Dr. H. Jungner (University of Helsinki, Finland) for the use of the thermoluminescence appratus.

6. References

1. Blasse, G. and Grabmaier, B.C. (1994) *Luminescent materials*, Springer, Berlin, 65 - 66.
2. Murayama, Y. (1999) Luminous paints, in Shionoya, S. and Yen, W.M. (eds.), *Phosphor handbook*, CRC Press, Boca Raton, FL, USA, 651 - 658.
3. Murayama, Y., Takeuchi, N. Aoki, Y., and Matsuzawa, T. (1995) Phosphorescent phosphor, *U.S. Patent* 5,424,006.

4. Palilla, F.C., Levine, A.K., and Tomkus, M.R. (1968) Fluorescent properties of alkaline earth aluminates of type MAl_2O_4 activated by divalent europium, *J. Electrochem. Soc.* **115**, 642 - 644.

5. Blasse G. and Bril, A. (1968) Fluorescence of Eu^{2+}-activated alkaline-earth aluminates, *Philips Res. Repts* **23**, 201 - 206.

6. Abbruscato, V. (1971) Optical and electrical properties of $SrAl_2O_4$:Eu^{2+}, *J. Electrochem. Soc.* **118**, 930 - 933.

7. Kodama, N., Sasaki, N., Yamaga, M., and Masui, Y. (2001) Long-lasting phosphorescence of Eu^{2+} melilite, *J. Lumin.* **94-95**, 19 - 22.

8. Yamaga, M., Tanii, Y., Kodama, N., Takahashi, T., and Honda, M. (2002) Mechanism of long-lasting phosphorescence of Ce^{3+}-doped $CaAl_2SiO_7$ melilite crystals, *Phys. Rev.* **B65**, 235108-1 – 235108-11.

9. Kodama, N., Tanii, Y., and Yamaga, M. (2000) Optical properties of long-lasting phosphorescent crystals Ce^{3+}-doped $Ca_2Al_2SiO_7$ and $CaYAl_3O_7$, *J. Lumin.* **87-89**, 1076 - 1078.

10. Kodama, N., Takahashi, T., Yamaga, M., Tanii, Y., Qiu, J., and Hirao, K. (1999) Long-lasting phosphorescence in Ce^{3+}-doped $Ca_2Al_2SiO_7$ and $CaYAl_3O_7$, *Appl. Phys. Lett.* **75**, 1715 - 1717.

11. Lin, Y., Zhang, Z., Tang, Z., Wang, X., Zhang, J., and Zheng, Z. (2001) Luminescent properties of a new long afterglow Eu^{2+} and Dy^{3+} activated $Ca_3MgSi_2O_8$, *J. Eur. Ceram. Soc.* **21**, 683 - 685.

12. Uheda, K., Maruyama, T., Takizawa, H., and Endo, T. (1997) Synthesis and long-period phosphorescence of $ZnGa_2O_4$:Mn^{2+} spinel, *J. Alloys Comp.* **262-263**, 60 - 64.

13. Qiu, J., Gaeta, A.L., and Hirao, K. (2001) Long-lasting phosphorescence in oxygen-deficient Ge-doped silica glasses at room temperature, *Chem. Phys. Lett.* **333**, 236 - 241.

14. Kinoshita, T. and Hosono, H. (2000) Materials design and example of long lasting phosphorescent glasses utilized electron trapped centers, *J. Non-Cryst. Solids* **274**, 257 - 263.

15. Qiu, J., Miura, K., Inouye, H., Fujiwara, S., Mitsuyu, T., and Hirao K. (1999) Blue emission induced in Eu^{2+}-doped glasses by an infrared femtosecond laser, *J. Non-Cryst. Solids* **244**, 185 - 188.

16. Kinoshita, T., Yamazaki, M., Kawazoe, H., and Hosono, H. (1999) Long lasting phosphorescence and photostimulated luminescence in Tb-ion-activated reduced calcium aluminate glasses, *J. Appl. Phys.* **86**, 3729 - 3733.

17. Hosono, H., Kinoshita, T., Kawazoe, H., Yamazaki, M., Yamamoto, Y., and Sawanobori, N. (1998) Long lasting phosphorescence properties of Tb^{3+}-activated reduced calcium aluminate glasses, *J. Phys. Condens. Matter* **10**, 9541 - 9547.

18. Qiu, J., Miura, K., Inouye, Y., Kondo, Y., Mitsuyu, T., and Hirao, K. (1998) Femtosecond laser-induced three-dimensional bright and long-lasting phosphorescence inside calcium aluminosilicate glasses doped with rare earth ions, *Appl. Phys. Lett.* **73**, 1763 - 1765.

19. Qiu, J., Kawasaki, M., Tanaka, K., Shimizugawa, Y., and Hirao, K. (1998) Phenomenon and mechanism of long-lasting phosphorescence in Eu^{2+}-doped aluminosilicate glasses, *J. Phys. Chem. Solids* **59**, 1521 - 1525.

20. Qiu, J., Kondo, Y., Miura, K., Mitsuyu, T., and Hirao, K. (1999) Infrared femtosecond laser induced visible long-lasting phosphorescence in Mn^{2+}-doped sodium borate glasses, *Jpn. J. Appl. Phys.* **38**, L649 – L651.

21. Qiu, J. and Hirao, K. (1998) Long lasting phosphorescence in Eu^{2+}-Doped calcium aluminoborate glasses, *Solid State Commun.* **106**, 795 - 798.

22. Yamazaki, M., Yamamoto, Y., Nagahama, S., Sawanobori, N., Mizuguchi, M., and Hosono, H. (1998) Long luminescent glass: Tb^{3+}-activated ZnO-B_2O-SiO_2 glass, *J. Non-Cryst. Solids* **241**, 71 - 73.

23. Lian, S.X., Ren, M., Lin, J.H., Gu, Z.N., and Su, M.Z. (2000) On the afterglow of the cerium doped silicate glasses, *J. Mater. Sci. Lett.* **19**, 1603 - 1605.

24. Matsuzawa, T., Aoki, Y., Takeuchi, N., and Murayama, Y. (1996) A new long phosphorescent phosphor with high brightness, $SrAl_2O_4$:Eu^{2+},Dy^{3+}, *J. Electrochem. Soc.* **143**, 2670 - 2673.

25. Katsumata, T., Nabae, T., Sasajima, K., Komuro, S., and Morikawa, T. (1997) Effects of composition on the long phosphorescent $SrAl_2O_4$:Eu^{2+},Dy^{3+} phosphor crystals, *J. Electrochem. Soc.* **144**, L243 – L245.

26. Katsumata, T., Nabae, T., Sasajima, T., and Matsuzawa, T. (1998) Growth and characteristics of long persistent $SrAl_2O_4$- and $CaAl_2O_4$-based phosphor crystals by floating zone technique, *J. Crystal Growth* **183**, 361 - 363.

27. Katsumata, T., Sasajima, K., Nabae, T., Komuro, S., and Morikawa T. (1998) Characteristics of strontium aluminate crystal used for long-duration phosphors, *J. Am. Ceram. Soc.* **81**, 413 - 416.

28. Sakai, R., Katsumata, T., Komuro, S., and Morikawa, T. (1999) Effect of composition on the phosphorescence from $BaAl_2O_4$:Eu^{2+},Dy^{3+} crystals, *J. Lumin.* **85**, 149 - 154.

29. Tsutai, I., Kamimura, T., Kato, K., Kaneko, F., Shinbo, K., Ohta, M., and Kawakami, T. (2000) Preparation of sputtered $SrAl_2O_4$:Eu films and their thermoluminescence properties, *Electr. Engin. Japan* **132**, 7 -14.

30. Nakazawa, E. and Mochida, T. (1997) Traps in $SrAl_2O_4:Eu^{2+}$ phosphors with rare-earth ion doping, *J. Lumin.* **72-74**, 236 - 237.

31. Yamamoto, H. and Matsuzawa, T. (1997) Mechanism of long phosphorescence of $SrAl_2O_4:Eu^{2+},Dy^{3+}$ and $CaAl_2O_4:Eu^{2+},Nd^{3+}$, *J. Lumin.* **72-74**, 287 - 289.

32. Jia, W., Yuan, H., Lu, L., Liu, H., and Yen, W.M. (1998) Phosphorescent dynamics in $SrAl_2O_4:Eu^{2+},Dy^{3+}$ single crystal fibers, *J. Lumin.* **76&77**, 424 - 428.

33. Jia, W., Yuan, H., Holmstrom, S., Liu, H., and Yen, W.M. (1999) Photo-stimulated luminescence in $SrAl_2O_4:Eu^{2+},Dy^{3+}$ single crystal fibers, *J. Lumin.* **83-84**, 465 - 469.

34. Kato, K., Tsutai, I., Kamimura, T., Kaneko, F., Shinbo, K., Ohta, M., and Kawakami, T. (1999) Thermoluminescence properties of $SrAl_2O_4:Eu$ sputtered films with long phosphorescence, *J. Lumin.* **82**, 213 - 220.

35. von Seggern, H. (1999) Photostimulable X-ray storage phosphors: a review of present understanding, *Braz. J. Phys.* **29**, 254 - 268.

36. Krupa, J.-C., Queffelec, M., Kirikova, N.Y., and Makhov, V.N. (1999) Rare earths in the luminescence of inorganic hosts excited in the VUV and XUV range, *Mater. Sci. Forum* **315-317**, 27 - 33.

37. Poort, S.H.M., Meijerink, A., and Blasse, G. (1997) Lifetime measurements in Eu^{2+}-doped host lattices, *J. Phys. Chem. Solids* **58**, 1451 - 1456.

38. Hörkner, W. and Müller-Buschbaum, H.K. (1976) Zür kristallstruktur von $CaAl_2O_4$, *J. Inorg. Nucl. Chem.* **38**, 983 - 984.

39. Schulze, A.-R. and Müller-Buschbaum, H.K. (1981) Zür struktur von monoklinem $SrAl_2O_4$, *Z. Anorg. Allg. Chem.* **475**, 205 - 210.

40. Shannon, R.D. (1976) Revised effective ionic-radii and systematic studies of interatomic distances in halides and chalcogenides, *Acta Cryst. A* **32**, 751 - 767.

41. Poort, S.H.M., Blockpoel, W.P., and Blasse, G. (1995) Luminescence of Eu^{2+} in barium and strontium aluminate and gallate, *Chem. Mater.* **7**, 1547 - 1551.

42. Chen, I.-C. and Chen, T.-M. (2001) Effect of host composition on the afterglow properties of phosphorescent strontium aluminate phosphors derived from the sol-gel method, *J. Mater. Res.* **16**, 1293 - 1300.

43. Zeng, Q., Pei, Z., Wang, S., and Su, Q. (1999) Luminescent properties of divalent samarium-doped strontium hexaborate, *Chem. Mater.* **11**, 605 - 611.

44. Meijerink, A. and Dirksen, G.J. (1995) Spectroscopy of divalent samarium in $LiBaF_3$, *J. Lumin.* **63**, 189 - 201.

45. Ellens, A., Zwaschka, F., Kummer, F., Meijerink, A., Raukas, M., and Mishra, K. (2001) Sm^{2+} in BAM: fluorescent probe for the number of luminescent sites of Eu^{2+} in BAM, *J. Lumin.* **93**, 147 - 153.

46. Gregorkiewicz, T., Thao, D.T.X., and Langer, J.M. (1999) Direct spectral probing of energy storage in Si:Er by a free-electron laser, *Appl. Phys. Lett.* **75**, 4121 - 4123.

47. Gregorkiewicz, T., Thao, D.T.X., Langer, J.M., Bekman, H.H.P.Th., Bresler, M.S., Michel, J., and Kimerling, L.C. (2000) Energy transfer between shallow centers and rare-earth ion cores: Er^{3+} ions in silicon, *Phys. Rev. B* **61**, 5369 - 5375.

RELAXATION OF MID- IR TRANSITIONS OF ND³⁺ IN LASER CRYSTALS WITH "SHORT" PHONON SPECTRA

T.T. BASIEV[a], YU.V. ORLOVSKII[a], I.N. VOROB'EV[a], L.N. DMITRUK[a], T.D. EFIMENKO[b], V.A. KONYUSHKIN[a], V.V. OSIKO[a]

[a]Laser Materials and Technology Research Center of General Physics Institute RAS, 38 Vavilov st., bldg. D, Moscow, 119991, GSP-1, Russia
[b]High Technologies Laboratory of Kuban State University, Stavropolskaya st., 149, Krasnodar, 350040, Russia

Abstract. The kinetics of fluorescence decay of the $^4G_{7/2}$ manifold of Nd^{3+} is directly measured in the row of fluorite type crystals like PbF_2, BaF_2, SrF_2, and CdF_2 as well in the $CaGa_2S_4$ and $PbCl_2$ crystals with "short" phonon spectra. Based on the measured kinetics the radiative and multiphonon relaxation rates of the $^4G_{7/2}$ manifold were determined. The lead fluoride crystals demonstrate the smallest nonradiative losses for the $^4G_{7/2} \rightarrow$ $^4G_{5/2}$; $^2G_{7/2}$ mid- IR transition of Nd^{3+} among the tested fluoride crystals. But much smaller nonradiative losses are found for $CaGa_2S_4$: Nd^{3+} and practically no losses are found for the $PbCl_2$: Nd^{3+} crystal. The increased rates of radiative decay in comparison with fluoride and oxide Nd^{3+} doped crystals were found in the sulfide and the chloride crystal. Based on these results the decay times of the $^4I_{15/2}$; $^4I_{13/2}$; and $^4I_{11/2}$ manifolds perspective as initial laser levels for mid- IR generation were estimated in all studied crystals.

1. Introduction

Information about multiphonon relaxation rate is of critical importance for searching of laser transitions in the 4- 6 μm spectral range where it may have a dominant contribution to the relaxation rate of optical excitation. For the Nd^{3+} ion three different optical transitions $^4I_{15/2}$ - $^4I_{13/2}$ (λ_{max} ~ 5.5 μm); $^4I_{13/2}$ - $^4I_{11/2}$ (λ_{max}~5 μm); $^4I_{11/2}$ - $^4I_{9/2}$ (λ_{max} ~5 μm) (Figure 1) reside in the desired spectral range. The lack of pulsed nanosecond laser excitation sources at 1.68 μm as well as photodetectors with nanosecond resolution in the 1.68 – 5 μm spectral range requires searching of alternative ways for relaxation rates prediction and estimation of above mentioned manifolds. One of possibilities is to measure and analyze the relaxation rate of a manifold having the energy in the visible spectral range and the energy gap ΔE to the next lower lying manifold equal to those for above mentioned mid IR transitions. The best candidate is the $^4G_{7/2}$ manifold (Figure 1) having for example for the pair M- center

J.-C. Krupa and N.A. Kulagin (eds.), Physics of Laser Crystals, 51–61.
© 2003 *Kluwer Academic Publishers. Printed in the Netherlands.*

in the CaF_2 crystal the energy gap between its lowest crystal –field level and the crystal –field levels of the next lower lying $^4G_{5/2}$; $^2G_{7/2}$ manifold in the range from 1555 to 1858 cm^{-1} [1]. These energies are very close to the values for the desired mid IR transitions (Tabl.1). The small number of phonons $p=3$ and consequently the fast multiphonon relaxation rate can be expected for the CaF_2: Nd^{3+} crystals with the effective longitudinal optical phonon frequency $\hbar\omega_{eff} = 474$ cm^{-1} (Tabl.2) [2]. The maximal number of phonons $p=7$ and, hence, the minimal nonradiative internal losses can be expected for the BaF_2: Nd^{3+} crystal [3] with $\hbar\omega_{eff} = 324$ cm^{-1}.

2. Experimental technique and results

Nd^{3+} doped fluoride PbF_2, BaF_2, SrF_2, CdF_2, and CaF_2 crystals and lanthanum co-doped and mixed crystals like PbF_2: $LaF_3(0.75\%)$: $Nd^{3+}(0.5\%)$, BaF_2: LaF_3 (10%): $Nd^{3+}(0.5\%)$, SrF_2: $LaF_3(1\%)$: Nd^{3+} (0.2%), CaF_2: YF_3 (6%): Nd^{3+} (1%), and CaF_2: LaF_3 (0.25%): Nd^{3+} (0.25%) were grown by modified Bridgman- Stockbarger technique in fluorinating atmosphere. $PbCl_2$ crystals doped with Nd^{3+} were grown by modified Bridgman technique in quartz ampoules. $CaGa_2S_4$ crystals doped with Nd^{3+} were grown by Bridgman- Stockbarger technique in quartz ampoules. The fluorescence kinetics of the $^4G_{7/2}$ manifold of Nd^{3+} in the row of fluorite type crystals (Fig.2-3) was monitored at the $^4G_{7/2} \rightarrow ^4I_{13/2}$ transition at 77K using correlated photon counting technique and copper vapor laser excitation at 510 nm. For the CaF_2: Nd^{3+}, CdF_2: Nd^{3+}, SrF_2: Nd^{3+} and BaF_2: Nd^{3+} crystals strongly non-exponential kinetics decay because of multi- site structure of these crystals was found. The results of double exponential decay curve fit for nonexponential decay are presented in Tabl.2. At the initial nanosecond stage of kinetics decay for the time interval 6- 8 ns strong scattering laser light is observed. The fastest decay is observed for the CaF_2: Nd^{3+} (0.6 m.%) crystal ($\tau = 7/25$ ns) (Figure 2) with the lowest cation mass (M(Ca)= 40) among fluorite type crystals and with the rare-earth fluorine distance $a = 2.36$ A. Very similar fluorescence kinetics decay is observed for the CdF_2: Nd^{3+} (0.3 m.%) crystal ($\tau = 13/27$ ns) with the larger cation mass (M(Cd)= 112) than in CaF_2 but with the shortest distance between a rare- earth ion and the nearest

TABLE 1. $4.5 – 7$ µm transitions for the pair M- center of Nd^{3+} in the CaF_2: Nd^{3+} crystal and their reduced matrix elements $(U^{(k)})^2$ ($k = 2, 4, 6$).

Transition	ΔE, $cm^{-1}/$ µm	n, number of phonons	Reduced matrix elements $(U^{(k)})^2$		
			k=2	k=4	k=6
$^4I_{11/2} \rightarrow ^4I_{9/2}$	1524- 2224/ 6.5 – 4.5	3	0.0195	0.1073	1.1652
$^4I_{13/2} \rightarrow ^4I_{11/2}$	1739–2107/ 5.8 – 4.7	4	0.0256	0.1353	1.2379
$^4I_{15/2} \rightarrow ^4I_{13/2}$	1442-1837/ 6.9- 5.4	3	0.0196	0.1189	1.4511
$^4G_{7/2} \rightarrow ^4G_{5/2}$;	1555-1858/ 6.4- 5.4	3	0	0.2246	0.0503
$^2G_{7/2}$			0.0002	0.1493	0.0874

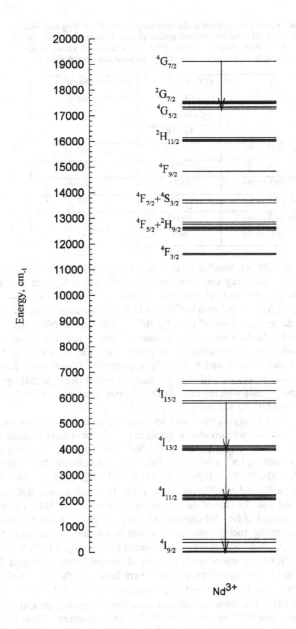

Figure 1. Energy level diagram of Nd³⁺ ions.

TABLE 2. The measured decay times at different temperatures of the $^4G_{7/2}$ manifold of the Nd^{3+} ion and effective longitudinal optical phonon frequencies in fluoride laser crystals with short phonon spectra. (The decay times of curve fit of nonexponential decay by double exponential decays are presented through slash.)

Crystal	$\tau(^4G_{7/2})$, ns ex. = 510 nm reg. = 660 nm T= 77K	Effective phonon frequency, cm^{-1} [2]
LaF$_3$:Nd^{3+} 0.75 m.%)	110 [5]	360
CaF$_2$: Nd^{3+}(0.6 m. %)	7/25	474
CdF$_2$:Nd^{3+}(0.3 m. %)	13/27	384
SrF$_2$:Nd^{3+}(0.5 m.%)	33/100	383
BaF$_2$:Nd^{3+}(1.5 m.%)	50/414	324
PbF$_2$:Nd^{3+}(0.75 m.%)	640	337

fluorine ion ($a(CdF_2)$ = 2.34 A). Smaller relaxation rate (τ = 6/33/100 ns) at the final stage of fluorescence kinetics decay was found in SrF$_2$: Nd^{3+} (0.5 m.%) with the cation mass (M(SrF$_2$)= 87) larger than in CaF$_2$ but smaller than in CdF$_2$. The cation – anion distance ($a(SrF_2)$ = 2.505 A) is larger than in CdF$_2$ and CaF$_2$. Even longer lifetime at the final stage of kinetics decay was observed in BaF$_2$: Nd^{3+} (1.5%) (τ = 8/50/414 ns).

This correlates with a highest rare- earth to nearest fluorine ion- ion distance (a= 2.68 A) and an increase of the cation mass in this crystal (M(Ba)= 137). For the PbF$_2$: NdF$_3$ (0.75%) crystal the slowest and practically exponential fluorescence kinetics decay with τ = 640 ns was measured. PbF$_2$ has the heaviest cation (M(Pb)= 207) among the studied fluorite type crystals with the rare- earth - fluorine distance (a = 2.57 A) lies in between for SrF$_2$ (a = 2.505 A) and BaF$_2$ (a = 2.68 A).

To prevent Nd- Nd ion aggregation and to make nonlinear quenching energy transfer and the cross- relaxation fluorescence decay in the cluster centers (like the pair M- and the quartet N- centers) negligible the CaF$_2$: Nd^{3+} and SrF$_2$: Nd^{3+} crystals were co- doped by optically inactive La^{3+} ions. Two - three times increase of the decay time was observed for CaF$_2$: LaF$_3$ (0.25%): NdF$_3$ (0.25%) (τ = 7.7/44/234 ns) and SrF$_2$: LaF$_3$ (1 %): NdF$_3$ (0.2 %) (τ = 99/265 ns) (Figure 3 and Table 3) in comparison with single doped crystals (Fig.2 and Tabl.2). One of the reasons is the formation of Nd-La clusters, so called M'- centers, instead of Nd- Nd clusters (M- centers). As the result the Nd- Nd cross- relaxational quenching energy transfer is absent inside the M'- centers. Also, slower decay than in CaF$_2$: Nd^{3+} (0.5 m.%) was measured in the CaF$_2$: YF$_3$ (6 %): NdF$_3$ (1 %) crystal (τ = 6/65/184 ns) where specific types of optical centers different from M'– centers are formed [4]. No significant changes were found for the BaF$_2$: LaF$_3$ (10 %): NdF$_3$ (0.5 %) crystal (τ = 13/73/359 ns) in comparison with the BaF$_2$: Nd^{3+} (1.5%) crystal (τ = 8/50/414 ns). This shows small concentrations of cluster optical centers (contains two, three, and four Nd^{3+} ions) in BaF$_2$: Nd^{3+}. Further increase of fluorescence decay time was found in the PbF$_2$: LaF$_3$ (0.75 %): NdF$_3$ (0.5%) crystal (τ = 779 ns) in

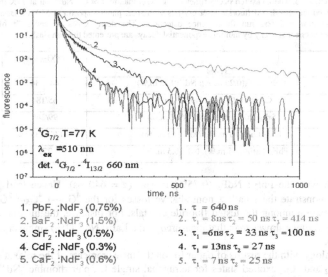

1. PbF$_2$:NdF$_3$ (0.75%)
2. BaF$_2$:NdF$_3$ (1.5%)
3. SrF$_2$:NdF$_3$ (0.5%)
4. CdF$_2$:NdF$_3$ (0.3%)
5. CaF$_2$:NdF$_3$ (0.6%)

1. τ = 640 ns
2. τ_1 = 8 ns τ_2 = 50 ns τ_3 = 414 ns
3. τ_1 = 6 ns τ_2 = 33 ns τ_3 = 100 ns
4. τ_1 = 13 ns τ_2 = 27 ns
5. τ_1 = 7 ns τ_2 = 25 ns

Figure 2. The kinetics of fluorescence decay of the $^4G_{7/2}$ manifold in the PbF$_2$: Nd^{3+}(0.75%) - 1, BaF$_2$: Nd^{3+}(1.5%) - 2, SrF$_2$: Nd^{3+}(0.5%) - 3, CdF$_2$: Nd^{3+}(0.3%)- 4, CaF$_2$: Nd^{3+}(0.6%)- 5 under direct laser excitation ($\lambda_{ex.}$ = 510 nm) and fluorescence detection at the $^4G_{7/2} \rightarrow {}^4I_{13/2}$ transition (λ_{det} = 660 nm) at 77K.

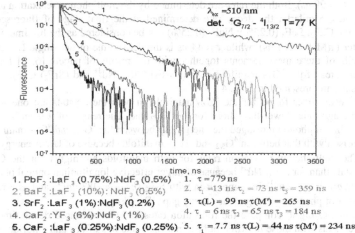

1. PbF$_2$:LaF$_3$ (0.75%):NdF$_3$ (0.5%)
2. BaF$_2$:LaF$_3$ (10%): NdF$_3$ (0.5%)
3. SrF$_2$:LaF$_3$ (1%):NdF$_3$ (0.2%)
4. CaF$_2$:YF$_3$ (6%):NdF$_3$ (1%)
5. CaF$_2$:LaF$_3$ (0.25%):NdF$_3$ (0.25%)

1. τ =779 ns
2. τ_1 =13 ns τ_2 = 73 ns τ_3 = 359 ns
3. $\tau(L)$ = 99 ns $\tau(M')$ = 265 ns
4. τ_1 = 6 ns τ_2 = 65 ns τ_3 = 184 ns
5. τ_1 = 7.7 ns $\tau(L)$ = 44 ns $\tau(M')$ = 234 ns

Figure 3. The kinetics of fluorescence decay of the $^4G_{7/2}$ manifold in the PbF$_2$: LaF$_3$(0.75%): Nd^{3+}(0.5%) - 1, BaF$_2$: LaF$_3$(10%): Nd^{3+}(0.5%) - 2, SrF$_2$:LaF$_3$(1%): Nd^{3+}(0.2%) - 3, CaF$_2$: YF$_3$(6%): Nd^{3+}(1%) - 4, CaF$_2$: LaF$_3$(0.25%): Nd^{3+}(0.25%) - 5 under direct laser excitation ($\lambda_{ex.}$ = 510 nm) and fluorescence detection at the $^4G_{7/2} \rightarrow {}^4I_{13/2}$ transition ($\lambda_{det.}$ = 660 nm) at 77K.

TABLE 3. The measured decay times of the $^4G_{7/2}$ manifold of the Nd^{3+} ion at 77K in mixed and co- doped fluoride laser crystals with short phonon spectra at 510 nm laser excitation and 660 nm fluorescence detection. (The decay times of curve fit of nonexponential decay by multiple exponential decays are presented through slash.)

Crystal	$\tau(^4G_{7/2})$, ns
CaF_2: LaF_3(0.25 m.%): Nd^{3+}(0.25 m.%)	44/ 234
CaF_2:YF_3(6 m.%):Nd^{3+} (1 m.%)	65/184
SrF_2:LaF_3(1 m.%):Nd^{3+} (0.2 m.%)	99/265
BaF_2:LaF_3(10 m. %):Nd^{3+} (0.5 m. %)	13/73/359
PbF_2: LaF_3(0.75 m.%): Nd^{3+}(0.5 m. %)	779

comparison with the PbF_2: NdF_3 (0.75%) crystal ($\tau = 640$ ns). Hence lead fluoride crystals demonstrate the smallest nonradiative losses for the $^4G_{7/2} \rightarrow {}^4G_{5/2}$; $^2G_{7/2}$ five-micron transition among the tested fluoride crystals. For instance, the measured decay times are 6- 8 times slower than that measured in the LaF_3: Nd^{3+} (1%) single crystal ($\tau = 110$ ns [5]).

By analogy with the results of paper [6] on the measurements of the relaxation rates of the $^4G_{5/2}$ and $^4F_{3/2}$ excited states for tetragonal single L-, pair rhombic (Nd^{3+}- Nd^{3+}) M- and pair (Nd^{3+}- La^{3+}) M'- optical centers in the Nd:SrF_2 and Nd:La:SrF_2 crystals one can consider the decay time determined at the final stage of fluorescence kinetics decay of the $^4G_{7/2}$ manifold in the SrF_2: LaF_3(1%): Nd^{3+}(0.2%) as the lifetime of the (Nd-La) M' - pair center (τ (M')= 265 ns). The decay time determined at the initial stage of fluorescence kinetics decay is considered as the lifetime of the single tetragonal L-center (τ(L)= 99 ns). Both the rates are determined by the multiphonon relaxation to the lowest states. Similarly, the decay rate determined at the final stage of fluorescence decay in the CaF_2: LaF_3 (0.25%): NdF_3 (0.25%) can be considered as the lifetime of the M'- center (τ(M') = 234 ns), while $\tau = 44$ ns is the decay time of the single L- center. The results of these measurements together with the results of papers [6, 7] for the relaxation rates of the $^4G_{5/2}$ manifold in the (SrF_2) CaF_2: Nd^{3+} and (SrF_2) CaF_2: Nd^{3+}: La^{3+} crystals are presented in Tabl. 4.

The decay times found for the $^4G_{7/2}$ manifold in both the crystals are one – two orders of magnitude slower than those for the $^4G_{5/2}$ one. The reason of it is an increase of the number of phonon p bridged the energy gap between the $^4G_{7/2}$ and $^4G_{5/2}$ manifolds in comparison with that between $^4G_{5/2}$ and $^2H_{11/2}$ manifolds because of larger energy gap ΔE_{min}. The multiphonon relaxation rates for both manifolds are higher for the CaF_2: Nd^{3+} crystals than for SrF_2: Nd^{3+} because of larger effective longitudinal optical phonon in CaF_2 ($\hbar\omega_{eff.} = 474$ cm^{-1}) in comparison with SrF_2 ($\hbar\omega_{eff.} = 383$ cm^{-1}) and as a result less number of phonons p bridged the energy gap $\Delta E = p\hbar\omega_{eff}$.

It is interesting to compare the relaxation rates and the measured decay times in fluoride crystals with those in different laser matrixes like in sulfides and chlorides with shorter phonon spectra and, therefore, larger number of phonon p bridged the same energy gap. The fluorescence kinetics decay of the $^4G_{7/2}$ manifold was measured in the $CaGa_2S_4$:Nd^{3+} (0.16 w.%) crystal at liquid nitrogen and room temperatures (Fig.4) using 2nd harmonics of YAG:Nd pulsed laser excitation and fluorescence detection at the $^4G_{7/2} \rightarrow {}^4I_{13/2}$ transition. The measured decay times are found almost equal for both the

TABLE 4. Multiphonon decay times of the $^4G_{5/2}$ and $^4G_{7/2}$ manifolds of Nd^{3+} in the CaF_2 and SrF_2 crystals

Crystal	$\hbar\omega_{eff},$ cm^{-1}	Type of center	$^4G_{5/2}$			$^4G_{7/2}$		
			ΔE, cm^{-1}	n	τ, ns	ΔE, cm^{-1}	n	τ, ns
CaF$_2$	474	L	948 [1]	2	<2[7]			44
		M'	1125 [1]	3	7.1[7]	1555 [1]	4	234
SrF$_2$	383	L	985 [1]	3	5.2[6]			99
		M'			32[6]			265

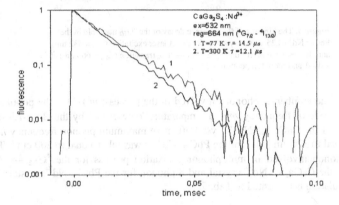

Figure 4. The kinetics of fluorescence decay of the $^4G_{7/2}$ manifold in the CaGa$_2$S$_4$: Nd^{3+} (0.16 w.%) crystal under direct laser excitation ($\lambda_{ex.}$ = 532 nm) and fluorescence detection at the $^4G_{7/2} \rightarrow {}^4I_{13/2}$ transition ($\lambda_{det.}$ = 664 nm) at 77K -1 and room temperature - 2.

temperatures (τ(77) = 14.5 µs and τ(300) = 12.1 µs). This is more than one order of magnitude slower than even for the PbF$_2$: LaF$_3$ (0.75 %): NdF$_3$ (0.5%) crystal with the longest lifetime among the fluoride crystals (τ = 779 ns).

For the PbCl$_2$:NdF$_3$ (0.25 w.%) crystal with shortest phonon spectrum among studied crystals additional 2 times increase of fluorescence life time is observed (Fig.5). The measured decay times reach 32.2 µs at 77K and 22.5 µs at room temperature.

In assumption that the radiative lifetime τ_R do not depend on the temperature in the measured temperature range and multiphonon relaxation rate is constant at cryogenic temperatures from liquid helium temperatures and up to 77K one can calculate the radiative relaxation rate and multiphonon relaxation rates W_{MR} for both the temperatures using the following equations:

58

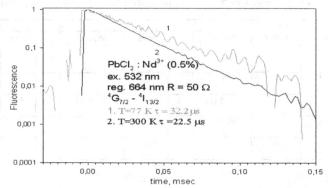

Figure 5. The kinetics of fluorescence decay of the $^4G_{7/2}$ manifold in the
PbCl$_2$: Nd^{3+}(0.25 w.%) crystal under direct laser excitation ($\lambda_{ex.}$ = 532 nm)
and fluorescence detection at the $^4G_{7/2} \rightarrow {}^4I_{13/2}$ transition ($\lambda_{det.}$ = 664 nm) at
77K -1 and room temperature - 2.

Here p is the number of phonons involved in the process; $n(\omega,T)$ is the population
of a phonon mode of frequency ω at a temperature T described by the Bose-Einstein
distribution. For the CaGa$_2$S$_4$: Nd^{3+} crystal effective maximum phonon frequency $h\omega_{eff}$
was taken equal to 320 cm^{-1} and for the PbCl$_2$: Nd^{3+} it was taken equal to 200 cm^{-1}. This
gives 5-6 phonon involved in multiphonon relaxation process for the $^4G_{7/2} \rightarrow {}^4G_{5/2}$
transition for the CaGa$_2$S$_4$: Nd^{3+} crystal and 8 phonon for the PbCl$_2$: Nd^{3+} crystal. The
result of calculation is presented in Tab. 5.

$$W_{meas.}(300K) = 1/\tau_R + W_{MR}(300) \qquad (1)$$

$$W_{meas.}(77K) = 1/\tau_R + W_{MR}(77) \qquad (2)$$

$$W_{MR}(300) = W_{MR}(77)(n(\omega,T)+1)^p \qquad (3)$$

$$n(\omega,T)=(\exp(h\omega_{eff}/kT)-1)^{-1} \qquad (4)$$

It is found that the contribution of multiphonon relaxation for the PbCl$_2$: Nd^{3+}
crystal at 77K is negligible. Rather fast radiative decay rates were found for both the
crystals - τ_R = 15.6 µs for CaGa$_2$S$_4$: Nd^{3+} and τ_R = 32.2 µs for PbCl$_2$: Nd^{3+}.

To check the high values of radiative relaxation probabilities in these crystals the
fluorescence kinetics decay of the metastable $^4F_{3/2}$ manifold was measured in the sulfide
and the chloride crystals under 532 nm laser excitation and fluorescence detection at the
$^4F_{3/2} \rightarrow {}^4I_{9/2}$ transition at room temperature (Fig. 6).

The measured fluorescence decay times for both the crystals are rather close (τ
(PbCl$_2$: Nd^{3+}) = 100 µs and τ(CaGa$_2$S$_4$: Nd^{3+}) = 83.5 µs). These measured decay times
are 3 - 10 times shorter than those for the fluoride and oxide crystals and indicate very

high rates of radiative decay for these chloride and sulfide crystals comparing to other well-known crystals.

TABLE 5. The measured decay times of the $^4G_{7/2}$ manifold of the Nd^{3+} ion at 77K and room temperature in the $CaGa_2S_4$: Nd^{3+} and $PbCl_2$: Nd^{3+} crystals with short phonon spectra at 532 nm laser excitation and 660 nm fluorescence detection as well as radiative and multiphonon relaxation rates determined.

Crystal	τ_{meas}, $^4G_{7/2}$		Radiative rate τ_R^{-1}, s^{-1}	Multiphonon relaxation rate W_{MR}, s^{-1}	
	77K	300K		77K	300K
$CaGa_2S_4$: Nd^{3+} (0.16 w.%)	14.5 µs	12.1 µs	$6.4\ 10^4$	$0.5\ 10^4$	$1.9\ 10^4$
$PbCl_2$: Nd^{3+}(0.25 w.%)	32.2 µs	22.5 µs	$3.1\ 10^4$	335	$1.4\ 10^4$

The obtained results allow estimate the decay times of the $^4I_{15/2}$, $^4I_{13/2}$, and $^4I_{11/2}$ manifolds of Nd^{3+} in fluoride crystals. Their values are mostly determined by multiphonon relaxation (MR) rates and are close to or even larger than the decay time of the $^4G_{7/2}$ manifold in each studied crystal. The values of multiphonon relaxation rates

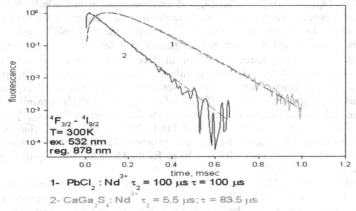

1- $PbCl_2$: Nd^{3+} $\tau_2 = 100$ µs $\tau = 100$ µs

2- $CaGa_2S_4$: Nd^{3+} $\tau_2 = 5.5$ µs; $\tau = 83.5$ µs

Figure 6. The kinetics of fluorescence decay of the $^4F_{3/2}$ manifold in the $PbCl_2$: Nd^{3+}(0.25 w.%) crystal -1 and $CaGa_2S_4$: Nd^{3+} (0.16 w.%) crystal - 2 under laser excitation into the $^4G_{7/2}$ manifold ($\lambda_{ex.} = 532$ nm) and fluorescence detection at the $^4F_{3/2} \rightarrow ^4I_{9/2}$ transition ($\lambda_{det.} = 878$ nm) at room temperature. The curve fit by the equation $I(t) = I_0 (1- \exp(-t/\tau_2)) \exp(-t/\tau)$ is given by dashed lines.

can be estimated using the nonlinear theory of multiphonon relaxation [8, 9]. According to this theory a MR rate of p phonon transition depends on the square of reduced matrix elements of corresponding electronic transition as

$$W(J \to J')(p) = \frac{1}{137}(2J+1)^{-1} \sum_{k=2,4,6} \Xi_k(p) \, (LSJ\|U^{(k)}\|L'S'J')^2 \, \eta^p \qquad (5)$$

The second rank reduced matrix elements $(U^{(2)})^2$ for the $^4G_{7/2} \to {}^4G_{5/2}$; $^2G_{7/2}$ transitions are practically zero. While for the other transitions ($^4I_{15/2} \to {}^4I_{13/2}$, $^4I_{13/2} \to {}^4I_{11/2}$, and $^4I_{11/2} \to {}^4I_{9/2}$) $(U^{(2)})^2)$ are of the order of 0.02. This is one order of magnitude smaller than the values of $(U^{(4)})^2)$ for all the transitions (see Table 1), but for the $^4G_{7/2} \to {}^4G_{5/2}$; $^2G_{7/2}$ transitions $(U^{(4)})^2)$ are larger than for the others. The matrix elements $(U^{(6)})^2$ are more than one order of magnitude larger for the desired mid IR transitions than for the $^4G_{7/2} \to {}^4G_{5/2}$; $^2G_{7/2}$ ones. But according to the result of Ref. [8] a contribution of the $\Xi_6(p)$ intensity parameter of multiphonon relaxation (Eq. (5)) to the relaxation rate for a number of phonon $p = 4$ and 5 is negligible and the MR rate is mostly determined by the $\Xi_2(p)$ and $\Xi_4(p)$ terms. Because $(U^{(4)})^2)$ is much larger than $(U^{(2)})^2)$ for all the transitions the contribution of the term with $k = 4$ is dominated. But this term is larger for the $^4G_{7/2} \to {}^4G_{5/2}$; $^2G_{7/2}$ transitions, and, also, the MR probabilities for two transitions from the $^4G_{7/2}$ manifold should be added. As the result the MR rate of the $^4G_{7/2}$ manifold will be larger than for the $^4I_{15/2}$, $^4I_{13/2}$, and $^4I_{11/2}$ ones.

Our estimation of radiative rates of initial laser levels of the $^4I_{15/2}$ - $^4I_{13/2}$ ($\lambda_{max} \sim 5.5$ μm); $^4I_{13/2}$ - $^4I_{11/2}$ ($\lambda_{max} \sim 5$ μm); $^4I_{11/2}$ - $^4I_{9/2}$ ($\lambda_{max} \sim 5$ μm) transitions in the CaGa$_2$S$_4$: Nd^{3+} and PbCl$_2$: Nd^{3+} crystals gives the values much smaller than the measured MR rates for the $^4G_{7/2}$ manifold. The only exception is the $^4I_{15/2}$ manifold in the PbCl$_2$: Nd^{3+} crystal at 77K where the radiative and MR rates are comparable. But for the later crystal the number of phonons $p = 8$ requires to account all three $\Xi_k(p)$ terms of Eq. (5) for estimation of MR rates of the interested manifolds. Our rough estimation without accounting of anharmonicity of crystal lattice vibrations in the PbCl$_2$: Nd^{3+} crystal shows that this may slightly increase the MR rates of the $^4I_{15/2}$, $^4I_{13/2}$, and $^4I_{11/2}$ manifolds in comparison with the $^4G_{7/2}$ manifold. As the result the decay times of the $^4I_{15/2}$, $^4I_{13/2}$, and $^4I_{11/2}$ manifolds in the CaGa$_2$S$_4$: Nd^{3+} and PbCl$_2$: Nd^{3+} crystals as for fluoride crystals will be mostly determined by the multiphonon relaxation rates. At room temperature this gives $\tau(^4I_J) = 1/W_{MR}(^4G_{7/2}) \sim 50$ μs for the CaGa$_2$S$_4$: Nd^{3+} crystal and ~ 70 μs for the PbCl$_2$: Nd^{3+} crystal.

For the fluoride and sulfide crystals the slowest decay times and minimal non-radiative losses are expected for the $^4I_{13/2}$ manifold having maximal energy gap to the next $^4I_{11/2}$ manifold below. Among the fluoride crystals the slowest decay time of this manifold and minimal non-radiative losses are expected for the PbF$_2$ based crystal matrixes. But much smaller non-radiative losses are expected for CaGa$_2$S$_4$: Nd^{3+} doped crystals and especially for the PbCl$_2$ Nd^{3+} doped crystals with shortest phonon spectrum.

3. Conclusion

Based on the measured kinetics of fluorescence decay of the $^4G_{7/2}$ manifold of Nd^{3+} in the row of fluorite type crystals like PbF$_2$, BaF$_2$, SrF$_2$, CdF$_2$, and CaF$_2$ as well as in the CaGa$_2$S$_4$ and PbCl$_2$ crystals with "short" phonon spectra the radiative and multiphonon relaxation rates of the $^4G_{7/2}$ manifold were found. The lead fluoride crystals demonstrate the smallest non-radiative losses for the $^4G_{7/2} \to {}^4G_{5/2}$; $^2G_{7/2}$ mid- IR transition of Nd^{3+} among the tested fluoride crystals. But much smaller non-radiative losses are found for

$CaGa_2S_4$: Nd^{3+} and practically no losses are found for the $PbCl_2$: Nd^{3+} crystal. The increased rates of radiative decay in comparison with fluoride and oxide Nd^{3+} doped crystals were found in the sulfide and the chloride crystal. Based on these results the decay times of the $^4I_{15/2}$; $^4I_{13/2}$; and $^4I_{11/2}$ manifolds perspective as initial laser levels for mid- IR generation were estimated in all studied crystals.

4. Acknowledgments

This work is partially supported by ISTC grant No 2022, RFBR grant No 00-02-17108, NSF grant ECS- 0140484, Collaborative Linkage NATO grant No PST.CLG.979125, grant of Russian Ministry of Industry, Science and Technology on Fundamental Spectroscopy No M-3-02, and the Program of the Government of Russian Federation for Integration of Science and High Education (project No IO821). We would like to thank Prof. A. Avanesov and Dr. V.Badikov for their kind help with crystal growth experiments and fruitful discussions.

5. References

1. Han, T.P.J., Jones, G.D., and Syme, R.W.G. (1993) Site- selective spectroscopy of Nd^{3+} centers in CaF_2: Nd^{3+} and SrF_2: Nd^{3+}, *Phys.Rev.*B **47**, 14706-14723
2. Denham, P., Field, G.R., Morse, P.L.B., and Wilkinson, G.R. (1970) Optical and dielectric properties and lattice dynamics of some fluorite structure ionic crystals, *Proc. Roy. Soc. Lond.* **A317**, 55-77
3. Orlovskii, Yu.V., Basiev, T.T., Pukhov, K.K., Vorob'ev, I.N., Papashvili, A.G., Pelle, F., Osiko, V.V. (2001) Multiphonon relaxation of mid IR transitions of rare- earth ions in the crystals with fluorite structure, *Journal of Lumin.*, **94/95**, 791-795
4. Basiev, T.T., Karasik, A.Ya., Shubochkin, R.L. (1995) Selective laser excitation and inhomogeneous band broadening of Nd^{3+} ions in disordered CaF_2- YF_3 crystals, *Journal of Lumin.*, **64**, 259-265
5. Basiev, T.T., Dergachev, A.Yu., Orlovskii, Yu.V., Prokhorov, A.M. (1994) Multiphonon nano- and subnanosecond relaxation from high- lying levels of Nd^{3+} ions in laser fluorides and oxides, *Proceedings of General Physics Institute*, Nauka, Moscow, **46**, 3-64
6. Orlovskii, Yu.V., Basiev, T.T., Vorob'ev, I.N., Osiko, V.V., Papashvili, A.G., Prokhorov, A.M. (1996) Site-selective measurements of $^4G_{5/2}$; $^2G_{7/2}$ nonradiative relaxation rate in Nd:SrF$_2$, Nd:La:SrF$_2$, and Nd:Sr:LaF$_3$ laser crystals, *Laser Physics International Journal*, **6**, 448-455
7. Orlovskii, Yu.V., Basiev, T.T., Abalakin, S.A., Vorob'ev, I.N., Alimov, O.K., Papashvili, A.G., Pukhov, K.K.(1998) Fluorescence quenching of the Nd^{3+} ions in different optical centers in fluorite-type crystals, *J.Lumin.*, **76/77**, 371-376
8. Orlovskii, Yu.V., Reeves, R.J., Powell, R.C., Basiev, T.T., Pukhov, K.K.(1994) Multiple-phonon nonradiative relaxation: Experimental rates in fluoride crystals doped with Er^{3+}and Nd^{3+} ions and a theoretical model, *Phys. Rev.*B **49**, 3821-3830
9. Basiev, T.T., Orlovskii, Yu.V., Pukhov, K.K., Sigachev, V.B., Doroshenko, M.E., Vorob'ev, I.N. (1996) Multiphonon relaxation rates measurements and theoretical calculations in the frame of non-linear and non-Coulomb model of a rare-earth ion-ligand interaction, *J. Lumin.*, **68**, 241-254

GROWTH AND CHARACTERIZATION OF Y-Lu-Gd ALUMINIUM PEROVSKITES

J.A. MARES[*1] , M. NIKL[1] , K. BLAZEK[2], P. MALY[2],
K. NEJEZCHLEB[2] C. D'AMBROSIO[3] , F. DE NOTARISTEFANI[4]

[1] *Institute of Physics, Academy of Sciences of the Czech Republic,*
Cukrovarnicka 10, 16253 Praha 6, Czech Republic
[2] *Crytur Ltd., Palackeho 175, 511 01 Turnov, Czech Republic*
[3] *CERN, EP Division, CH-1211 Geneve 23, Switzerland*
[4] *INFN Section Rome, 3rd University of Rome, Italy*

Abstract. Y-Lu-Gd pure, mixed and doped aluminium perovskite crystals belong to efficient and durable crystals used in different laser and scintillation applications. This paper will review methods of their growth (the Czochralski, Bridgman, EFG ones and their modifications), their overall growth yield and present performances. Various kinds of spectroscopies are used to characterize the crystals in the range $\approx$ 0.5 eV to $\approx$ 1 MeV. Finally, status and possible development of Y-Lu-Gd aluminium perovskite crystals will be evaluated with emphasis to their use in modern applications.

1. Introduction

Now, pure, mixed and doped (mainly by rare earth or transition metal ions) $AAlO_3$ or $A_xB_{1-x}AlO_3$ perovskite single crystals (orthorhombic structure) are used in different modern applications [1-13]. These applications are mainly concentrated to either lasing or scintillating ones but also in special ones (science, industry and medicine) [2-7]. Nd^{3+}-doped perovskite crystals such as $YAlO_3$, $SrTiO_3$ etc. are used in different solid state lasers [1,9,10,14-16]. From other rare earth ions Ce^{3+} is widely used both for scintillation and selected laser applications (first of all for near UV laser generation around 310 nm [4, 6, 11, 17-22]). Ce^{3+} UV laser generation was mainly obtained with complex fluoride crystals such as $LiCaAlF_6$, $LiLuF_4$ etc. [17-22].

Here, we will summarize, compare and review the spectroscopic, scintillation and laser properties of pure or doped $AAlO_3$ (A = Y, Lu, Gd) or mixed $A_xB_{1-x}AlO_3$ (A, B = Y, Lu, Gd) aluminium perovskite crystals. Nd^{3+} is the most used lasing ion radiating at 1.079 nm in $YAlO_3$ ($^4F_{3/2} \rightarrow {}^4I_{11/2}$ laser transition) and this laser crystal is known for more than 30 years and is used both in pulsed or cw laser regime

---

* Corresponding author: amares@fzu.cz

J.-C. Krupa and N.A. Kulagin (eds.), Physics of Laser Crystals, 63–74.

[14,16]. Another important dopant for aluminium perovskites is Ce^{3+}. This rare earth ion is characterized by allowed and fast $4f \leftrightarrow 5d$ transitions and Ce^{3+}-doped perovskites belong to the most important, efficient, fast and medium or heavy scintillators. Some of the most recent Ce^{3+} ion applications include also experimental UV laser systems (fixed or partly tunable in the near UV) [17-22]. Generally, Nd^{3+}- or Ce^{3+}-doped aluminium perovskites have suitable properties for lasing, scintillation or other applications because they are durable, chemically and mechanically inert with good luminescence and scintillation properties [2-9]. Pure or doped $AAlO_3$ or mixed $A_xB_{1-x}AlO_3$ perovskites are crystals characterized by high melting point (around 2000 ^0C) and complicated phase diagrams which result in rather difficult preparation of these crystals.

Nd^{3+} and Ce^{3+} ions are the most important lasing or scintillating ones from group of rare earth ions but other doping ions as Er^{3+}, Ho^{3+}, Pr^{3+} and Tm^{3+} are used. In the pure or mixed Y-Lu-Gd aluminium perovskites the dopants replace mainly Y^{3+}, Lu^{3+} or Gd^{3+} lattice ions but their ionic radii are a bit lower compared with those of Nd^{3+} and Ce^{3+} [2,14] ($r_{ion}(Nd^{3+})$ = 1.12 Å, $r_{ion}(Ce^{3+})$ = 1.14 Å, $r_{ion}(Y^{3+})$ = 1.02 Å, $r_{ion}(Lu^{3+})$ = 0.97 Å and $r_{ion}(Gd^{3+})$ = 1.06 Å). This reflects in relatively low distribution coefficients in these high temperature alumimium perovskites. Generally, the lasing and scintillating perovskite crystals can contain around 1 at. % of doping ions related to A, A_x or B_{1-x} lattice ions. The aim of this paper is to characterize and compare performances, spectroscopic and scintillation properties and overall efficiency of the pure and mixed Y-Lu-Gd aluminium perovskites together with characterization and comparison of different methods of their growth.

2. Characterization of Y-Lu-Gd Aluminium Perovskites

Such properties of Y-Lu-Gd aluminium perovskites as emission, excitation and absorption spectra, scintillation efficiency and decays are those from which we can determine their laser and scintillation parameters and efficiencies. These properties and parameters can be studied by various kinds of spectroscopic techniques as (i) the classical spectroscopy (near UV, visible and near IR from ~ 0.5 to 6 eV), (ii) VUV spectroscopy (from ~ 6 to 1 keV) and (iii) X- or γ-ray spectroscopy (in the energy range above 1 keV). Results presented here were obtained mainly with (i) and (iii) spectroscopies. The classical UV, visible and near IR spectroscopy can determine energy level structure and properties of doping ions in crystals from the following measurements:

(i) emission spectra;

(ii) excitation spectra;

(iii) absorption spectra, and

(iv) decays and decay kinetics.

Classical spectroscopy can characterize the basic luminescence properties of impurity ions (Nd^{3+}, Ce^{3+}, Pr^{3+}, Er^{3+}, Ho^{3+}, Cr^{3+} and similar ones) used for laser generation or scintillation.

VUV spectroscopy uses either special VUV monochromators [24] or large and unique synchrotron radiation sources [25]. This spectroscopy measures the same characteristics as the classical one and can determine such properties as interband energy structure of the crystals, excitons and their levels, bandgap absorption etc.

The third and last spectroscopy is the X and γ-ray spectroscopy, which uses radiation with energies above ≈ 1 keV. Especially, this radiation is used for excitation of scintillation emission spectra, scintillation decays and kinetics [2, 3, 7]. This is not continuous spectroscopy as the classical and VUV ones because only some lines with fixed energies are usable for measurements. As X- or γ-ray excitation lines either those of the individual radioisotopes or by $_{26}^{55}Fe$ 6.47 keV or $_{95}^{241}Am$ 59.5 keV excited multiple X and γ-ray sources [2] can be used. E.g. multi-line source generated by $_{95}^{241}Am$ 59.5 keV line radiates at the following X- and γ-ray lines: 8.16 keV (Cu), 13.6 keV (Rb), 17.8 keV (Mo), 22.6 keV (Ag), 32.9 keV (Ba) and 44.5 keV (Tb). This source is sometimes called as the small synchrotron. Other important X- or γ-ray lines used for measurements are mainly those of $_{27}^{57}Co$ (122 keV), $_{11}^{22}Na$ (511 keV) and $_{55}^{137}Cs$ (662 keV).

3. Methods of Growth of Y-Lu-Gd Aluminium Perovskites

Various crystal growth methods (either classical as the Czochralski and Bridgman ones or their modifications) are used to grow Y-Lu-Gd aluminium perovskite crystals [6,11,13,16,24,26-32]. Various raw materials are used, mainly those of powder oxides of Y, Lu, Gd, Al, Ce, Nd and other ones of 4N or 5N purity. Some purification methods can improve the purity of these raw materials. The purified powder oxides are pressed, homogenized and annealed. The sintered tables are placed into appropriate crucibles (manufactured mainly from Mo or Ir) and these crucibles with raw materials are put into furnaces. Generally, as was mentioned the growth of Y-Lu-Gd perovskites is difficult either due to their high melting points (T_{melt} is between 1850 and 2000 °C) or high prices of some oxide raw materials, especially high expensive Lu_2O_3 [26,30]).

Growth of Y-Lu-Gd aluminium perovskite crystals is also difficult from another reason: due to very limited temperature range where the stable perovskite phase can appear and grow. The garnet phase arises more easily and sometimes grows instead of the perovskite phase. Besides the perovskite phase (orthorhombic structure) even a cubic phase or oxides can also appear [26,27].

3.1. THE CZOCHRALSKI METHOD

The Czochralski method is often used for growth of Y-Lu-Gd aluminium perovskite crystals. It is successfully used for growth of selected crystals because its main advantage is homogeneity of the as-grown crystals. Generally, the Czochralski crystal growth method is carried out in an inert atmosphere and the post-growth annealing is used for stabilizing and maximizing content and valency of doping ions as e.g. Nd^{3+}, Ce^{3+}, Er^{3+} and others. The Czochralski grown crystals have roughly homogeneous distribution of impurities and e.g. large and efficient YAP:Nd and YAP:Ce laser and scintillation crystals of 40 mm in diameter and up to 150 in length were grown by this method [28]. Unfortunately, up to now almost no pure LuAP crystals were grown by this method [9,29] but the mixed $Lu_xY_{1-x}AP:Ce$ crystals were grown more successfully ($Lu_{0.3}Y_{0.7}AP:Ce$ crystals of 15 mm in diameter and 50 mm long were grown by this method [26,28]). Recent results presented at the

CCC meetings this year [30] have shown that $Lu_{0.7}Y_{0.3}AP:Ce$ crystals of 150 mm long and 15 mm in diameter were grown successfully by this method.

3.2. THE BRIDGMAN AND BRIDGMAN-STOCKBARGER METHODS

The Bridgman method of crystal growth was successfully used by Petrosyan et al. [13, 24, 31,32] to grow LuAP:Ce crystal. It was also used for growth of the mixed
$Lu_xY_{1-x}AP:Ce$ crystals and last results have shown that crystals with x = 0.8 were prepared [30]. A disadvantage of this method is inhomogeneous distribution of doping ions along crystal growth axis. Now, the Bridgman method is the only one, which allows to grow of reproducible LuAP:Ce or LuAP:Nd crystals [31,32].

The Bridgman-Stockbarger method is the modification of the Bridgman one. Now $Lu_{0.8}Y_{0.2}AP:Ce$ crystals of polycrystalline form were grown by it [2,27,28]. Compared with the classical Bridgman method it seems that the Bridgman-Stockbarger method could also be used for successful growth of the mixed or even pure Y-Lu-Gd perovskites.

3.3. EFG METHOD

EFG (Edge-defined Film Fed method) method is the special one to be used for growth of various small and specific profiles (e.g. cylinders, wires, tubes of dimensions up to 10 mm in diameters or sides). However, this method has some drawbacks and no good quality crystals were grown by it. The main drawback of the EFG method is connected with temperature fluctuation at the border between solid and melting phase and this can results into inhomogeneous growth. By this method the garnet phase can arise easily and also various defects appear. Generally, the EFG method has lower growth yield (= a number of perfectly grown crystals to the number of the whole crystal growth attempts) compared with the Czochralski and Bridgman methods [2].

The modification of EFG method is the micro-pulling-down one (μ-PD) [23]. This method was successfully used for growth of almost homogeneous micro-crystals of $LiNbO_3$ (of very thin rod-like needles of 300 - 800 μm in diameter).

4. Review of Spectroscopic Properties of Y-Lu-Gd Aluminium Perovskites

4.1. CLASSICAL SPECTROSCOPY

Classical spectroscopic studies of Y-Lu-Gd aluminium perovskites were carried out in the spectral range from ≈ 0.5 eV to 6 eV (near IR, visible and near UV ranges). For pure or mixed and impurity-doped crystals this spectroscopy presents their emission and excitation properties, e.g. those of Ce^{3+}, Nd^{3+}, Er^{3+} and other impurities. We can also obtain the emission and other properties of different radiating defects as e.g. color centers [26, 33]. Fig. 1 presents the emission and excitation spectra of Ce^{3+} ions in pure YAP:Ce and LuAP:Ce crystals at 80 K.

Generally, Ce^{3+} ions in Y - Lu – Gd aluminium mixed perovskites have almost the same emission properties as those in YAP:Ce and LuAP:Ce [2, 26]. A comparison of Ce^{3+}-doped Y-Lu-Gd perovskites with garnets (YAG:Ce and LuAG:Ce) shows that the crystal field splitting in perovskites is smaller than that in the garnets. Emission properties of Ce^{3+} ions are given by $5d^1 \rightarrow 4f^0$

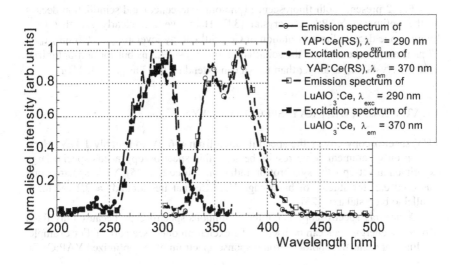

Figure 1. Classical spectroscopy - excitation and emission spectra of YAP:Ce and LuAP:Ce at T= 80 K   (parameters of measurements are given in the inset top right).

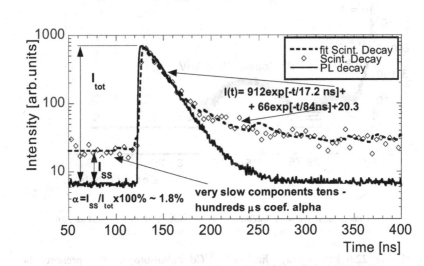

Figure 2. Scintillation and photoluminescence decay curves of the optimized YAP:Ce,Zr crystal at RT (scintillation decay was fitted and the fit is described above the curve).

68

allowed transitions characterized by fast decays (see Fig. 2). Nd^{3+} doping ions behave quite differently [16, 28] because their transitions arise from $4f^3$ electronic configuration (forbidden transitions).

Fig. 2 presents both fluorescence (photoluminescence) and scintillation decays of the optimized YAP:Ce,Zr crystal [34]. Here, we can clearly see that Ce^{3+} fluorescence decay (direct excitation of Ce^{3+} $5d^1$ energy level at $\lambda_{ex} = 290$ nm) $\tau_{uv} = 17$ ns is almost the same as the scintillation one $\tau_{sc} = 17.2$ ns but for non-optimized YAP:Ce crystals τ_{sc} is longer than τ_{uv} (it can extend up to 27 ns [11, 24, 34]).

4.1. VUV AND X OR γ-RAY SPECTROSCOPY

VUV spectroscopy covers the spectral range from ≈ 6 eV to roughly 1 keV. Here, we can only comment some results because this spectroscopy needs special large experimental set-up - the synchrotron radiation source [24, 25]. VUV spectroscopy was used e.g. for studies of band edge YAP:Ce that lies around ≈ 7.2 eV for E parallel to b crystal axis [25].

X or γ-ray spectroscopy is not the continuous spectroscopy because only fixed X or γ-ray energy lines can be used as the excitation ones (see part 2). For example, Fig.3 presents the photoelectron response spectrum of the optimized YAP:Ce:Zr

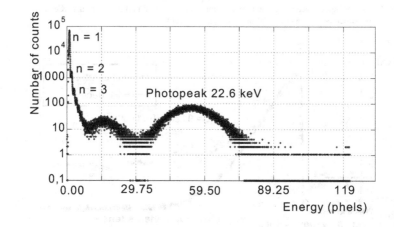

Figure 3. Photoelectron response spectrum of the optimized YAP:Ce,Zr crystal for 22.6 keV energy of ^{109}Cd radioisotope (description of measurement - see [2,7])

crystal for 22.6 keV γ-ray line of ^{109}Cd radioisotope. Fig.4 presents the scintillation photoelectron response curves of various perovskites and LSO crystal.

Their scintillation photoelectron responses depend linearly on energy of γ-ray lines roughly up to 1 MeV [2]. The slopes of the curves which characterize the individual scintillation photoelectron responses of the crystals are the following: $\Delta N_{phels}(LuAP:Ce) = 1.79$ phels/keV, $\Delta N_{phels}(Lu_{0.3}Y_{0.7}AP:Ce) = 3.52$ phels/keV and $\Delta N_{phels}(LSO:Ce) = 5.52$ phels/keV. Other Y-Lu-Gd Ce^{3+}-doped aluminium perovskites exhibit also linear scintillation photoelectron responses up to 1 MeV and their actual $\Delta N_{phels}(E)$ values are similar

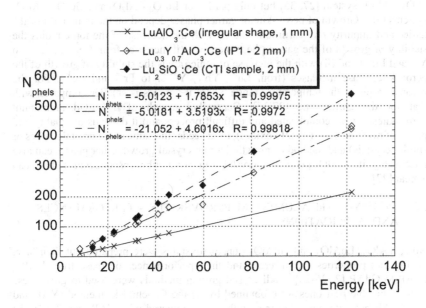

Figure 4. N_{phls} photoelectron number versus values of the energies of the total absorption peaks of $LuAP:Ce$, $Lu_{0.3}Y_{0.7}AP:Ce$ and LSO in the range 5 to 122 keV.

5. Overview-Growth Possibilities, Performances and Characteristics of Y-Lu-Gd Aluminium Perovskites

5.1. GROWTH POSSIBILITIES OF Y-Lu-Gd PEROVSKITES

Growth of Y-Lu-Gd pure or mixed aluminium perovskites is difficult task, which is not only connected with their high melting points (between 1850 - 2050 ^0C). Generally, growth of perovskite crystals (or perovskite phase) is more difficult compared with the growth of garnet crystals because the garnet phase is more stable compared with the perovskite one [26-28]. Growth possibilities of pure or mixed Y-Lu-Gd aluminium perovskites are mainly limited by a thermodynamic instability of this phase [27]. Especially, if temperature gradient is low the perovskite phase disappears to the more stable garnet one and rare earth oxides [27, 28]. If the temperature gradient is high than defect growth and twinning appear due to internal stresses in crystals during growth. Large efforts by Kvapil et al. [27] shown that growth of the perovskite crystals improves if the matrix is doped by some additional rare earth ions as e.g. by Gd^{3+}, Y^{3+} and La^{3+}. Their presence can increase the temperature range of the stability of the perovskite phase a bit.

Growth of crystals depends on knowledge of the phase diagram between raw materials from, which are prepared [35]. Now, phase diagram is well known for Y_2O_3 - Al_2O_3 system [27, 35] but only partly for Lu_2O_3 - Al_2O_3 and Gd_2O_3 - Al_2O_3 systems [30]. Growth of perovskite or garnet phases depend on the ionic radii of the lattice and impurity rare earth ions and (i) with the decrease of the ionic radius the stability of growth of the garnet phase increases (it increases from Ce^{3+}, Pr^{3+}, ... to Y^{3+} and Lu^{3+}) and (ii) with the increase of ionic radius the stability of growth of the perovskite phase increases (from Lu^{3+}, Yb^{3+}, Y^{3+}... to Pr^{3+} and Ce^{3+}). For the perovskite phase the temperature range where it appears is smaller compared with that for the garnet one [27,35]. Generally, growth of Y-Lu-Gd aluminium perovskites is very complicated and difficult process and it depends on quality and purity of raw materials (mainly oxides), used seeds (the same small crystals, wires or similar crystals) and method of growth. During crystal growth other phases can also arise as e.g. if $GdAlO_3$ perovskite crystal is grown a cubic (pseudocubic) phase can appear [27].

5.2. PURE Y-Lu-Gd ALUMINIUM PEROVSKITES - PERFORMANCES AND APPLICATIONS

Now, $YAlO_3$, $LuAlO_3$ and $GdAlO_3$ pure perovskites were prepared but their size, quality and properties are different and their performances decrease from YAP - LuAP to GdAP [2, 26-28]. All crystal growth methods were used to grow these crystals but the best ones were obtained by (i) the Czochralski method (YAP and also LuAP but with much lower growth yield compared with YAP one [2, 9]), (ii) the Bridgman and Bridgman-Stockbarger methods (mainly LuAP [24, 31-32]) and (iii) EFG method ($Lu_xY_{1-x}AP$:Ce and GdAP as thin rods or small plates [2, 27]). From application point of view we can exclude GdAP crystal because its performance is not sufficient (inhomogeneous samples with various defects [2]). LuAP (grown by the Bridgman) and mainly YAP Czochralski grown crystals were prepared in sufficient performance for applications. Their major applications include medical imaging, radiation detection and lasers [2-8]. These applications demand either doping by scintillating (Ce^{3+}, Pr^{3+}) or lasing (Nd^{3+}, Er^{3+}, Ho^{3+} and Tm^{3+}) ions, respectively. This is not large problem because these doping ions replace "3+" rare earth lattice ions mainly and almost no charge compensation defects are necessary.

From the above-described pure Y-Lu-Gd aluminium perovskites the highest level of performance reached YAP crystal grown by the Czochralski method [2, 27, 28]. By Ce^{3+} or Nd^{3+} doped YAP crystals were grown of up to 40 mm in diameter and 150 mm in length [28]. Their scintillation applications are mainly radiation detectors of ionizing radiation where are used as plates or multidetectors (for position sensitive measurements) [2, 6, 26]. YAP:Ce gamma imaging camera can reach up to submilimeter resolution [36], small systems for high resolution PET are constructed (small animal PET systems [30]), YAP:Ce window is used as the photocathode one of the HPMT (31 mm in diameter) [7] and scintillators are used to test the ISPA system [37]. YAP:Ce can also be used for identification and separation of different particles and ions (protons, α-particles, ^{16}O ions etc.) in the range up to 30 MeV [38, 39] and as X-ray imager can resolve and identify them by comparison of Ce^{3+} fast and slow decay components and their intensities [40]. Further scintillation applications of YAP:Ce include also various lines of standards or special detectors [28], detectors for electron microscopy and light guides. Nd^{3+}-doped YAP crystals are used in lasers [16, 28] and this laser radiation is partly tunable because Nd^{3+} laser wavelength in YAP depends on the crystallographic orientation [28].

5.3. MIXED Y-Lu-Gd ALUMINIUM PEROVSKITES - PERFORMANCES AND APPLICATIONS

Different growth methods were used to grow the mixed Y-Lu-Gd aluminium perovskites [26, 27]. Ce^{3+}-doped crystals containing Lu or Gd are characterized by higher density and high effective Z_{eff} compared with YAP. Now, large efforts are devoted to grow the mixed crystals, especially Ce^{3+}- doped $Lu_xY_{1-x}AP$ ones [26, 31].

Detailed results of growth of $Lu_xY_{1-x}AP$:Ce were described in [24, 26, 31]. The Czochralski grown ones (up to x = 0.3 [26]) have structure and properties very similar to those of YAP:Ce and LuAP:Ce. Last results presented at CCC collaboration this year shown that larger and heavier $Lu_{0.7}Y_{0.3}AP$:Ce crystals were prepared (15 mm in diameter and 50 mm long). However, these crystals exhibit rather intense slow Ce^{3+} decay components (around 50 % of the whole decay intensity) [30]. It was also possible to grow polycrystalline $Lu_{0.8}Y_{0.2}AP$:Ce crystals [27, 30] consisting from both garnet and perovskite phases but also rare earth oxides were indicated.

$Lu_xGd_{1-x}AP$:Ce crystals were grown by the Czochralski method [26, 27] but their performances are worse compared with those of $Lu_xY_{1-x}AP$. Especially, these crystals exhibit Ce^{3+} slow decay components due to $Gd^{3+} \rightarrow Ce^{3+}$ energy transfer [26].

5.4. FUTURE DEVELOPMENT OF Y-Lu-Gd ALUMINIUM PEROVSKITES OR OTHER POSSIBILITIES

Now, it seems that from pure and mixed Y-Lu-Gd aluminium perovskites the highest level of performance reached YAP:Ce and YAP:Nd crystals [34,41] scintillation and laser crystals, respectively. From other Y-Lu-Gd aluminium perovskites the average level of performance reached $Lu_xY_{1-x}AP$:Ce crystals, especially those with x up to 0.3 or 0.7 which parameters are close to the parameters of average YAP:Ce crystals. Y-Lu-Gd mixed aluminium perovskites could be used in medical imaging similarly as YAP:Ce in gamma camera with 140 keV γ-ray radiation of ^{99}Tc radioisotope [2, 36]. Ce^{3+} cannot be used as only the scintillating ion but probably also as lasing one as like as in $LiLuF_4$, $LiCaAlF_6$ or similar crystals [17, 20]. Growth of pure LuAP:Ce is very difficult task with extremely low growth yield. A disadvantage of this crystal is its large self-absorption [24]. Promising applications of the mixed $Lu_xY_{1-x}AP$:Ce crystals could be connected with the HPMT tubes where these crystals and YAP:Ce can be used instead of the quartz photocathode window [7]. They can be also used in small gamma cameras or PET systems [2,6,30,36]. Technology achievement in the future should reach not only improvement of YAP:Ce but also other pure or mixed Y-Lu-Gd aluminium perovskites. Nd^{3+} is used as laser ion for YAP but it could be interesting to try Ce^{3+} lasing in the near UV. Generally, the improvement of the methods of growth of these crystals should result in more reliable and reproducible growth of these crystals.

6. Acknowledgements

Support of NATO SfP 973510 - Scintillators and GA CR 202/01/0753 projects are gratefully acknowledged along with the support of Ministry of Industry and Trade of the Czech Republic, project N°. FF-P/125.

7. References

1. Kulagin, N.A. (2000) Blue intensive luminescence of unperfect perovskite crystals, *Proc. of Abstracts of the 2nd Intern. Symp. on Laser, Scintillator and Nonlinear Optical Materials*, Lyon, May 28-31, 2000, poster No 23.

2. Mares, J.A., Nikl, M., Maly, P., Bartos, K., Nejezchleb, K., Blazek, K., de Notaristefani,F., D'Ambrosio, C., Puertolas, D. and Rosso, E. (2002) Growth and properties of Ce^{3+}-doped Lu$_x$(RE^{3+})$_{1-x}$AP scintillators, *Opt. Mat.* **19**, 117 - 122.

3. Nikl, M. (2000) Wide band gap scintillation materials, *phys. stat. sol. (a) 178*, 595 - 620.

4. van Eijk, C.W.E. (2002) Inorganic-scintillator development, *NIM Phys. Res. A 460*, 1 - 14.

5. van Eijk, C.W.E. (2002) Inorganic scintillators in medical imaging, *Phys. Med. Biol. 47*, R85 - R106.

6. Baccaro, S., Blazek, K., de. Notaristefani F., Maly, P., Mares, J.A., Pani, R., Pellegrini, R., Soluri A., (1995) Scintillation properties of YAP:Ce, *NIM. Phys. Res A* **361**, 209 - 215.

7. D' Ambrosio, C., de Notaristefani, F., Leutz H., Puertolas, D. and Rosso, E. (2000) X-ray detection with a Scintillating YAP-window hybrid photomultiplier tube *IEEE Trans. Nucl. Sci.* **47**, 6 - 12.

8. Zhang, L., Madej, C., Pedrini, C., Moine, B., Dujardin, C., Petrosyan, A.G. and Belsky, A. (1997) Elaboration and spectroscopic properties of new dense cerium doped lutetium based scintillator materials, *Chem. Phys. Lett. 268*, 408 -412.

9. Lempicki, A., Randles, M.H., Wisniewski, D., Balcerzyk, M., Brecher, C. and Wojtowicz, A.J. (1995) LuAlO$_3$:Ce and other aluminate scintillators, *IEEE Trans. Nucl. Sci. 42*, 280 -284.

10. Weber, M.J. (1973) Optical spectra of Ce^{3+} and Ce^{3+}-sensitized fluorescence in YAlO$_3$, *J. Appl. Phys. 44*, 3205 - 3208.

11. Autrata, R., Schauer, P., Kvapil, Jo. and Kvapil, Ji. (1983) Cathodoluminescent efficiency of Y$_3$Al$_5$O$_{12}$ and YAlO$_3$ single crystals in dependence on Ce^{3+} and other dopant concentration, *Cryst. Res. Technol. 18*, 907 - 913.

12. Okada, F., Togawa, S., Ohta, K. and Koda, S. (1994) Solid-state ultraviolet tunable lasers -a Ce^{3+} doped LiYF$_4$ crystal, *J. Appl. Phys. 75*, 49 - 53.

13. Kaminskii, A.A., Petrosyan, A.G., Markosyan, A.A. and Shironyan, G.O. (1991) Bridgman growth and stimulated-emission spectroscopy of orthorhombic LuAlO$_3$ single crystals doped with Nd^{3+} and Pr^{3+} ions, *phys. stat.sol. (a) 125*, 353 - 361.

14. Weber, M.J., Bass, M., Andringa, K., Mochamp, R.R. and Comperchio, E. (1969) Czochralski growth and properties of YAlO$_3$ laser crystal, *Appl. Phys. Lett. 15*, 342 - 345.

15. Mares, J.A., Chval, J., Nikl M. and Boulon, G. (1993) Identification of trace impurities in pure and doped YAlO$_3$ and Y$_3$Al$_5$O$_{12}$ crystals by their fluorescence and by EMA method, *Czech. J. Phys. 43*, 683 - 696.

16. Kvapil, J., Kvapil, Jos., Perner, B.and Hamal, K. (1988) Luminescence sensitization in Nd-Cr-Ce doped yttrium aluminates, *Czech. J. Phys.* B**37**, 1277 -1287.

17. Sarukura, N., Liu, Z., Izumida, S., Dubinskii, M.A., Abdulsabirov, R.Yu. and Korableva S.L. (1998) All-solid-state tunable ultraviolet subnanosecond laser with direct pumping by the fifth harmonic of Nd:YAG laser, *Appl. Opt.* **37**, 6446 - 6448.

18. Liu, Z., Ohtake, H., Sarukura, N., Dubinskii, M.A., Semashko, V.V., Naumov, A.K., Korableva, S.L. and Abdulsabirov, R.Yu. (1997) Subnanosecond tunable ultraviolet pulse generation from a low-Q, short-cavity Ce:LiCAF laser, *Jpn. J. Appl. Phys.* **36**, L1384 - L1386.

19. Liu, Z., Sarukura, N., Dubinskii, M.A., Semashko, V.S., Naumov, A.K., Korableva, S.L. and Abdulsabirov, R.Yu. (1998) Tunable ultraviolet short-pulse generation from Ce:LiCAF laser amplifier systém and its sum-frequency mixing with an Nd:YAG laser, *Jpn. J. Appl.Phys.* **37**, L36 - L38.

20. Liu, Z., Izumida, S., Ohtake, H., Sarukura, N. Shimamura, K., Mujilatu, N., Baldochi, S.L. and Fukuda, T. (1998) High-pulse-energy, all-solid-state, ultraviolet laser oscillator using large Czochralski-grown Ce:LiCAF crystal, *Jpn. J. Appl. Phys.* **37**, L1318 - L1319.

21. Liu, Z., Shimamura, K., Nakano, K., Mujilatu, N., Fukuda, T., Kozeki, T., Ohtake, H. and Sarukura, N. (2000) Direct generation of 27-mJ, 309-nm pulses from a Ce^{3+}:LiLuF$_4$ oscilator using a large-size Ce^{3+}:LiLuF$_4$ crystal, *Jpn. J. Appl. Phys.* **39**, L88 - L89.

22. Shimamura, K., Mujilatu, N., Nakano, K., Baldochi, S.L., Liu, Z., Ohtake, H., Sarukura, N. and Fukuda, T. (1999) Growth and characterization of Ce-doped LiCaAlF$_6$ single crystals J. *Cryst. Growth* **197**, 896 - 900.

23. Yoon, Dae-Ho and Fukuda, T. (1994) Characterization of LiNbO$_3$ micro single crystals grown by the micro-pulling-down method, *J. Cryst. Growth* **144**, 201 - 206.

24. Dujardin, C., Pedrini, C., Blanc W., Gacon, J.C., van't Spijker, J.C., Frijns, O.W.V., van Eijk, C.W.E., Dorenbos, R., Chen, R., Fremount, A., Tallouf, F., Tavernier, S., Bruyndonckx, P. and

Petrosyan, A.G. (1998) Optical and scintillation properties of large $LuAlO_3:Ce^{3+}$ crystals, J. *Phys.C: Condens. Matter* **10**, 3061 - 3073.

25. Tomiko, T., Fukudome, F., Kaminao, M., Fujisawa, M. and Tanahara, Y. (1986) Optical spectra of $Y_3Al_5O_{12}$ and $YAlO_3$ in VUV, J. *Phys. Soc. Japan* **55**, 2090 - 2091.

26. Chval, J., Clement, D., Giba, J., Hybler, J., Loude, J.-F., Mares, J.A., Mihokova, E., Morel, C., Nejezchleb, K., Nikl, M., Vedda, A. and Zaidi, H. (2000) Development of new mixed $Lu_x(RE^{3+})_{1-x}AP:Ce$ scintillators ($RE^{3+} = Y^{3+}$ or Gd^{3+}): Comparison with other Ce- doped or intrinsic scintillating crystals, *NIM Phys. Res. A* **443**, 331 - 341.

27. Kvapil, J., and Kvapil, Jos., private communications. Crytur Ltd., Palackeho 175, 511 01 Turnov, Czech Republic - pages http://www.crytur.cz

28. Mares, J.A., Nikl, M., Chval, J., Dafinei, I., Lecoq, P. and Kvapil, J. (1995) Fluorescence and scintillation properties of $LuAlO_3:Ce$ crystal, *Chem. Phys. Lett.* **241**, 311 - 316.

29. Materials from General meetings of CCC collaboration, CERN, Geneve, 2002.

30. Petrosyan, A.G., Shirinyan, G.O., Ovanesyan, K.L., Pedrini, C. and Dujardin, C. (1999) Bridgman single crystal growth of Ce-doped $(Lu_{1-x}Y_x)AlO_3$, J. *Cryst. Growth* **198/199**, 492 - 496.

31. Petrosyan, A.G., Shirinyan, G.O., Pedrini, C., Dujardin C., Ovanesyan, K.L., Manucharyan, R.G., Butaeva, T.A. and Derzyan, M.V. (1998) Bridgman growth and characterization of $LuAlO_3-Ce^{3+}$ scintillator crystals, *Cryst. Res. Technol.* **33**, 241 - 248.

32. Mares, J.A., Cechova, N., Nikl, M., Kvapil, J., Kratky, R. and Pospisil, J. (1998) J. *Alloys Comp.* **275-277**, 200 - 204.

33. Nikl, M., Mares, J.A., Chval, J., Mihokova E., Solovieva, N., Martini, M., Vedda, A., Blazek, K., Maly, P., Nejezchleb, K., Fabeni, P., Pazzi, G.P., Babin, V., Kalder, K., Krasnikov, A., Zazubovich, S. and D'Ambrosio, C. An effect of Zr^{4+} co-doping of YAP:Ce scintillator, *NIM Phys. Res. A* **486/1, 2**, 250 - 253.

34. *Phase Diagrams for Ceramics* (various supplements published from 60-ties), Eds. E.M. Lewin and H.F. Mc Murde, The American Ceramic Society Inc., published continuosly from 1960.

35. Pani, R., de Notaristefani, F., Blazek, K., Maly, P., Pellegrini, R., Pergola, A., Soluri, A. and Scopinario, A. (1994) Multi-crystal YAP:Ce detector system for position sensitive measurements, *NIM Phys. Res. A* **348**, 551 - 558.

36. Puertolas, D., Piedigrossi, D., Leutz, H., Gys T., de Notaristefani F. and D'Ambrosio, C. (1997) An ISPA-camera for gamma rays with improved energy resolution, *IEEE Trans. Nucl. Sci.* **44**, 747 - 1752.

37. Slunga, E., Cederwall, B., Ideguchi, E., Kerek A., Klamra, W., van der Marel, J., Novak, D. and Norlin, L.-O. (2001) Scintillation response of BaF_2 and $YAlO_3:Ce$ (YAP:Ce) to energetic ions, *NIM Phys. Res. A* **469**, 70 - 76.

38. Westman, S., Kerek, A., Klamra, W., Norlin, L.-O. and Novak, D. (2002) Heavy ion detection at extreme high vacuum by means of a YAP:Ce scintillator, *NIM Phys. Res. A* **481**, 655 - 660.

39. Suzuki, M., Toyokawa, H., Mizumaki, M., Ohashi, Y., Yagi, N., Kamitsubo, H., Kuroda, K., Gorin, A., Manouilov, I., Riazantsuev, A., Nomachi, M., Yosoi, M., Ishikawa, T., Morimoto, K. and Tokanai, F. (2001) A YAP(Ce) imager operated in high energy X-ray region, *NIM Phys Res. A* **467-468**, 1121 - 1124.

40. Moszynski, M., Kapusta, M., Wolski, D., Klamra, W. and Cederwall, B. (1998) Properties of the YAP:Ce scintillator, *NIM Phys. Res. A* **404**, 157 - 165.

WHAT KIND OF INFORMATION IS POSSIBLE TO OBTAIN FROM ^{2S+1}L TERMS OF $4f^N$ CONFIGURATIONS?

E. ANTIC-FIDANCEV
Laboratoire de Chimie Appliquée de l'État Solide, CNRS, UMR-C7574, ENSCP, 11, Rue Pierre et Marie Curie, F-75231 Paris Cedex 05, France
e-mail: antic@ext.jussieu.fr

Abstract:

The optical spectra of f-f transitions of trivalent rare earth (R^{3+}) ions in solids are very useful for different purposes. First of all, to analyse the purity and the local environment about the rare earth site in various compounds. Then, to follow the evolution of the barycenter positions of $^{2S+1}L_J$ levels inside the $4f^N$ configurations. This evolution is linear and is separated in two parts. i) The shift of excited manifolds following ionic – covalent character of R^{3+} ion and ii) The shift of the ground multiplet states which describes the crystal field strength of the R^{3+} ion. As well, we shall demonstrate that the absorption / emission lines of R^{3+} ion embedded in an isostructural family display a linear relation with the ionic radii of the cation host in the lattice along the lanthanide series. It is then possible to predict the phase limit existence following the evolution of Stark components of some isolated crystal field levels. It is clearly shown on the rare earth garnet family.

Keywords: Rare earths; Optical analysis; Barycenter curves; Crystal field strength; Phase limit.

J.-C. Krupa and N.A. Kulagin (eds.), Physics of Laser Crystals, 75–91.

1. Introduction

The spectroscopy of rare earth elements is widely investigated for divers purposes: as luminescent materials, for laser pumping, for second harmonic generation oscillators, e.g.. The rare earth (R^{3+}) ions embedded in a variety of crystalline host lattices display very complex absorption / emission spectra corresponding to f-f transitions situated from VUV to far IR. This complexity is due to the degeneracy removal which is in relation with the local environment around the R^{3+} ion in solid. The number of states increases with increasing number of electrons (or holes) in the f shell. So, from lanthanum to gadolinium (lutetium to gadolinium) the number of states is: 1, 14, 91, 364, 1001, 2002, 3003 and 3432.

According to the local symmetry around the rare earth ion in the host lattice a degeneracy is partly or completely lifted. This degeneracy is first broken by electrostatic interaction between electrons, expressed as:

$$H_{el} = \sum_{ij} \frac{e^2}{r_{ij}}$$

by spin-orbit interaction:

$$H_{SO} = \sum_{i=1}^{N} \zeta_i (l_i s_i)$$

and finely by the crystal field, as follows:

$$H_{CF} = \sum_{k,q} B_q^k \sum_i C_q^k(i)$$

The calculation of the energy levels in zero-order approximation is given by Dieke [1, and references cited herein]. Garcia and Faucher gave recently an overview on the methods used for *ab initio* calculation as well as the fitting of phenomenological parameters [2].

The identification of the crystal field levels is not simple, especially when some of them are situated in a very near spectral domain. Due to an even or an odd number of electrons in the 4f shell, 2J+1 or J+1/2 Stark components are expected for each $^{2S+1}L_J$ level, respectively. If the compound is single phased and if the rare earth ion possess one local environment in the structure, only one line is detected in the absorption / emission spectrum for J = 0 or 1/2. Therefore, the energy position of these single lines along the nephelauxetic scale can be related to the shift of different $^{2S+1}L_J$ level barycenters (*bc*) for a given R^{3+} ion. It is evident from following observations that $^{2S+1}L_J$ level barycenter situated far from the slope underlines a wrong identification of the wavenumber of the level [2]. But we shall show that the barycenter energy values of any two $^{2S+1}L_J$ levels for a given rare earth ion in a host lattice can be connected. The only restriction is to connect the excited states with the excited states and the ground sates with the ground states.

By following the smooth evolution of a particular R^{3+} crystal field level in function of the ionic radius of the host cation in an isostructural family along the lanthanide series, it is possible to predict the limit of the phase existence of that family [4]. The rare earth gallium garnet family is a good example for this design. The comparison is given with yttrium aluminium garnet doped with rare earth ions.

We shall present some experimental data in view to underline the utility of the rare earth optical spectroscopy.

2. Results and Discussion

2.1. RARE EARTH IONS AS THE SPECTROSCOPIC PROBES

The rare earth ions are commonly used as the spectroscopic probes, namely praseodymium, neodymium and europium ions. These rare earth ions possess the levels situated in accessible spectral domain, which are isolated from other electronic transitions and are also non degenerated by the crystal field: $^3P_0$ (Pr^{3+}), $^2P_{1/2}$ (Nd^{3+}) and $^5D_0$ (Eu^{3+}). The maximal number of lines for each J level depends on the local symmetry around the rare earth ion in the structure. Europium is most suitable due to the lowest ^7F ground level with J=0. Therefore, the absorption or emission spectrum even at room temperature can give us enough information.

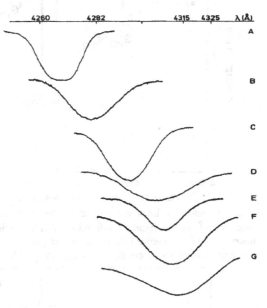

Figure 1: $^4I_{9/2} \rightarrow {}^2P_{1/2}$ transition of Nd^{3+} ion in several glasses: A and B fluorides, C phosphate, D silico-borate, E silico-aluminate, F silicate, G aluminate; T = 4.2 K.

In glasses the spectral lines are broad (Figure 1), contrary to the crystals in which very sharp and numerous lines are observed (Figure 2) [5, 6]. As we are mostly interested in crystals in following paragraphs we report the optical data on the rare earth ions embedded in crystalline environment.

On Figure 3 is presented the emission spectrum of Eu^{3+} doped lanthanum chromate. According to the crystal structure the rare earth ion exhibits two different environments in that host lattice. Under 457.9 nm argon ion laser excitation (the lower trace on the Figure) the emission spectrum is obtained [7]. From this spectra it is evident that the optical and the crystallographic data are in agreement: two lines for $^5D_0 \rightarrow {}^7F_0$ transition are the confirmation of the existence of two different local environments around the rare earth ion. And the selective excitation in these two 0-0 lines, noted "a" and "b", allows the separation of two sites for Eu^{3+} ion. It is clear from $^5D_0 \rightarrow {}^7F_1$ transition (a and b spectra on the figure) that two local environments around the rare earth ion are different. In fact, one can say that the crystal field parameters (*cfp*) of rank 2 are higher for "a" polyhedron than in the case of the "b" one. Moreover, it suggests some higher symmetry for the "a" polyhedron.

In Figure 4 is presented the emission spectrum of two Gd^{3+} sites in the cubic $C-Y_2O_3$.

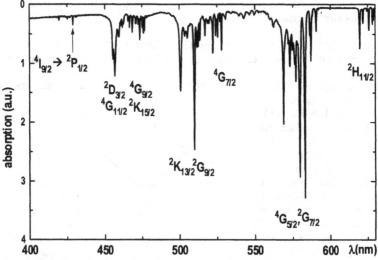

Figure 2:: A part of the absorption spectra of $NdBO_3$; T = 9 K.

Four low intensity lines correspond to the rare earth site which possess the C_{3i} symmetry (↑ in the figure) conforming with the selections rules imposed by group theory due to the presence of inversion centre, and the other 4 lines to C_2 symmetry site. The crystal field study on the rare earth in the C_2 symmetry site has been already

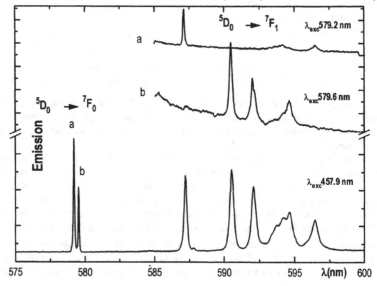

Figure 3: A part of the emission spectra of Eu^{3+} doped lanthanum chromate; T = 77 K.

reported, namely Eu^{3+} [8], but nothing about C_{3i} site. From the spectrum in Figure 4 it has been possible to estimate for the first time the *cfp* of rank 2 for the C_{3i} site [9].

2.2. BARYCENTER CURVES

In this paragraph we shall present the evolution of $^{2S+1}L_J$ level barycenter curves for several rare earth ions, namely praseodymium, neodymium, europium, gadolinium and ytterbium, in various crystalline environments. The first part treats the excited electronic levels settled above the ground manifold and in the second part only the data concerning the ground manifold are presented.

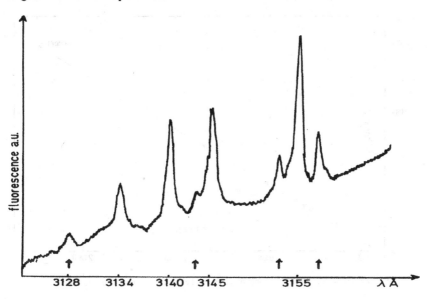

Figure 4: Emission spectrum of Gd^{3+} doped C-type Y_2O_3 corresponding to $^6P_{7/2} \rightarrow {}^8S_{7/2}$ transition.

2.2.1. *Excited states*

4f^2 configuration

Praseodymium is quite suitable for our purpose. There are many experimental data concerning the optical spectra of Pr^{3+} embedded in various crystalline host lattices. From [10, 11] and our own experimental measurements we established the nephelauxetic scale for various energy levels of Pr^{3+}. Pr^{3+} ion possesses one electronic level which is situated in the visible spectral domain and is isolated from other excited sates. So, we consider this level, $^3P_0$ level, and the evolution of the energy level barycenter of $^3P_1$ in function of $^3P_0$ level in praseodymium pure or Pr^{3+} doped host lattices. In low temperature absorption spectra of praseodymium ion $^3H_4 \rightarrow {}^3P_1$ transition is often mixed in some crystalline compounds with another electronic transition situated in the same energy interval, it means $^3H_4 \rightarrow {}^1I_6$ transition. In that case the following assertion can be helpful to predict the correct position of $^3P_1$ barycenter, only if $^3P_0$ is unambiguously determined. Moreover, the barycenter position is also the indication of energy positions of respective Stark components of this level. For exemple, if two of three expected $^3P_1$ Stark components are observed in the spectrum (in the case of low symmetry site) and one component is missing, it is possible to find its value from the barycenter curve.

The relation of $^3P_1$ level barycenter as a function of $^3P_0$ level energy is presented in Figure 5. Almost all the experimental data are located on the straight line. There is one exception. It is evident from this curve that the $^3P_1$ level barycenter in $LiYO_2 : Pr^{3+}$ is badly assigned [12]. The $^3P_1$ calculated level barycenter in $LiYO_2 : Pr^{3+}$ is also wrong. It

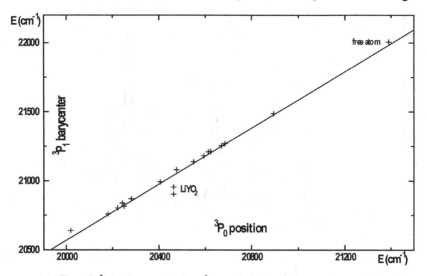

Figure 5: $^3P_1$ level bc as a function of $^3P_0$ level along the Pr^{3+} nephelauxetic scale.

indicates that the phenomenological set of the free ion and / or crystal field parameters must be improved in this case. It is also possible to consider only $^3P_0$ / $^3P_1$ barycenter energy difference, which is about 580 cm^{-1}, and to check these two levels.

In the same way one can draw the barycenter curves for other levels, but the simplest way is this first indication on the respective energy differences to be respected. For example, $^3P_2$ barycenter is at about 1800 cm^{-1} above the $^3P_0$ level and $^1D_2$ barycenter at about 3800 cm^{-1} under $^3P_0$ level.

The correct position of $^1I_6$ level is a real problem in the spectroscopic study of praseodymium ion in solids. As there are no good experimental data on this level (vibronics present associated to $^3P_{0, 1}$ levels) the only value to be considered is in the free atom. Therefore, one can suppose that the $^1I_6$ barycenter is at about 850 cm^{-1} above the $^3P_0$ level. But up to now nobody was able to verify this experimentally.

$4f^3$ configuration

We assume the same approach in the case of neodymium ion. Nd^{3+} possess also one $^{2S+1}L_J$ level which is far from other excited levels, situated in the visible area and un-splitted by the crystal field. This is $^2P_{1/2}$ level (J = ½) located in 22700-23500 cm^{-1}

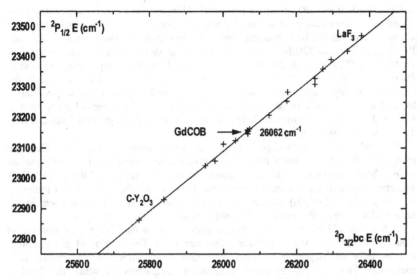

Figure 6 : $^2P_{3/2}$ level bc as a function of $^2P_{1/2}$ level along the Nd^{3+} nephelauxetic scale.

spectral interval. Like in example of praseodymium, it is possible to establish a linear relation between this $^2P_{1/2}$ level and the barycenter energy position of any excited manifold correctly assigned. $^2P_{3/2}$ barycenter in function of $^2P_{1/2}$ level is shown in Figure 6. To illustrate the potentiality of these barycenter curves, Nd^{3+} ion doped

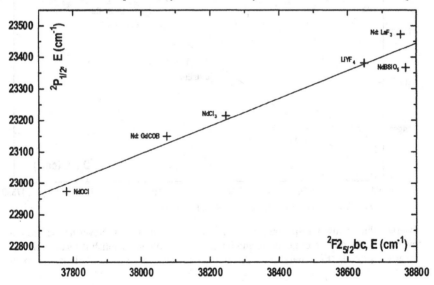

Figure 7: $^2F_{5/2}$ level bc as a function of $^2P_{1/2}$ level along the Nd^{3+} nephelauxetic scale.

$Ca_4GdO(BO_3)_3$ (GdCOB) crystal is presented. This host lattice is actually intensively studied as efficient non-linear optical materials. Consequently, the correct energy level scheme of Nd^{3+} in the wide spectral domain is requested. We shall consider $^2P_{1/2}$ and $^2P_{3/2}$ levels. For GdCOB:Nd^{3+} one intense line is detected in the low absorption spectrum for $^2P_{1/2}$ level: at 23150 cm^{-1}. In 5 % Nd^{3+} doped sample only one line is observed for $^2P_{3/2}$ level, at 25964 cm^{-1}, instead of two expected lines. From the corresponding $^2P_{3/2}$ barycenter in function of $^2P_{1/2}$ level (Figure 6) the second line is deduced at 26160 cm^{-1}. In fact, in 7 % Nd^{3+} doped crystal the second component is found at 26151 cm^{-1}. A difference between the experimental and predicted value is only 10 cm^{-1}, what is really negligible.

Two other neodymium doped compounds are presented in Figure 6. They are neodymium doped lanthanum fluoride and neodymium doped cubic C-type yttrium oxide. These two compounds are situated on two extremities of the slope. Their respective positions on this energy nephelauxetic scale are the indication of ionicity / covalency (LaF$_3$ / C-Y_2O_3) character of these compounds as well as the coordination number of the rare earth cation in these host lattices.

$^2F2_{5/2}$ manifold of $4f^3$ configuration is situated far in the UV spectral domain, at about 38000 cm^{-1}, and is isolated from other manifolds. The position of this manifold, or even only its barycenter position, is helpful in the free ion and crystal field calculation. There are not so many neodymium compounds for which this level is

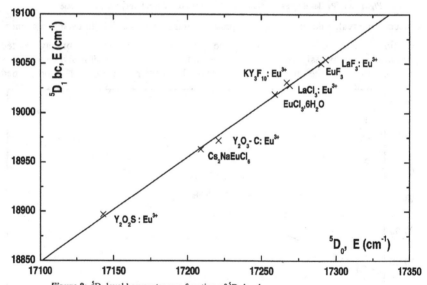

Figure 8: $^5D_1$ level barycenter as a function of $^5D_0$ level.

observed due to low transparency of the majority of host lattices. Nevertheless, from a few data we found in the literature, and ours, it is possible to establish the same relation already reported, $^2F2_{5/2}$ barycenter position in function of $^2P_{1/2}$ level. It is shown in Figure 7.

4f⁶ configuration

This advisement can be developed for other rare earth ions, for example, for europium ion. The emission of Eu^{3+} is intensely studied and this ion is frequently used as the local structural probe for identification of the crystallographic phases, the presence of impurity in compounds as well as for the determination of the crystal field (cf) parameters. The europium ion, Eu^{3+} ($4f^6$ configuration), possess the ground level manifold, ^7F_J, and the first excited level, $^5D_0$, situated at 12000 cm⁻¹ above the highest ground ^7F_J level.

The relationship of $^5D_0$ level with $^5D_1$ barycenter is presented in Figure 8. One can see quite linear dependence between these two levels, the first two excited levels of $4f^6$ configuration. In addition to the result presented in Figure 8, it is possible to link $^5D_0$ level with $^5D_2$ or $^5D_3$ level barycenters e.g., or any other barycenter of excited levels of Eu^{3+} ion.

4f⁷ configuration

The barycenter curves for gadolinium ion follow the same trend as presented previously for Pr^{3+}, Nd^{3+} and Eu^{3+} excited states. It has been verified for ^6P term and also for $^6D_{1/2}$ level situated between 40000 and 40500 cm⁻¹. On Figure 9 is presented $^6P_{5/2}$ barycenter energy position in function of $^6P_{7/2}$ bc for Gd^{3+} ion in different host lattices. As for other $4f^N$ configurations presented above, a fluoride is situated at higher energy value of the nephelauxetic scale and oxides at lower.

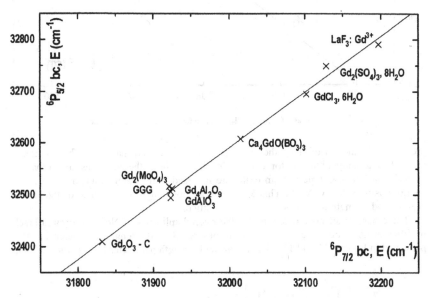

Figure 9: $^6P_{5/2}$ level bc as a function of $^6P_{7/2}$ level bc.

2.2.2. Ground manifold

The energy level barycenter values of different $^{2S+1}L_J$ levels in the ground manifold of $4f^N$ configurations are connected with the crystal field strength around the rare earth ion embedded in various crystalline environments. It is evidenced for $4f^2$, $4f^6$, and $4f^{13}$

configurations, respectively. Linear slopes are obtained for the ground level barycenter as a function of the upper level barycenters of the ground manifold.

$4f^3$ configuration

The nephelauxetic scale, previously obtained for the excited states of neodymium in various compounds classifying them in ionic or covalent ones [5, 13] is not found in the case of the ground term. The effect of the chemical liaison is overpass by the crystal field effect. It is clearly observed in Figure 10 where the $^4I_{9/2}$ ground level barycenter is

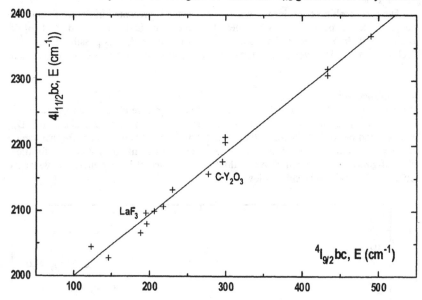

Figure 10:; $^4I_{11/2}$ level bc as a function of $^4I_{9/2}$ level bc for various Nd^{3+} compounds.

presented in function of $^4I_{11/2}$, the upper level of the ground manifold. The relation is linear like already obtained for excited levels, but here the positions of lanthanum fluoride and cubic C-type yttrium oxide are inversed, LaF_3 being situated at the lower energy position than C-Y_2O_3. This lower energy position is an evidence of the lower crystal field strength in the case of LaF_3 host lattice.

These barycenter positions, but also the overall splittings of Nd^{3+} ion ground levels in different host lattices, can be related to the crystal field strength parameter, N_v. It is presented in Figures 11 and 12. The crystal field strength parameter N_v is given by [14]:

$$N_v = \left[\sum_{kq} \left(\frac{4\pi}{2\kappa + 1} \right) \left| B_q^k \right|^2 \right]^{1/2}$$

If we present the overall splitting of the ground level $^4I_{9/2}$ of Nd^{3+} ion in different host lattices in function of N_v, the linear evolution is observed. It has been also reported by F. Auzel [15 - 17]

As previously shown in Figure 6, LaF_3 is also situated below $C-Y_2O_3$ in Figures 11 and 12. This lower position of LaF_3: Nd^{3+} on the slope is the confirmation of the lower crystal field strength in LaF_3 host lattice in respect to the cubic yttrium oxide.

Some ambiguity persists concerning the crystal field strength of Nd^{3+} ion in fluoroapatite $Ca_5(PO_4)F$ host lattice, (FAP). According to $^4I_{9/2}$ overall splitting, the

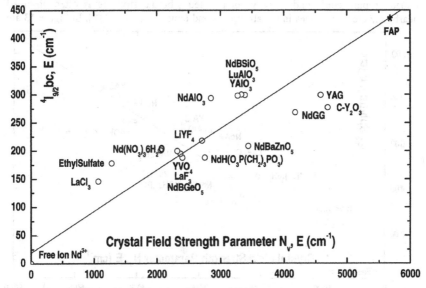

Figure 11 :$^4I_{9/2}$ level bc as a function of N_v, crystal field strength parameter.

crystal field strength parameter N_v should be drastically lower for FAP in Figure 11 than in Figure 12 with the respect to $^4I_{9/2}$ barycenter position. But on our knowledge, there are no full N_v data for Nd^{3+} ion embedded in FAP host lattice, so no clear answer can be done at the moment.

$4f^6$ configuration

Ground term of trivalent europium ion, $4f^6$ configuration, is another example we consider in this study. ^7F term is far from the first excited ^5D term, about 12000 cm^{-1} separates the highest $^7F_6$ level and the lowest $^5D_0$ level. One can consider then ^7F_J levels as quite pure. The position of $^7F_1$ barycenter is very important to be determined exactly, especially for the compounds for which the $^5D_0 \rightarrow {}^7F_0$ transition is forbidden, due to group theory rules, or not observed, due to low intensity of 0-0 line. Then, taking into account this $^7F_1$ bc position and the energy value of $^5D_0 \rightarrow {}^7F_1$ transition, it is possible to predict the exact energy position of $^5D_0$ electronic line [3].

$4f^{13}$ configuration

As it has been already reported on excited manifolds, some missing sublevels (or wrongly assigned) of a given $^{2S+1}L_J$ level of the ground manifold can be extrapolated on the basis of the corresponding barycenter positions deduced from slopes, too. It is particularly convenient for ytterbium trivalent ion, $4f^{13}$ configuration. In ytterbium doped or pure compounds a very complicated vibronic structure is often observed nearby the electronic lines in the absorption and emission spectra [18]. In Figure 13 are

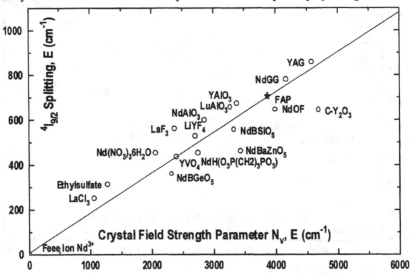

Figure 12 : $^4I_{9/2}$ level splitting as a function of N_v, crystal field strength parameter.

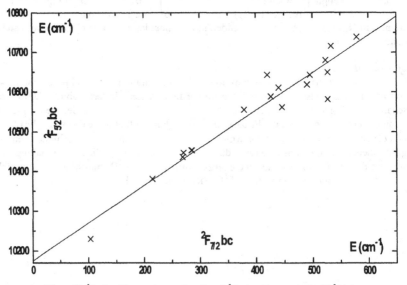

Figure 13: $^2F_{5/2}$ level barycenter as a function of $^2F_{7/2}$ level barycenter for Yb^{3+} ion.

shown the barycenter energy positions for two manifolds, $^2F_{5/2}$ and $^2F_{7/2}$, of Yb^{3+} ion doped several host lattices. The relation is almost linear but some points are dispersed. It, perhaps, suggests the wrong attribution of some crystal field levels inside the manifolds.

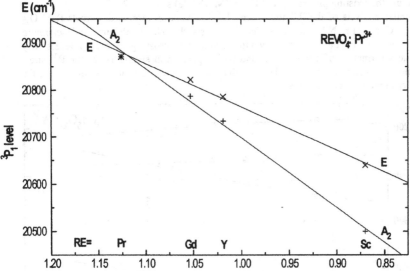

Figure 14: Evolution of two Stark components of $^2P_1$ level of Pr^{3+} ion in vanadates.

2.3. RARE EARTH IONS IN AN ISOSTRUCTURAL FAMILY

Absorption / emission lines of a given rare earth ion in an isostructural family exhibit a linear relation with the ionic radii of the rare earth cation of the host lattice along the lanthanide series. On Figure 14 are shown two Stark components of $^3P_1$ level of Pr^{3+} ion doped zircon-type rare earth orthovanadate along the lanthanide series [4]. $REVO_4$ (RE = La [19], Ce - Lu, Y and Sc included) belong to an isostructural family and the evolution of the spectral lines is linear from praseodymium to scandium. The identification of some lines in one of the compounds in the isomorphous series is sometimes impossible, or the line is not observed in the emission / absorption spectrum, then the extrapolation can be used.

In one isostructural family the evolution of some f-f levels along the lanthanide series suggests the phase limit existence. The rare earth gallium garnet family is a good example for this design.

2.4. PHASE LIMIT

To find a phase limit in one isomorphic family across the lanthanide series one can use semi-empirical methods based *e.g.* on the comparison of the actual - or extrapolated - bond lengths to those in the stable rare earth compounds as done by the bond valence model [20].

The extent of the strains supported by the lattice before the structural change occurs is very difficult to predict by this model, however. Otherwise, it seems possible to predict wherein the structural change may occur on the basis of the optical spectroscopic data.

The energy level scheme deduced from the electronic spectra of R^{3+} ion in the host lattice is very closely related to its structure and has been shown to be of the utility in predicting the existing phases in the MR_2O_4 (M=Sr, Ba) systems [21, 22].

Similar analysis of the R^{3+} ion energy level scheme in rare earth garnets $R_3M_5O_{12}$ (M=Al, Ga; R=Pr, Nd, Eu-Lu) systems suggests that the lanthanum and scandium compounds are beyond the stability range. Since the crossing of energy levels may indicate a formation of a new compound with higher point symmetry for the R^{3+} site - and thus a transformation of the lattice - this observation can be used in predicting the stability range of R compounds.

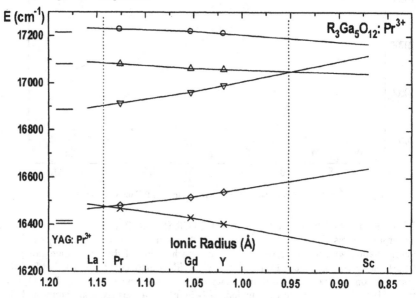

Figure 15: $^1D_2$ level in $R_3Ga_5O_{12}$: Pr^{3+} family as a function of R^{3+} ionic radii..

It is clear from curves presented in Figures 15 and 16 from which it is possible to predict the phase limit existence in gallium garnet systems for rare earth ionic radii between 1.14< IR >0.95 (Å). The same is obtained for Nd^{3+} ion in the same matrix.

3. Conclusion

In several examples presented in previous paragraphs one can see the advantages of the optical spectroscopy of rare earth ions in solids. First of all, in connection with the structure, in crystal field analysis and for identification of spectral lines in an isostructural family along the lanthanide series. Moreover, the utility of the "barycenter curves" in the assignment of electronic levels for a given rare earth in any compound is underlined.

We present a novel approach in the apprehension of electronic levels of f-f transitions, which could be applied for all studied compounds of a given rare earth ion.

And it holds for any rare earth ion. The linear evolution of $^{2S+1}L_J$ barycenters can be useful as a validity test in the assignment of absorption / emission lines. From the excited states an ionic-covalent character of the compound is apprehended (see Figure 2, LaF$_3$ and C-Y$_2$O$_3$) and from the ground multiplet one can appreciate the crystal field strength (see Figure 4, LaF$_3$ and C-Y$_2$O$_3$)-higher bc position means higher crystal field strength parameter N$_v$.

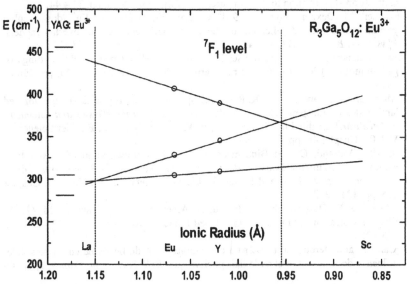

Figure 16 : ^{7}F$_1$ level in R$_3$Ga$_5$O$_{12}$: Eu^{3+} family as a function of R^{3+} ionic radii..

Phase limit existence in a rare earth isomorphic family can be predicted following the evolution of R^{3+} optical spectra along the lanthanide series. Rare earth gallium garnets doped with Pr^{3+} and Eu^{3+} (Nd^{3+} too) clearly illustrate the phase limit of an isostructural family.

4. References

1. Dieke, G.H. (1968) *Spectra and Energy Levels of Rare Earth Ions in Crystals,* Editors Crosswhite, H.M., and Crosswhite, H., Interscience Publishers, John Wiley&Sons, New York-London-Sydney-Toronto, Chapters 6 and 7, pp. 46-103.
2. Garcia, D. and Faucher, M. (1995) Crystal field in non-metallic (rare earth) compounds in *Handbook on the Physics and Chemistry of Rare Earths,* Edited by Gschneidner, Jr. K.A. and Eyring, L. Elsevier Science B. V., Vol. 21, Chapter 144, pp. 263-304.
3. Antic-Fidancev, E. (2000) Simple way to test the validity of $^{2S+1}L_J$ barycenters of rare earth ions (e.g. 4f^2, 4f^3 and 4f^6 configurations), *J. Alloys Comp.*, **300-301**, pp. 2 - 10
4. Antic-Fidancev, E. Lemaître-Blaise, M. and Porcher, P. (1998) Optical study of praseodymium 3+ in zircon-type orthovanadate phases, *Spectrochimica Acta* A **54**, pp. 2151 - 2156.

5. Antic-Fidancev, E. Lemaître-Blaise, M. and Caro, P. (1987) The lanthanides' nephelauxetic effect revisited, *New J. Chem.*, (France) **11**, pp. 467 - 472

6. Antic-Fidancev, E. Aride, J. Chaminade, J.P. Lemaître-Blaise, M. and Porcher, P. (1992) The aragonite-type neodymium borate $NdBO_3$: energy levels, crystal field and paramagnetic susceptibility calculations, *J. Solid State Chem.*, **97**, pp. 74 - 81.

7. Antic-Fidancev, E. Lemaître-Blaise, M. Parada, C. (1996) Spectroscopic study of rare earth chromates: relation to the structure, *Acta Phys. Pol.* **A 90**, pp. 33 – 44.

8. Faucher, M. Dexpert-Ghys, J. (1981) Crystal-field analysis of Eu^{3+} doped cubic yttrium sesquioxide. Application of the electrostatic and angular overlap models, *Phys. Rev. B* **24**, pp. 3138 - 3144.

9. Antic-Fidancev, E. Lemaître-Blaise, M. Caro, P. (1982) Crystal field splittings of gadolinium ($4f^7$) levels in C-type rare earth oxides, *J. Chem. Phys.*, **76**, pp. 2906 - 2913.

10. Morrison, C. A. and Leavitt, R. P. (1982) Spectroscopic properties of triply ionized lanthanides in transparent host crystals in *Handbook on the Physics and Chemistry of Rare Earths*, Edited by Gschneidner, Jr. K.A. and Eyring, L. Elsevier Science B. Vol. 05, Chapter 46, pp. 461 - 692.

11. Görller-Walrand, C. and Binnemans, K. (1996) Rationalization of crystal field parametrization in *Handbook on the Physics and Chemistry of Rare Earths*, Edited by Gschneidner, Jr. K.A. and Eyring, L. Elsevier Science B. V., Vol. 23, Chapter 155, pp. 121 - 283.

12. Moon, O. K. Dexpert-Ghys, J. Piriou, B. Alves, M. G. and Faucher, M. (1998) Electronic structure of Pr^{3+} and Tm^{3+} doped $LiYO_2$, *J. Alloys Compds.*, **275-277**, pp. 258 - 263.

13. Caro, P. and Derouet, J. (1972) La configuration $4f^3$ du néodyme en phase solide: influence de la structure et de la liaison chimique, *Bull. Soc. Chim. France*, **1**, pp. 46 - 54.

14. Auzel, F. and Malta, O.L. (1983) A scalar crystal-field strength parameter for rare earth ions: meaning and usefulness, *J. Phys.* **44**, pp. 201- 206.

15. Auzel, F. (1979) L'auto-extinction de Nd^{3+}: son mécanisme fondamental et un critère prédictif simple pour les matériaux minilaser, *Mater. Res. Bull.* **14**, pp. 223 - 231.

16. Auzel, F. (2001) On the maximum splitting of the $(^2F_{7/2})$ ground state in Yb^{3+} - doped solid state laser materials, *J. Lumin.* **93**, pp. 129 - 135.

17. Auzel, F. (2002) A relationship for crystal field strength along the lanthanide series; application to the prediction of the $(^2F_{7/2})$ Yb^{3+} maximum splitting, *Opt.. Mater. 19*, pp. 89 - 94.

18. Haumesser, P.H. Gaumé, R. Viana, B. Antic-Fidancev, E. and Vivien, D. (2001) Spectroscopic and crystal field analysis of new Yb-doped laser materials, *J. Phys.: Condens. Matter*, **13**, pp. 5427 - 5447.

19. Antic-Fidancev, E. Lemaître-Blaise, M. Porcher, P. (1998) Optical evidence of the zircon-type lanthanum orthovanadate, *Ann. Chim.*, (Paris) **23**, pp. 273 - 276.

20. Brown, I. D. (1987) Recent Developments in the Bond Valence Model of Inorganic Bonding, *Phys. Chem. Miner.*, **15**, pp. 30 - 34.

21. Taibi, M. Antic-Fidancev, E. Aride, J. Lemaître-Blaise, M. and Porcher, P. (1993) Spectroscopic properties of europium (3+) doped $SrRE_2O_4$; (RE Eu, Gd, Y and In): crystal-field analysis and paramagnetic susceptibility measurements, *J. Phys.: Condens. Matter*, **5**, pp. 5201 - 5208.

22. Taibi, M. Aride, J. Antic-Fidancev, E. Lemaître-Blaise M. and Porcher, P. (1994) Fluorescence of Eu^{3+} in $BaRE_2O_4$ (RE = Gd, Y). Determination of the crystal field parameters, *Phys. Status Solidi (a)*, **144**, 453 - 459.

LOCATION OF CHARGE-COMPENSATING VACANCY IN IONIC CRYSTALS DOPED WITH RARE EARTH IONS:

In Case of Cubic Perovskite KMgF$_3$ Doped with Eu^{2+} Ions

T. TSUBOI

Faculty of Engineering, Kyoto Sangyo University,
Kamigamo, Kita-ku, Kyoto 603-8555, Japan
E-mail: tsuboi@cc.kyoto-su.ac.jp

Abstract. Optical study on Eu^{2+} ions in KMgF$_3$ crystal is reviewed. Eu^{2+} ions are substituted for monovalent K$^+$ ions in KMgF$_3$ crystal, resulting in creation of charge-compensating K$^+$ vacancies. Spectroscopic study and electron paramagnetic resonance study have been undertaken to determine the position of Eu^{2+} ions and vacancies. Two-photon excitation spectroscopy indicates coexistence of Eu^{2+} ions with site symmetries of cubic, C$_{3v}$, C$_{4v}$ and C$_{2v}$, which are created depending on location of vacancy. It is suggested that, of three positively-charged ions K$^+$, Eu^{2+} and Mg^{2+} which attract the negatively charged vacancy, the vacancy has Coulomb interaction with not only Eu^{2+} but also the same divalent Mg^{2+}, giving rise to frustration to vacancy in selecting the location and resulting in various locations of vacancy. This suggestion is confirmed to be reasonable by comparing with the cases of Eu^{2+} doped KCl and Ce^{3+} doped KMgF$_3$.

1. Introduction

It is important to determine the electronic energy levels and locations of luminescent and laser-active ions in host lattice sites to achieve high lasing and luminescence efficiencies [1]. Like Nd^{3+} doped YAG, rare-earth ions are used as laser-active ions. Doping of divalent and trivalent rare-earth ions creates charge compensating vacancies in ionic crystals, which change the optical characteristics of rare earth ion. In Eu^{2+}-doped KMgF$_3$ crystals, it is conceivable that Eu^{2+} is substituted for positively charged ion Mg^{2+} or K$^+$.

KMgF$_3$ crystal has a cubic perovskite crystal structure as shown in Fig.1. Mg^{2+} ion is surrounded by six nearest neighbor F$^-$ ions and eight next-nearest neighbor K$^+$ ions, while K$^+$ ion is surrounded by twelve nearest neighbor F$^-$ ions and eight next-nearest neighbor Mg^{2+} ions. Firstly we have to answer a question "At which place Eu^{2+} ion is located in the KMgF$_3$ lattice when Eu^{2+} ions are doped ? ". Eu^{2+} ion has ionic radius of 1.12 Å, while K$^+$ and Mg^{2+} have radii of 1.33 and 0.65 Å, respectively. Mg^{2+} ion site is too small to substitute Eu^{2+} ion although they are the same divalent ions, while K$^+$ ion site is enough space to substitute Eu^{2+} ion. Therefore it is assumed that the Eu^{2+} substitutes for the singly-charged K$^+$ with a

93

J.-C. Krupa and N.A. Kulagin (eds.), Physics of Laser Crystals, 93–107.
© 2003 *Kluwer Academic Publishers. Printed in the Netherlands.*

twelve-fold F coordination in KMgF$_3$ lattice.

Spectroscopic study (measurements of one-photon absorption and emission spectra and emission decay time) and electron paramagnetic resonance study have been undertaken to confirm the position of Eu^{2+} ions and vacancies [2]. Recently more accurate method has been made using two-photon excitation spectroscopy (i.e. measurement of excitation spectra for luminescence produced by excitation using two photons).

Here we review the previous studies on Eu^{2+} doped KMgF$_3$ including recent study by the present author [2, 3] and describe how the presence and location of vacancy in Eu^{2+} doped KMgF$_3$ have been determined. Additionally we describe what kind of element is responsible for the location of vacancy in the KMgF$_3$ lattice. To confirm whether the suggested element is reasonable and check whether it is applicable to other rare earth ions, we examine a case of Ce^{3+} ion in KMgF$_3$.

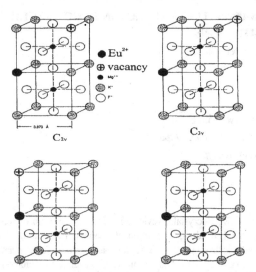

● Eu^{2+}
⊕ vacancy
● Mg^{++}
● K$^+$
○ F$^-$

C$_{2v}$ C$_{3v}$

Figure 1. Crystal structure of KMgF$_3$ with Eu^{2+} ion, showing various possible locations of charge-compensating K$^+$ ion vacancy.

2. Energy Levels of Eu^{2+} and Gd^{3+} Ions

Eu^{2+} ion has the electronic configuration 4f^{7}5s^{2}5p^6 (named f^7 hereafter) in the outer shell in the ground state. The next high-energy electronic configuration is 4f^{6}5s^{2}5p^{6}5d (named f^{6}d). Gd^{3+} ion has the same electron configuration as Eu^{2+}ion, i.e. Gd^{3+} is isoelectronic to Eu^{2+}. The f^7 configuration gives rise to the ground state ^{8}S$_{7/2}$ and the excited state multiplets ^{6}P$_J$ (J = 7/2, 5/2, 3/2 in order of increasing energy), ^{6}I$_J$ (J = 7/2, 9/2, 17/2, 11/2, 15/2, 13/2) and ^{6}D$_J$ (J = 9/2, 1/2, 7/2, 3/2, 5/2) with increasing energy by Coulomb interaction among the seven 4f-electrons [4]. These multiplets split into several sublevels by spin-orbit interaction and furthermore by crystal field although the amount of crystal field splitting is much smaller than that

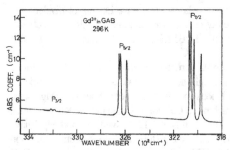

Figure 2. Absorption bands due to the ^{8}S$_{7/2}$ → ^{6}P$_{7/2}$, ^{6}P$_{5/2}$ and ^{6}P$_{3/2}$ transitions of Gd^{3+} ions in GdAl$_3$(BO$_3$)$_4$ (named GAB) crystal at 296 K [6]. ABS. COEFF. : absorption coefficient

of the spin-orbit splitting. These excited states are located at 32000-41000 cm^{-1} (corresponding wavelength is 312.5-243.9 nm) above the ground state [4, 5]. The positions and order of the multiplets are almost the same in various materials (even if they are aqueous solutions) since the $4f^7$- electrons are shielded by the outer $5s^25p^6$- electrons from the surrounding ligand ions of Gd^{3+} and therefore the energies of their states are not modified by the ligand field. The sharp absorption lines due to the electronic transition from the ground state $^8S_{7/2}$ to all the excited state multiplets ^6P_J, ^6I_J and ^6D_J are observed in various Gd^{3+} doped materials [5-13]. In addition to these excited state, the other $4f^7$- state (e.g. ^6F_J and ^4H_J) and the $4f^65d$- states are located in the vacuum-ultraviolet region, which have been studied in LiYF$_4$ [14] and CaF$_2$ [15], respectively.

An example of absorption spectrum due to the $^8S_{7/2} \rightarrow ^6P_J$ (J = 7/2, 5/2, 3/2) transitions in Gd^{3+} doped GdAl$_3$(BO$_3$)$_4$ is shown in Fig.2 [6]. Four, three and two absorption lines are observed for the $^8S_{7/2} \rightarrow ^6P_J$ (J = 7/2, 5/2, 3/2) transitions, respectively. Gd^{3+} ion has a hexagonal site symmetry (precisely speaking, D$_{3h}$ symmetry) in GdAl$_3$(BO$_3$)$_4$ crystal. The $^6P_{7/2}$, $^6P_{5/2}$ and $^6P_{3/2}$ states split into four, three and two levels in the hexagonal crystal field, respectively. Therefore the electronic transitions from the non-degenerated $^8S_{7/2}$ state to these states give rise to the absorption spectrum as shown in Fig. 2. Bouazaoui observed three and two lines in the excitation spectra related with the $^8S_{7/2} \rightarrow ^6P_{7/2}$ and $^6P_{5/2}$ transitions of Gd^{3+} in Cs$_2$NaGdCl$_6$ crystal [16]. The Gd^{3+} ion has an octahedral site symmetry in this crystal, thus the $^6P_{7/2}$ and $^6P_{5/2}$ states split into three and two levels, respectively. Like this we understand the observed the absorption and excitation spectra of Gd^{3+}. In the case of Gd^{3+} in Cs$_2$NaGdCl$_6$, the same numbers of lines were observed in the excitation spectra for emission produced by optical excitation to the $^6P_{7/2}$ and $^6P_{5/2}$ states by two-photon absorption, i.e. in the two-photon excitation spectra [16].

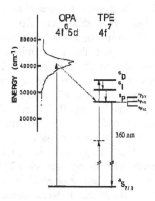
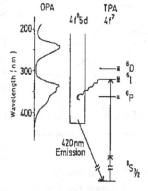

Figure 3. Schematic diagram of the low-energy levels of Eu^{2+} in KMgF$_3$ crystal showing two-photon excitation (TPE), relaxation and luminescence [28]. The one-photon absorption (OPA) is drawn at the left side.

Figure 4. Schematic diagram of the low-energy levels of Eu^{2+} in KCl crystal showing TPE, relaxation and luminescence [29]. OPA is drawn at the left side. Alkali ion is surrounded by six halide ions.

Unlike the case of Gd^{3+}, sharp absorption lines have not observed in Eu^{2+} doped crystals. Intense and broad absorption bands have been observed in e.g. KMgF$_3$ and KCl crystals as shown at the left side of Figs.3 and 4, respectively. These broad bands are caused by not the intra-configurational $f^7(^8S_{7/2}) \rightarrow f^7$ transition but inter-configurational electronic $f^7(^8S_{7/2}) \rightarrow f^6d$ transition. The $f^7 \leftrightarrow f^7$ transitions are parity-forbidden and the electric dipole transition is not allowed, while the $f^7 \rightarrow f^6d$ transition is parity-allowed, resulting in much intense absorption than the $f^7 \leftrightarrow f^7$ absorption bands [1, 17]. Contrary to the case of Eu^{2+} doped crystals, the f^6d states lie far above the f^7 states in Gd^{3+} doped crystals. Therefore we can observe the absorption bands due to the $f^7 \leftrightarrow f^7$ transitions in ultraviolet spectral region without interference by absorption bands due to the $f^7 \rightarrow f^6d$ transition.

The energy of the f^6d states depends on the crystal field, because, unlike the 4f electrons, the 5d electron is in the outer shell and not shielded by the $5s^25p^6$ - electrons. The 5d energy level is split by crystal field into two to five components. The amount of the splitting depends on the strength of crystal field. Fig.5 shows the schematic energy level diagram of the f^7- and f^6d- configurations where the splitting of 5d- level by crystal field is also illustrated. The Eu^{2+} emission is caused by electronic transition from the lowest excited state to the f^7 ($^8S_{7/2}$) state. When the lowest excited state $^6P_{7/2}$ of the f^7- configuration is located below the lowest state of f^6d- configuration, sharp emission line is observed. In the reverse case, a broad emission band is observed by the transition from the lowest f^6d- state. Thus the emission wavelength depends on the strength of crystal field around Eu^{2+} ion, i.e. it depends on the host material. e.g. sharp 360 nm emission line is observed in $KMgF_3$ [3], $RbMgF_3$ [18], $SrAlF_5$ [19,20], a broad emission band with peak at 390 nm in $BaFBr$ [21], at 400 nm in BaF_2 and $BaCl_2$ [22], at 410 nm in SrF_2 and $SrCl_2$ [22], at 425 nm in CaF_2 [22,23], at 430 nm in $CaCl_2$ [22], and broad violet, blue, green, yellow and red emission bands in $Sr_2P_2O_7$, $BaAl_{12}O_{19}$, $SrAl_2O_4$, Ba_3SiO_5 and CaS, respectively [24], as shown in Fig.5.

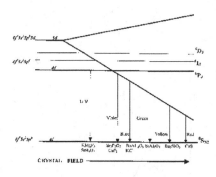

Figure 5. Energy level diagrams of the f^7- and f^6d – configurations of Eu^{2+} ion as a function of the crystal field, with Eu^{2+} emission observed in various crystals.

It should be noted that, when in Eu^{2+} doped $Ca_xSr_{2-x}MgSi_2O_7$ the concentration ratio x of Ca^{2+} and Sr^{2+} is changed, the peak of the Eu^{2+} broad emission band shifts from 470 nm to 550 nm with increasing x value from 0 to 2 [25], indicating that the Eu^{2+} emission is sensitive to the strength of ligand field. Like this, either sharp emission line or broad emission band is observed depending on crystal. In SrB_4O_7 [26], $BaFCl$ and $SrFCl$ [22], however, both sharp 360 nm emission line and broad 390 nm emission band are observed. The sharp line is observed predominantly at low temperatures below 77 K, while the broad band is observed at high temperatures like 293 K. The broad emission band is induced by thermal activation from the low-energy-lying $^6P_{7/2}$ state with increasing temperature, i.e. by crossover from the $^6P_{7/2}$ level to the lowest f^6d- level in a configuration coordinate diagram [1, 22, 26].

In $KMgF_3$, CaF_2 and alkali halide crystals, the f^6d energy states are located close to the f^7- state, i.e. the optical absorption bands due to the $f^7 \rightarrow f^7$ transition of Eu^{2+} appear to lie quite close to the bands due to the $f^7 \rightarrow f^6 d$ transition. Therefore it is difficult to obtain clear evidence of the $f^7 \leftrightarrow f^7$ absorption bands because their intensity is considerably weaker than the $f^7 \rightarrow f^6 d$ intensity. So far, weak absorption bands have been reported as the $f^7(^8S_{7/2}) \rightarrow f^7$ absorption bands in Eu^{2+} doped $SrAlF_5$ [19], without the detailed description and analysis. The excited f^7- states of the Eu^{2+} ion in $KMgF_3$ are expected to lie at much lower energy with respect to the $f^6 d$ states than in other ionic crystals such as CaF_2 and alkali halides [17]. Therefore it should be possible to observe the $f^7 \leftrightarrow f^7$ absorption spectrum separated from the intense $f^7 \rightarrow f^6 d$ absorption spectrum clearly in $KMgF_3$. We tried to find the $f^7(^8S_{7/2}) \rightarrow f^7$ absorption bands in $KMgF_3$.

An intense absorption band due to $f^7 \rightarrow f^6 d$ dipole-allowed transition is observed in the region 30000 - 45000 cm^{-1} as shown in Fig.6 [2]. Additionally two very weak absorption bands are observed at the low energy side of the intense band. Their peaks are at 27830 and 28251 cm^{-1} (the corresponding wavelengths are about 359.3 and 353.9 nm, respectively) at 296 K. The 27830 cm^{-1} band is more intense than the 28251 cm^{-1} band. From comparison with the isoeletronic Gd^{3+} ion spectrum shown in Fig.2 and the previous studies of luminescence [27, 28],

the weak bands are attributed to the $f^7 \rightarrow f^7$ absorption : the lower-energy band is attributed to the $^8S_{7/2} \rightarrow {}^6P_{7/2}$ transition, while the higher-energy one to the $^8S_{7/2} \rightarrow {}^6P_{5/2}$ transition. Contrary to the case of Gd^{3+} in $GdAl_3(BO_3)_4$ crystal (Fig.2), the $^8S_{7/2} \rightarrow {}^6P_{7/2}$ band consists of three lines at 27827, 27837 and 27846 cm^{-1} in $KMgF_3$, while the $^8S_{7/2} \rightarrow {}^6P_{5/2}$ band consists of two lines at 28262 and 28270 cm^{-1}. Their enlarged spectra are shown in upper part of Fig.6. Absorption bands due to the $^8S_{7/2} \rightarrow {}^6P_{3/2}$, ^6I_J and ^6D_J transitions were not found because they were immersed under the $f^7 \rightarrow f^6 d$ absorption bands. These bands are observed not by one-photon absorption but they are observed by luminescence using one-photon or two-photon excitation spectroscopy.

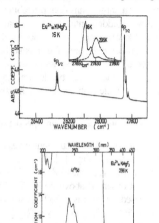

Figure 6. Absorption spectra of $KMgF_3:Eu^{2+}$

3. Two-Photon Excitation Spectroscopy

The two-photon excitation spectroscopy is undertaken using two photons with same photon energy (i.e. same light wavelength) [28-33]. The first photon gives rise to impurity ions such as Eu^{2+} to excite to an intermediate state (which does not correspond to any f^7 (or $f^6 d$) state) from the ground state G by electric dipole transition, and immediately (i.e. simultaneously) the second photon gives rise to excite the impurity ion at the intermediate state to the upper excited state K by the electric dipole transition. By such a simultaneous absorption of two photons, the impurity a simultaneous absorption of two photons, the impurity ion is excited to the upper state K although the electric dipole transition from the ground state G to the excited state K is not allowed by one photon absorption. For example, the transition from the ground state $f^7(^8S_{7/2})$ to the excited state $f^7(^6P_{7/2})$ at 27840 cm^{-1} of Eu^{2+} [3] is parity- and electric-dipole forbidden but magnetic-dipole allowed, i.e. the excitation to the f^7 $(^6P_{7/2})$ state from the ground state is not undertaken in $KMgF_3$ by electric-dipole transition by absorption of one photon with energy of

27840 cm⁻¹ (i.e. wavelength of 359.2 nm). However, its transition is allowed by two photons, each of which has the same photon wavenumber of 13920 cm⁻¹ (i.e. half the energy 27840 cm⁻¹, the corresponding wavelength is 718.4 nm), i.e. the transition from the ground state $^8S_{7/2}$ to the excited state $^6P_{7/2}$ is undertaken by intense light of 718.4 nm wavelength. This is a non-linear optical process. Eu^{2+} ions excited to the f⁷ ($^6P_{7/2}$) state come back to the ground state, emitting photon by magnetic dipole transition. Thus the 359.1 nm luminescence is observed in Eu^{2+} doped $KMgF_3$. We can measure the two-photon excitation spectrum for the 359.1 nm emission by varying the wavelength (i.e. photon energy) of infrared light. In this way we can observe all the Stark levels of the $^6P_{7/2}$ state without being interfered by the intense f⁷→f⁶d absorption bands. The stark levels are revealed for not only the $^6P_{7/2}$ state but also the other high energy states such as $^6P_{5/2}$ state by monitoring the 359.1 nm $^6P_{7/2}$ emission as shown at the right side of Fig.2. Emission from the lowest f⁶d state is used as the monitored emission in case of alkali halides as shown at the right side of Fig.3.

Unlike the case of usual absorption process by one-photon transition, such an intraconfigurational transition (by the two-photon absorption) is electric-dipole allowed, giving rise to a large transition probability. As a result we obtain intense signal for the originally parity-forbidden intraconfigurational transition. In summary, (1) the excited f⁶d state lies at much lower energy in Eu^{2+} ion, it completely overlaps even the lowest sharp f⁷ excited states, (2) consequently only the intense broad, electric dipole allowed f⁷→f⁶d absorption bands are observed in the one-photon absorption spectra, (3) by two-photon excitation spectroscopy, however, the broad bands are suppressed by the selection rule of two-photon transition, resulting in clear observation of the sharp f⁷ states.

4. Two-Photon Excitation Spectra of Eu^{2+} in $KMgF_3$

Optical excitation of Eu^{2+} ions gives rise to sharp emission bands at 359 and 322 nm in $KMgF_3$ at 10 K as seen in Fig.7 [3]. The 322 nm emission is considerably weaker than the 359 nm emission (its intensity is about 1/68 of the 359 nm emission intensity) and it is observed at low temperatures but not observed above 20 K. The 359 nm emission shifts to low energy side as temperature is increased, its peak is at 360 nm at room temperature. The 359 and 322 nm emission are attributed to the $^6P_{7/2}$→ $^8S_{7/2}$ and $^6I_{7/2}$→ $^8S_{7/2}$ [3, 27]. The decay time of 359 nm emission is 6.9 ms, while the 322nm emission from the $^6I_{7/2}$ nm emission is 37 s. state takes much shorter decay time is as follows. The energy gap between the $^6I_{7/2}$ state and lower $^6P_{3/2}$ state is about 2600 cm⁻¹ which is much bigger than the gap between the $^6P_{7/2}$ state and the lower $^8S_{7/2}$ state (i.e. the gap is 27840 cm⁻¹). Therefore the non-radiative multi-phonon relaxation occurs from the $^6I_{7/2}$ state to the $^6P_{3/2}$ state because the highest phonon energy of $KMgF_3$ is 530 cm⁻¹, while the non-radiative multi-phonon relaxation from the $^6P_{7/2}$

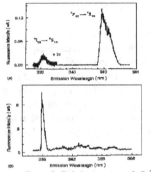

Figure 7. Emission spectrum of Eu^{2+} in $KMgF_3$ crystal which was excited by two-photon absorption with 581.58 nm laser radiation at 10 K [3].

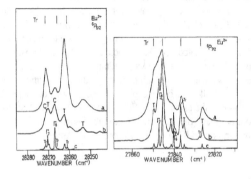

Figure 8. $^8S_{7/2}$ → $^6P_{7/2}$ transition of Eu^{2+} ions in $KMgF_3$ at 16 K

state to the $^8S_{7/2}$ state is impossible. As a result, the lifetime of the $^6I_{7/2}$ state becomes shorter than that of the $^6P_{7/2}$ state, giving rise to shorten the decay time of the 322 nm emission due to the $^6I_{7/2} \rightarrow {}^8S_{7/2}$ transition.

On Fig.8, the excitation spectra for the 359 nm emission by one-photon and two-photon absorption processes are shown together with one-photon absorption spectra [2]. The right side of figure shows a fine structure of the one-photon absorption band (Curve a) due to the $^8S_{7/2} \rightarrow$ $^6P_{7/2}$ transition of Eu^{2+} ions in $KMgF_3$ at 16 K, compared with the one-photon excitation spectrum in the $^8S_{7/2} \rightarrow {}^6P_{7/2}$ spectral range (Curve b, obtained by Ellens et al [27]) for the luminescence due to $^6P_{7/2} \rightarrow {}^8S_{7/2}$ transition at 4.2 K and compared with the two-photon excitation spectrum (curve c, obtained by Francini et al [28]) for the luminescence at 15 K. Γ_6 means the Γ_6 symmetry absorption band due to Eu^{2+} at cubic lattice site. Tr means the absorption bands due to trigonal Eu^{2+} ion. The left side shows a fine structure of the absorption band (curve a) due to the $^8S_{7/2} \rightarrow {}^6P_{5/2}$ transition of Eu^{2+} ions in $KMgF_3$ at 16 K, compared with the one-photon excitation spectrum in the $^8S_{7/2} \rightarrow {}^6P_{5/2}$ spectral range (Curve b, obtained by Altshuler et al [34]) at 77 K and compared with the two-photon excitation spectrum (Curve c, obtained by Francini et al [28]) for the luminescence at 359 nm which is due to $^6P_{7/2} \rightarrow {}^8S_{7/2}$ transition at 15 K.

It is observed that the one - photon absorption spectrum due to the $^8S_{7/2} \rightarrow {}^6P_{7/2}$ transition is quite similar (in the lineshape and peak positions) to the corresponding one- and two- photon excitation spectra for the luminescence transition. The excitation for lines for the two-photon-excited 359 nm narrower than the absorption lines, making possible to reveal several additional lines related the $^8S_{7/2} \rightarrow {}^6P_{7/2}$ and $^6P_{5/2}$ transition. The excitation lines for one-photon-excited 359 nm emission is similar to those for two-photon-excited emission, but the former is broader than the latter and some of them are overlapped with each other. Two-photon allowed transition takes place between states weakly coupled to the crystal lattice and lattice vibration, resulting in unusually narrow excitation lines with well resolved Stark components [32]. Therefore the two-photon excitation spectra show much more lines in both the $^8S_{7/2} \rightarrow$ $^6P_{7/2}$ and $^6P_{5/2}$ regions than the bands observed by one-photon absorption spectra

It is assumed that the Eu^{2+} substitutes for the singly-charged K^+ with a twelve-fold F coordination in $KMgF_3$ lattice as mentioned above. In this case the charge-compensating positive ion vacancy is necessary and it is expected to be located at one of nearest K^+ sites around the Eu^{2+} ion. If the vacancy is present at one of the nearest-neighbour K^+ sites in the <001> axis (shown at left side in the lower part of Fig.1), the site symmetry of Eu^{2+} is tetragonal C_{4v}, and if the vacancy is located at one of the third or second nearest-neighbour K^+ site in the <111> or <110> axis (shown at right or left side in the upper part of Fig.1), respectively, the site symmetry is trigonal C_{3v} or orthorhombic C_{2v}. Francini et al. have attributed three relatively intense lines (at 27840.3, 27845.8 and 27847.3 cm^{-1}) observed in the $^8S_{7/2} \rightarrow {}^6P_{7/2}$ two-photon spectrum to the Eu^{2+} ion at cubic site [28]. The $^6P_{7/2}$ state is split into three components Γ_6, Γ_8 and Γ_7 states with increasing energy by the cubic crystal field. Therefore the 27840.3, 27845.8 and 27847.3 cm^{-1} lines are assigned to the Γ_7, Γ_8 and Γ_6 states by crystal field calculation, respectively, as shown in Fig.8. Francini et al. have attributed four weak lines (at 27835.0, 27836.8, 27841.5 and 27849.8 cm^{-1}) to the Eu^{2+} at trigonal site from the crystal-field calculation, while two weak lines (at 28267.0 and 28270.5 cm^{-1}) observed in the $^8S_{7/2} \rightarrow {}^6P_{5/2}$ two-photon spectrum to the cubic Eu^{2+} and three lines (at 28261.0, 28262.3 and 28269.7 cm^{-1}) to the trigonal Eu^{2+} [28]. Besides these lines, there are several intense lines (at 27827, 27843, 28253 and 28271 cm^{-1} lines), which have not been assigned yet, these are shown by arrows in Fig.8. By the electron spin resonance experiment, Eu^{2+} ion at tetragonal site has been observed in addition to the cubic Eu^{2+} [35]. Therefore it is suggested that the unassigned lines are due to the tetragonal Eu^{2+}. The tetragonal Eu^{2+} gives rise to four and three absorption lines in the $^8S_{7/2} \rightarrow ^6P_{7/2}$ and $^8S_{7/2} \rightarrow ^6P_{5/2}$ transitions, respectively. All of these lines are not observed in Fig.8 clearly since it seems that some lines are overlapped with the lines due to the cubic and trigonal Eu^{2+} ions.

Two-photon excitation spectra for the 359 and 322 nm emission is also observed for the higher excited states ^6I_J (J = 7/2, 9/2, 17/2, 11/2, 15/2, 13/2) and ^6D_J(J = 9/2, 1/2, 7/2, 3/2, 5/2) as shown in Fig.9 [3]. Two-photon excitation spectra due to the $^8S_{7/2} \rightarrow ^6D_J$ and ^6I_J regions of Eu^{2+} ions in KMgF$_3$ crystal at 10 K [3]. The lower curve in each figure is obtained for the 322 nm $^6I_{7/2}$ emission, while the upper curve is obtained for 359.1 nm $^6P_{7/2}$ emission.

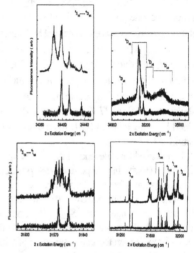

The excitation lines are much narrower for the 322 nm $^6I_{7/2}$ emission than for the 359 nm $^6P_{7/2}$ emission, and number of lines is smaller for the former emission than for the latter emission. This is explained as follows. The $^6I_{7/2}$ level lies below the lowest level of f^6d –configuration for Eu^{2+} ion at cubic site (i.e. Eu^{2+} which is not accompanied by vacancy at neighboring site), while it lies above the f^6d level for non-cubic site Eu^{2+} ions

Figure 9. Two-photon excitation spectra due to the $^8S_{7/2} \rightarrow ^6D_J$ and ^6I_J regions of regions of Eu^{2+} ions in

(i.e. Eu^{2+} ions which are accompanied by vacancy at neighbors). When the non-cubic Eu^{2+} ions are excited to the ^6I_J and ^6D_J states, they relax to the lowest f^6d level and then to the $^6P_{7/2}$ state, giving rise to only the 359 nm emission, while when the cubic Eu^{2+} ions are excited to the ^6I_J and ^6D_J states, they relax to the lowest f^6d level and then to the $^6I_{7/2}$ state, giving rise to the 322 nm emission, and simultaneously they relax from the $^6I_{7/2}$ state to the $^6P_{7/2}$ state non-radiatively, giving the additional 359 nm emission as shown in Fig.10. Two-photon excitation, relaxation and emission processes are shown by upward, wave-like and downward arrows, respectively. Δ E means energy gap between the lowest f^6d level and $^6I_{7/2}$ level [3]. Since the effect of crystal field is different between

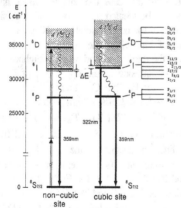

Figure 10. Energy level diagram of cubic- and non-cubic sites of Eu^{2+} ions in KMgF$_3$ crystal

cubic and non-cubic Eu^{2+} ions, such a consideration is possible. Non-cubic Eu^{2+} ions consist of Eu^{2+} with different site symmetry therefore they give rise to a broad excitation lines as observed (because several lines due to different site Eu^{2+}ions are located close to each other).

In the following chapters we describe the f^6d states of Eu^{2+} ions in other crystals and compare them with the case of $KMgF_3$ to understand the observed two-photon excitation spectra of Fig.8.

5. One-Photon Excitation Spectra of LiBaF$_3$

$LiBaF_3$ has the same crystal structure as cubic perovskite $KMgF_3$. Similar to the case of $KMgF_3$, a narrow sharp emission band is observed at 359 nm at low temperatures below 150 K. On the other hand, in addition to the narrow emission band, a broad emission band with a peak at about 420 nm is observed above 150 K [36]. The 359 nm emission is caused by the $^6P_{7/2} \rightarrow ^8S_{7/2}$ transition, while the 420 nm broad emission band is caused by the transition from the lowest f^6d- state to the $f^7(^8S_{7/2})$ state. Since theenergy difference between the $^6P_{7/2}$ state with lowest f^6d- state is 800 cm^{-1} [36], the f^6d- state is feed up by thermal population from the $^6P_{7/2}$ state with increasing temperatures. Unlike the case of $KMgF_3$, only two narrow one-photon excitation lines were observed for the 359 nm emission in both the $^8S_{7/2} \rightarrow ^6P_{7/2}$ and $^8S_{7/2} \rightarrow ^6P_{5/2}$ regions by Altshuler et al at 77 K [37]. Meijerink investigated the excitation spectra by high-resolution luminescence spectroscopy at low temperature, and revealed that the high energy line of the two lines in the $^8S_{7/2} \rightarrow ^6P_{7/2}$ region consists of two lines at 4.2 K which are located close to each other [36]. Thus, three and two narrow one-photon excitation lines are observed in the $^8S_{7/2} \rightarrow ^6P_{7/2}$ and $^8S_{7/2} \rightarrow ^6P_{5/2}$ regions, respectively. This is understood as follows.

Li^+ and Ba^{2+} have radii of 0.70 and 1.40 A, respectively. Ba^{2+} ion site is enough space to substitute Eu^{2+} ion (1.12 Å). Thus, it is believed that Eu^{2+} ion is substituted for the same divalent Ba^{2+} ion. In this case no charge-compensating vacancy is necessary. Eu^{2+} ion has a cubic site symmetry in $LiBaF_3$ lattice. No other site symmetry is possible for Eu^{2+} in $LiBaF_3$. The $^6P_{7/2}$ state splits into three levels in the cubic crystal field, while the $^6P_{5/2}$ state splits into two levels. Therefore we understand the reason why three and two excitation lines are observed in the $^8S_{7/2} \rightarrow ^6P_{7/2}$ and $^8S_{7/2} \rightarrow ^6P_{5/2}$ regions, respectively. In case of Eu^{2+}doped $KMgF_3$, the cubic-site Eu^{2+} is substituted for K^+ without vacancy at neighbor in $LiBaF_3$. Separation between the cubic-field Γ_7 and Γ_6 states of the $^6P_{7/2}$ state is 4.7 cm^{-1} in $LiBaF_3$ [36], while it is 7.0 cm^{-1} in $KMgF_3$ [28]. The ionic radius of Ba^{2+} is not so different from that of K^+. We have a question why the crystal field splitting is much smaller in $LiBaF_3$ than in $KMgF_3$. This is explained as follows. In the latter crystal the divalent Eu^{2+} is substituted for monovalent K^+, while in the latter it is substituted for the same divalent Ba^{2+}. Substitution by Eu^{2+} with different valency induces a strong attractive force to twelve F^- nearest neighbor ions in $KMgF_3$, giving rise to strong crystal field for the central Eu^{2+} ion. On the other hand, it is suggested that ligand field by six F^- nearest neighbor ions in $LiBaF_3$ is not so strong for Eu^{2+}. Such a suggestion is consistent with a theoretical crystal field calculation for the two crystals [37].

6. Two-Photon Excitation Spectra of Alkali Halides

Eu^{2+} ion is present in alkali halide crystals with NaCl structure. As shown in Fig. 4. Eu^{2+} absorption bands are broad and two bands appear at 410-320 nm and at 270-210 nm in KCl. Almost same absorption spectrum is observed in other alkali halides such as NaCl and KBr [17]. The broad absorption bands are due to the transition from the $^8S_{7/2}(f^7)$ state to the f^6d-levels and therefore sharp absorption lines due to the intraconfigurational $f^7 \rightarrow f^7$ transition is immersed under the intense $f^7 \rightarrow f^6d$ bands because the excited f^7 states lie above the lowest f^6d level. Unlike the case of $KMgF_3$, Eu^{2+} ion in alkali halide shows a broad emission band, with a peak at about 420 nm for KCl, NaCl and KBr. This emission due to the transition from the lowest f^6d- level to the ground state $^8S_{7/2}$ and observed when Eu^{2+} ions are excited to the f^6d state by one-photon absorption.

When the high-energy f^7 states such as $^6P_{7/2}$ and ^6I_J states are excited by two-photon absorption, the broad 420 nm emission band is also observed [29, 30, 38]. However, two-photon excitation spectrum has not been observed for the $^8S_{7/2} \rightarrow ^6P_{5/2}$ region in alkali halides [30, 38], presumably because of too small signal. The reason is unknown. Eu^{2+} excited into the $^6P_{7/2}$ and ^6I_J states relaxes to the lowest f^6d level non-radiatively, resulting in the 420 nm emission (see Fig.4). Figure 11 shows the excitation spectra for two-photon excited 420 nm emission for KCl, NaCl and

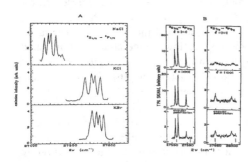

Figure 11. **A**: TPE spectra due to the $^8S_{7/2} \rightarrow ^6P_{7/2}$ transition of Eu^{2+} ions in NaCl, KCl and KBr at 77 K [30]. **B**: TPE spectra due to the $^8S_{7/2} \rightarrow ^6P_{7/2}$ (left side) and $^8S_{7/2} \rightarrow$ (right side) transitions of Eu^{2+} ions in CaF_2 [31].

KBr [29, 30]. Same result is also obtained for KI crystal [38]. Unlike the case of $KMgF_3$, only four lines are observed as the all the crystals. These lines are associated with the $^8S_{7/2} \rightarrow ^6P_{7/2}$ transition. The observed four lines are understood as follows.

The divalent Eu^{2+} ion enters the lattice substitutionally at monovalent alkali ion site which possesses octahedral symmetry. The substitutional Eu^{2+} is accompanied by a charge-compensating cation vacancy in one of the nearby alkali ion sites. If the vacancy is at one of the next-nearest- neighbour sites along the <110> directions (i.e. one of the nearest neighbour alkali ion sites), the symmetry of the Eu^{2+} environment is C_{2v}. The $^6P_{7/2}$ state splits into three levels in the cubic crystal field and four levels in the orthorhombic (e.g. C_{2v}) crystal fields. Therefore, taking into account that the ground state $^8S_{7/2}$ state is non-degenerate, it is suggested that the observed four absorption lines due to C_{2v} symmetry Eu^{2+} ion. It is noted that neither absorption lines due to octahedral Eu^{2+} (where vacancy is not located at the neighboring site but it is located far from the Eu^{2+} site) no absorption lines due to other low-symmetry Eu^{2+} (e.g. Eu^{2+} ions accompanied by a vacancy at second or third nearest neighbor cation site) are observed. This indicates that, unlike the case of $KMgF_3$ where even the cubic-site Eu^{2+} is present together with the C_{3v} and C_{4v} site Eu^{2+} ions, only the C_{2v} symmetry Eu^{2+} ion (with a vacancy at the nearest-neighbor cation site) is present in alkali halides. This is reasonable, since the C_{2v} site-symmetry of Eu^{2+} has a stronger coupling with the vacancy than that of Eu^{2+} with the other site symmetries because of the shortened distance between vacancy and Eu^{2+}. Therefore we suggest the vacancy chooses its location to achieve the most strong coupling with Eu^{2+}.

7. Two-Photon Excitation Spectra of CaF_2

CaF_2 has a cubic CsCl type crystal structure. Like the cases of alkali halides, any $f^7 \rightarrow f^7$ absorption band has not been observed and but $f^7 \rightarrow f^6 d$ absorption bands have been observed because the lowest f^7 state $^6P_{7/2}$ is above the lowest $f^6 d$ state in Eu^{2+} doped CaF_2. A broad emission band with a peak at about 413 nm is observed [31]. This emission is caused by the transition from the lowest $f^6 d$ state to the $^8S_{7/2}$ ground state. By the two-photon excitation spectroscopy, well-resolved three and two lines have been observed for the two-photon excited 413 nm emission in the $^8S_{7/2} \rightarrow ^6P_{7/2}$ and $^8S_{7/2} \rightarrow ^6P_{5/2}$ regions by Downer et al, respectively, as shown in Fig.11B [31]. They also observed the excitation lines in high energy $^8S_{7/2} \rightarrow ^6I_J$ (J = 7/2, 9/2, 11/2, 13/2, 15/2, 17/2) and $^8S_{7/2} \rightarrow ^6D_J$ (J = 1/2, 3/2, 5/2, 7/2) regions.

Eu^{2+} ion enters the lattice substitutionally at positive ion site Ca^{2+} which possesses cubic symmetry surrounded by eight F^- ions. Both the Eu^{2+} and Ca^{2+} ions are divalent ones, no charge-compensating vacancy is needed. Ca^{2+} ion has an ionic radius of 0.99 A. Thus, Eu^{2+} ion (1.12 Å) can be substituted for Ca^{2+} ion with the cubic symmetry. The $^6P_{7/2}$ state splits into three levels in the cubic crystal field, while the $^6P_{5/2}$ state splits into two levels. Therefore we understand the reason why three and two excitation lines are observed in the $^8S_{7/2} \rightarrow ^6P_{7/2}$ and $^8S_{7/2} \rightarrow ^6P_{5/2}$ regions, respectively.

Same three excitation lines have been also for the $^8S_{7/2} \rightarrow ^6P_{7/2}$ region in SrF_2 and BaF_2 crystals with the same crystal structure as CaF_2 [39]. This indicates that Eu^{2+} ion is substituted for the alkali-earth ion in these fluorides. The highest, middle and lowest energy lines are attributed to the cubic Γ_7, Γ_8 and Γ_6 states, respectively, as the case of $KMgF_3$ shown at the right figure of Fig.8. The amount of splitting decreases in order of CaF_2, SrF_2 and BaF_2, i.e. the separation between the Γ_7 and Γ_6 states is 26, 16 and 11 cm^{-1}, respectively [39]. Taking into account that the lattice constant is 5.463, 5.800 and 6.200 A for CaF_2, SrF_2 and BaF_2, respectively [17], it is suggested that the distance between Eu^{2+} and the ligand F^- ions becomes large and therefore crystal field becomes weak in order of CaF_2, SrF_2 and BaF_2. In this way we understand the difference of the splitting in these fluorides.

8. Why Various Locations are Possible for Vacancy in Eu^{2+} Doped $KMgF_3$ Crystal ?

In CaF_2 and $LiBaF_3$ crystals where Eu^{2+} is substituted for cubic-symmetry Ca^{2+} and Ba^{2+} ions, respectively, and no charge-compensating vacancy is necessary. Spectroscopic study supports that there is no vacancy around Eu^{2+} ion in these crystals, because three and two Eu^{2+} lines, which are expected for cubic site Eu^{2+}, have been observed in the $^8S_{7/2} \rightarrow ^6P_{7/2}$ and $^6P_{5/2}$ band spectra [31, 36, 39]. The numbers of the excitation lines are quite different from the case of $KMgF_3$ where at least eleven and seven lines are observed in the $^8S_{7/2} \rightarrow ^6P_{7/2}$ and $^6P_{5/2}$ band spectra, respectively. It is suggested as mentioned in Chapt.4 that the Eu^{2+} spectra observed in $KMgF_3$ consist of absorption lines due to Eu^{2+} ions with different site-symmetries, e.g. cubic, trigonal and tetragonal ones. In the case of alkali halides, vacancy is present as in $KMgF_3$.

However, unlike the case of KMgF₃, only one lattice place is allowed for the vacancy as mentioned in Chapt.6. KMgF₃ allows presence and coexistence of various types of Eu^{2+} ions, with and without vacancy at various neighboring K^+ sites. In the following we try to answer question why such a difference occurs between these crystals.

A possible explanation is that the location of the vacancy is not critical in KMgF₃. The vacancy is not necessarily located at one of special nearest-neighbor sites, and therefore Eu^{2+} ions with various kinds of local site-symmetries can be present. If the vacancy is far from the Eu^{2+} site, the site symmetry can be assumed to be cubic although it is not exactly so. In case of alkali halides, the positive ions which attract alkali-ion vacancy (with a negative charge effectively) are Eu^{2+} and alkali ion. The vacancy has a stronger attractive force with divalent Eu^{2+} than monovalent alkali ion. Therefore the vacancy chooses its location at nearest alkali lattice site, resulting in formation of Eu^{2+} with C_{2v} site symmetry. Thus, it seems that in monovalent crystals, such as alkali halides, the charge compensating vacancy is requested to be located considerably close to the impurity to form a stable Eu^{2+} vacancy dipole. On the other hand, formation of such a Eu^{2+} vacancy dipole is not necessary in the KMgF₃ crystal, leading to the coexistence of various Eu^{2+} ions with different site symmetries.

The reason why, unlike the case of alkali halides, the vacancy is allowed to locate at various K^+ sites in KMgF₃ is suggested as follows. In KMgF₃, besides the dopant Eu^{2+} ion, Mg^{2+} is also present in the crystal, and as a consequence the positive ion vacancy is expected to be attracted by these two divalent ions with almost the same electric force when crystal is grown. Therefore it should be deduced that this competition gives rise to a frustration to the vacancy in selecting the lattice place for location, and the result of the competition is that the vacancy is not necessarily located at the nearest-neighbour Eu^{2+} site, which produces the strongest vacancy-Eu^{2+} coupling. In this way a mixture of vacancies at different locations around Eu^{2+} is easily generated.

9. Ce^{3+} Ions in KMgF₃ Crystal

We shall check whether the above-mentioned deduction is reasonable or not. For this purpose we examine the case of Ce^{3+} doped KMgF₃. Like the case of Eu^{2+}, trivalent Ce^{3+} ion (ionic radius: 1.03 Å) is substituted for not smaller-size Mg^{2+} ion (0.65 A) but larger monovalent K^+ ion (1.33 Å) in KMgF₃ and therefore two charge-compensating K^+ ion vacancies are necessary. On Fig.12 are shown the structure of KMgF₃ with Ce^{3+} ion, showing possible locations of charge-compensating K^+ ion vacancy [40]. Ce^{3+} ion is substituted for K^+ ion. (a) Two K^+-ion vacancies on the C_{4v} axis around Ce^{3+} which is substituted for a K^+ ion, (b) no vacancy around Ce^{3+}. If the vacancies are located at K^+ site of [100] direction and close to Ce^{3+} as shown in Fig.12(b), the site symmetry of Ce^{3+} is C_{4v} [40]. If the vacancies are located far from Ce^{3+} as shown in Fig.12(a), the site symmetry is cubic, and Ce^{3+} is surrounded by six F nearest neighboring ions and eight K^+ next nearest neighboring ions. Ce^{3+} ion has the electronic configuration $4f5s^25p^6$ (named 4f hereafter) in the outer shell in the ground state and $5s^25p^65d$ (named 5d) in the first excited state. Therefore, compared with the other rare-earth ions,

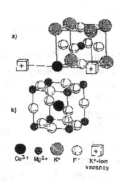

Figure 12. Structure of KMgF₃:Ce^{3+}.

the energy-level diagram is relatively simple. 4f level splits into the $^2F_{5/2}$ and $^2F_{7/2}$ manifolds separated by about 2000cm^{-1}. The ground state is the $^2F_{5/2}$ state, while the crystal-field-dependent 5d state is about 51000 cm^{-1} above the ground in KMgF$_3$ crystal. The 5d level splits into two states the $^2D_{5/2}$ and $^2D_{3/2}$ in the cubic crystal field and into five states in the C$_{4v}$ crystal field.

Five absorption bands due to the f→d transition in Ce^{3+} have been observed at 271, 255, 245, 234 and 227 nm by one-photon absorption spectroscopy [40]. It is observed that excitation into the latter two bands gives the 265 nm emission, while excitation into the first three bands gives the 350 nm emission. From the absorption spectra of differently grown crystals and the emission spectra and emission decay time, it has been suggested that the first three bands are due to Ce^{3+} ions with C$_{4v}$ site symmetry, while the last two absorption bands are due to the Ce^{3+} ions with cubic symmetry. This is confirmed by the two-photon excitation spectroscopy as follows [40].

The f→d transition is parity-allowed and electric-dipole allowed in one-photon transition, while it is parity forbidden in two-photon transition. This is true for Ce^{3+} ions with strictly cubic symmetry. However, this is not true for Ce^{3+} with the lower C$_{4v}$ site symmetry. The C$_{4v}$ non-centrosymmetry gives rise to mixing of the even parity 5d- state with odd parity states. As a result, the two-photon transition is allowed for the f→d transition for non-centrosymmetric Ce^{3+} ions. This means that the non-centrosymmetric Ce^{3+} ions give rise to both the one-photon absorption and two-photon absorption where the former intensity is much stronger than the latter. The two-photon excitation peaks are observed at 271, 255 and 245 nm but not at 234 and 227 nm. This indicates that the first three bands observed in the one-photon absorption spectra are certainly caused by the Ce^{3+} ions with C$_{4v}$ symmetry, while the latter two bands by Ce^{3+} ions with cubic symmetry.

Unlike the case of Eu^{2+} doped KMgF$_3$ where charge-compensating K$^+$ vacancies are located at several lattice sites, two vacancies are located at critical positions in the case of Ce^{3+}-doped KMgF$_3$ crystal. This is understood as follows. In Eu^{2+} doped KMgF$_3$, the vacancy is attracted by two divalent ions Eu^{2+} and Mg^{2+} with almost the same force. Therefore the vacancy is not necessarily located at the nearest-neighbor site, and a mixture of vacancies at different locations around Eu^{2+} is easily generated. In the case of Ce^{3+}, however, the vacancy is attracted more strongly to trivalent Ce^{3+} than the divalent ions Mg^{2+}. As a result the vacancies are located at the nearest-neighbor sites to produce the strong vacancy-Ce^{3+} coupling, forming only C$_{4v}$ site Ce^{3+} ion with two vacancies at critical nearest neighbors, not forming C$_{3v}$ and C$_{2v}$ site Ce^{3+} ions (where vacancy is at K$^+$ site of [111] and [110] axes, respectively) because the vacancy- Ce^{3+} distance is much shorter in C$_{4v}$ site Ce^{3+} ion than the C$_{3v}$ and C$_{2v}$ site Ce^{3+} ions. Presence of such a Ce^{3+} is consistent with the experimental result of electron spin resonance [41].

10. Concluding Remarks

In conclusion the location of charge compensating vacancy is determined by the electric attractive force to divalent Eu^{2+} and Mg^{2+} ions in KMgF$_3$ crystal. This is confirmed by comparing with the cases of Eu^{2+} doped KCl and Ce^{3+} doped KMgF$_3$. Next problem is to determine the concentration ratio among the cubic, C$_{4v}$, C$_{3v}$ and C$_{2v}$ site Eu^{2+} ions. From two-photon excitation spectra of Fig.8, it is suggested that the cubic site Eu^{2+} ions are more predominantly present than the other non-cubic Eu^{2+} ions because of the intense excitation lines.

However, we can not give the precise ratio at this moment. It should be reminded that it depends on the condition of crystal growth strongly. Another question is why cubic-site Eu^{2+} ions, which have no vacancy near Eu^{2+}, are predominantly present. A detailed crystal field calculation is necessary to answer to this question.

11. Acknowledgments

The author thanks   Prof. A. Scacco, University of Rome "La Sapienza", and Prof. H.J. Seo, Pukyong National University, Korea for useful discussion.   The present work is a part of joint work with them.   This work was partially supported by the Grant-in-Aid for Scientific Research   (C) of   Japan Society for Promotion of Science.

12. References

1. Blasse, G. and Grabmaier, B.C (1994) *Luminescent Materials*, Springer, Heidelberg.
2. Tsuboi, T. and Scacco, A. (2000) Observation of absorption bands due to the $4f^7 \rightarrow 4f^7$ parity-forbidden transitions of Eu^{2+} in $KMgF_3$ crystals, *J. Phys. Cond. Matter* **10**, 7259-7266.
3. Seo, H.J., Moon, B.K. and Tsuboi, T. (2000) Two-photon excitation spectroscopy of $4f^7 \rightarrow 4f^7$ transitions   in Eu^{2+} ions doped in $KMgF_3$ crystal, *Phys. Rev. B* **62**, 12688-12695.
4. Crosswhite, H.M. and Moos, H.W. (1967) *Optical Properties of Ions in Crystals*, Interscience Publishers, N.Y. vii., Huefner, S. (1978) *Optical Spectra of Transparent Rare Earth Compounds*, Academic Press, N.Y.
5. Carnall, W.T., Fields, P.R. and Rajnak, K. (1968) Electronic energy levels of the trivalent lanthanide aquo ions. II. Gd^{3+}, *J. Chem. Phys.* **49**, 4443-4446.
6. Tsuboi, T. (1998) Absorption spectra due to the $4f^7 \leftrightarrow 4f^7$ transitions of Gd^{3+} ions in $GdAl_3(BO_3)_4$ crystals, *J. Phys. Cond. Matter* **10**, 9155-9159.
7. Kundu, L., Banerjee, A.K. and Chowdhury, M. (1991) Two-photon absorption spectrum of gadolinium elpasolite, *Chem. Phys. Lett.* **181**, 569-574.
8. Schwiesow, R.L. and Crosswhite, H.M. (1969) Energy levels of Gd^{3+} in five hexagonal crystals, *J. Opt. Soc. Am.* **59**, 592-602.
9. Schwiesow, R.L. and Crosswhite, H.M. (1969) Energy levels of Gd^{3+} in LaF_3, *J. Opt. Soc. Am.* **59**, 602-610.
10. Piksis, A.H., Dieke, G.H. and Crosswhite, H.M. (1967)   Energy levels and crystal field of $LaCl_3:Gd^{3+}$, *J. Chem. Phys.* **47**, 5083-5089.
11. Detrio, J.S., Ferralli, M.W. and Yaney, P.P. (1970) Concentration study determination of the ^6P, ^6I, and ^6D energy levels of Gd^{3+} in SrF_2 at a C_{4v} site, *J. Chem. Phys.* **53**, 4372-4377.
12. Downer, M.C. and A. Bivas, A. (1983) Third- and fourth-order analysis of the intensities and polarization dependence of two-photon absorption lines of Gd^{3+} in LaF_3 and aqueous solution, *Phys. Rev. B* **28**, 3677-3696.
13. Bouazaoui, M., Jacquier, B., Linares, C. and Strek, W. (1991) Two-photon transitions of Gd^{3+} in cubic $Cs_3NaGdCl_6$, *J. Phys.: Cond. Matter* **3**, 921-926.
14. Wegh, R.T., Donker, H., Meijerink, A., Lamminmaki, R.J. and Holsa, J. (1997) Vacuum-ultraviolet spectroscopy and quantum cutting for Gd^{3+} in $LiYF_4$, *Phys. Rev. B* **56**, 13841-13848.
15. Szczurek, T. and Schlesinger, M. (1984) *Rare Earths Spectroscopy*, in B. Jezowska- Trzebiatowska, J. Legendziewicz and W. Strek (eds), World Scientific Publ., Singapore, 309.
16. Bouazaoui, M., Jacquier, B., Linares, C. and Strek, W. (1991) Two-photon transitions of Gd^{3+} in cubic $Cs_2NaGdCl_6$, *J. Phys.: Cond. Matter* **3**, 921-926.
17. Rubio, O.J. (1991) Doubly-valent rare-earth ions in halide crystals, *J. Phys. Chem. Solids* **52**, 101-174.
18. Alcala, A., Sardar, D.K. and Sibley, W.A. (1982) Optical transitions of Eu^{2+} ions in $RbMgF_3$ crystals, *J. Lumin.* **27**, 273-284.
19. Meehan, J.P. and Wilson, E.J. (1972) Single crystal growth and characterization of $SrAlF_3$ and $SrEuAlF_5$, *J. Cryst. Growth* **15**, 141-147.

20. Hewesand, R.A. and Hoffman, M.V. (1971) *J. Lumin.* **3**, 261.
21. Spoonhower, J.P. and Burberry, M.S. (1989) Time-resolved spectroscopy of BaFBr:Eu^{2+}, *J. Lumin.* **43**, 221-226.
22. Kobayashi, T. , Mroczkowski, S., Owen, J.F. and Brixner, L.H. (1980) Fluorescence lifetime and quantum efficiency for 5d→4f transitions in Eu^{2+} doped chloride abd fluoride crystals, *J. Lumin.* **21**, 247-257.
23. Tsuboi, T. and Silfsten, P. (1991) The lifetime of Eu^{2+}- fluorescence in CaF$_2$ crystals, *J. Phys.: Cond. Matter* **3**, 9163-9167.
24. Blasse,G. (1978) *Luminescence of Inorganic Solids*, in B. Di Bartolo (ed.), Plenum Press, N.Y. , 457-472.
25. Poort, S.H.M. and Blasse, G. (1997) The influence of the host lattice on the luminescence of divalent europium, *J. lumin.* **72-74**, 247-249.
26. Meijerink, A., Nuyten, J., and Blasse, G. (1989) Luminescence and energy migrtion in (Sr,Eu)B$_4$O$_7$, *J. Lumin.* **44**, 19-31.
27. Ellens, A., Meijerink, A., and Blasse, G. (1994) ^{6}I emission and vibraonic trasnsition of Eu^{2+} in KMgF$_3$, *J. Lumin.* **59**, 293-301; Ellens, A., Meijerink, A. and Blasse, G. (1994) The first observation of ^{6}I → ^{8}S emission from Eu^{2+} in KMgF$_3$, *J. Lumin.* **60/61**, 70-73.
28. Francini, R. Grassano, U.M., Tomini, M., Boiko, S., Tarasov, G.G. and Scacco, A. (1997), *Phys. Rev. B* **55**, 7579-7595.
29. Casalboni, M., Francini, R., Grassano, U.M., and Pizzoferrato, R. (1986) Two-photon spectroscopy in KCl:Eu^{2+}, *Phys. Rev. B* **34**, 2936-2938.
30. Casalboni, M., Francini, R., Grassano, U.M., and Pizzoferrato, R. (1987) Two-photon spectroscopy in Eu^{2+} doped alkali halides, *Cryst. Lattice Deffect Amorph. Mat.* **16**, 261-267.
31. Downer, M.C. , Cordero-Montalvo, C.D. and Crosswhite, H. (1983) Study of new 4f^7 levels of Eu^{2+} in CaF$_2$ and SrF$_2$ using two-photon absorption spectroscopy, *Phys. Rev. B* **28**, 4931-4943.
32. Francini, R. (1997) Spectroscopy of rare earth ions in Insulating materials, *SPIE Proc.* **3176**, 2-11.
33. Bellatreccia, M., Casalboni, M., Francini, R., and Grassano,U.M. (1991) Even-parity excited states of Ag$^-$ centers in alkali halides,*Phys. Rev. B* **43**, 2334-2338.
34. Altshuler, N.S., Livanova, L.D., and Stolov, A.L. (1974) Spectra of f-f transitions of the Eu^{2+} ion in KMgF$_3$, *Opt. Spectrosc.* **36**, 72-75.
35. Altshuler, N.A., Ivoilova, E.Kh., Livanova, L.D., Stepanov, V.G., and Stolov, A.L. (1974) Many-center structure of the ESR spectra of KMgF$_3$ and KZnF$_3$ crystals activated with Eu^{2+} and Gd^{3+} ions, *Phys. Solid State.* **15**, 1973-1975.
36. Meijerink, A. (1993) Spectroscopy and vibronic transitions of divalent europium in LiBaF$_3$, *J. Lumin.* **55**, 125-138.
37. Altshuler, N.S., Korableva, S.L., Livanova, L.D. and Stolov, A.L. (1974) ESR and optical spectra of Eu^{2+} ion in LiBaF$_3$, *Phys. Solid State* **15**, 2155-2157.
38. Nunes, L.A.O., Matinaga, F.M. and Castro, J.C. (1985) Two-photon spectroscopy in Eu^{2+} ions in KCl and KI, *Phys. Rev. B* **32**, 8356-8360.
39. Dujardin,C., Moine, B. and Pedrini, C. (1993) One- and two-photon spectroscopy of f→d and f→f transitions of Eu^{2+} ions in M$_{1-x}$N$_x$F$_2$ mixed fluoride crystals (M,N = Ba, Sr, Ca), *J. Lumin.* **54**, 259-270.
40. Francini, R., Grassano, U.M., Landi, L., Scacco, A. D'Elena, M., Nikl, M., Cechova, N. and Zema, N. (1997) Ce^{3+} luminescent centers of different symmetries in KMgF$_3$ single crystals, *Phys. Rev. B* **56**, 15109-15114.
41. Yamaga,M., Honda, M., Kawamata, N., Fujita, T., Shimamura, K., and Fukuda, T. (2001) Site symmetry and crystal field of Ce^{3+} luminescent centers in KMgF$_3$, *J. Phys. Cond. Matter* **13**, 3461-3473.

FLUORESCENCE DYNAMICS OF Er^{3+} IONS IN MBE-GROWN GaN-THIN FILMS

F. PELLE[1], F. AUZEL[1], J.M. ZAVADA[2], U. HÖMMERICH[3],
D.S. LEE[4], A.J. STECKL[4]

[1]·UMR7574-CNRS 1, Place Aristide Briand, F-92195 Meudon cedex
[2]·US Army Research Office, Electronics Division, Durham,
   NC 27709, USA
[3]·Hampton University, Department of Physics, Hampton,
   VA 23668, USA
[4]·University of Cincinnati, Nanoelectronics Laboratory,
   Cincinnati, OH 45221, USA

Abstract. Spectroscopic properties of erbium ions in GaN thin films are reported. Solid state MBE growth allows doping the semiconductor with rare earth concentrations as high as 10^{21} ions/cm^{-3}. The site selective excitation spectra allow to identify three types of sites for Er^{3+} ions in GaN. The majority of Er^{3+} ions are substituted on the Ga sub-lattice while the two other centers are ascribed to Er^{3+} centers in interstitial positions in the lattice and are assigned to Er-related defects. The $^4S_{3/2}$ fluorescence lifetime of the main center is strongly quenched with increasing Er concentration. A non-exponential time behavior for the $^4S_{3/2}$ decay profile is observed for the main center and is interpreted in a diffusion-limited model. The microscopic parameters of the interaction are determined.

1. Introduction

Rare earth (RE) doped wide gap semiconductors have recently appeared as a promising new class of materials with potential applications in the field of electroluminescent (EL) displays and as electrically activated laser sources. Up to now, infrared spectroscopic properties have mostly been studied for Erbium in GaAs [1-5] and Ytterbium in InP semiconductors [6,7] which have band gaps at a low energy (less than 8000 cm^{-1}). III-V nitrides are wide bandgap semiconductors known to have many advantages over II-VI semiconductors due to their larger strong bonding and robust nature. Furthermore, it has been demonstrated [8] that thermal quenching in Er –doped semiconductors decreases with increasing band gap. Therefore, wide band gap semiconductors (WBGS) appear to be very attractive for optoelectronic and photonic applications. However, few studies on RE visible optical properties in such materials have been reported [9,10] to date. No clear correlation between the efficiency of the visible RE^{3+} luminescence and the interaction of the rare earth ion with the semiconductor host has been established, especially concerning the RE excitation and deactivation pathways. A full

J.-C. Krupa and N.A. Kulagin (eds.), Physics of Laser Crystals, 109–124.
© 2003 Kluwer Academic Publishers. Printed in the Netherlands.

understanding of the properties of the rare earth dopants in the semiconductors and their relationship to growth processes is still incomplete. Among WBGS, GaN has been mainly studied to realize electroluminescent devices since it is transparent to visible RE^{3+} emission and it is chemically and thermally stable. Recent use of nonequilibrium growth techniques (ion implantation or thin-film deposition by Molecular Beam Epitaxy or Metallorganic Chemical Vapor Deposition) allows one to increase the solubility threshold of the RE in such materials and to dope the semiconductor with concentrations as high as 10^{21} cm^{-3}. This opens the opportunity to realize laser sources based on RE^{3+} doped III-V nitride semiconductors. EL device operation based on RE^{3+} doped GaN has recently been reported [11]. Emission of the three primary colors has been accomplished using Eu^{3+} for the red, Er^{3+} for the green and Tm^{3+} for the blue.

In order to optimize the performance of current RE^{3+} doped GaN devices it is important to obtain a better understanding of the RE^{3+} doping process, excitation schemes and luminescence efficiency. In this paper, new results on the spectroscopy of Er^{3+} ions in GaN thin films are reported.

2. Experimental

The Er^{3+} doped GaN samples were synthesized using the Solid Source Molecular Beam Epitaxy (SSMBE). The films are grown on a p-type (111) Si substrate after deposition of an AlN buffer layer. The growth of Er^{3+} doped GaN proceeds using Ga and Er solid sources in conjunction with a rf-plasma source supplying atomic nitrogen. The Ga cell temperature is kept constant and the Er^{3+} concentration in the film is controlled by adjusting the Er-cell temperature. The films are grown for 1h resulting in film thickness less than 1μm. The synthesis procedure is detailed in [12]. The Er^{3+} concentration measured using Rutherford Back Scattering and Secondary Ion Mass Spectroscopy ranges from 0.025 to 11.2 at.% in the studied thin films [12]. The crystal structure of the films was studied using X ray diffraction (XRD). The (0002) characteristic peak of GaN is observed to broaden and decreases in intensity as the Er^{3+} concentration increases. Beyond about 2 at.%, a new phase, assigned to ErN, appears for the highest Er^{3+} concentration (11.2%) [12].

The Er^{3+} fluorescence has been excited either by an OPO pumped by the third harmonic frequency of a pulsed Nd:YAG laser (501-DNS,720 BM Industrie) or using Ar^+ lines (514.5 and 496.5 nm) (Coherent Innova 300). The fluorescence was dispersed through a HR460 Jobin-Yvon monochromator, the signal was detected by a EMI 9558 QBM photomultiplier in the visible range and a InGaAs (Hamamatsu) photodiode for infrared luminescence. Time resolved spectra and fluorescence transients were digitized and averaged by an oscilloscope (TDS350 Tektronix) with data acquisition on a microcomputer. Low temperature measurements were performed by cooling the samples in a closed-cycle CTI cooling system.

3. Spectroscopic data : Results and discussion

3.1. VISIBLE EMISSION

Under excitation in the $^4F_{7/2}$ multiplet, the observed fluorescence is dominated by the emission from the $^2H_{11/2}$, $^4S_{3/2}$ multiplets, which are thermalized at room temperature due to the small energy gap between them. Radiative transitions are also observed from lower multiplet ($^4F_{9/2}$, $^4I_{9/2}$) with a very weak intensity (Fig.1). Surprisingly, emission from the $^4I_{13/2}$ multiplet exhibits a rather high intensity compared to that observed from the $^4F_{9/2}$ and $^4I_{9/2}$ multiplets . At low temperature, emission lines recorded in the visible range arise from the $^4S_{3/2}$ state exhibit the same relative intensity and are observed at the same wavelength whatever the Er^{3+} concentration in the samples (Fig.2). This seems to indicate that the environment of Er^{3+} remains the same, no other centers are observed in the visible range with increasing Er concentration . Energies of the Stark components and the overall splitting of the multiplets are very similar to that observed in insulators.

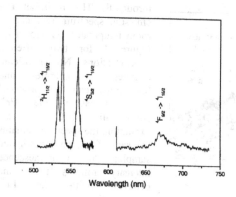

Figure1. Visible fluorescence spectrum recorded at Room temperature under resonant excitation in the $^4F_{7/2}$ multiplet ($\lambda_{excitation}$= 498 nm).

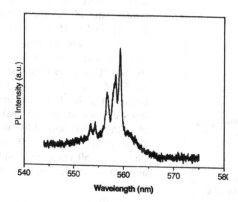

Figure 2. $^4S_{3/2} \rightarrow {}^4I_{15/2}$ emission spectrum recorded at 10K with excitation in the $^2H_{11/2}$ multiplet (GaN :Er^{3+} 0.17 at.%).

The fluorescence intensity dependence with the concentration exhibits a common behavior: it increases with the concentration up to a maximum (0.6 at.%) and then decreases due to interactions between nearby Er^{3+} ions leading to other de-excitation pathways.

At room temperature, the decay of the $^4S_{3/2}$ state is quite short and exhibits a complex behavior with increasing Er^{3+} concentration, as shown on Fig.3. The relative high intensity of the emission from the $^4I_{13/2}$ state compared to those from higher levels indicates that the optical relaxation involves energy transfer processes among Er^{3+} ions. The detailed study of the energy transfer process is developed in Part 4.

3.2. INFRARED EMISSION

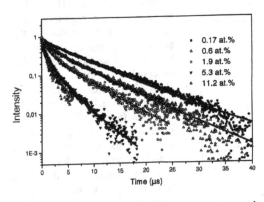

Figure 3. $^4S_{3/2}$ decay profile as a function of Er^{3+} concentration (T=300K) (Symbols: experimental data; full lines: theoretical curves calculated using Eq(4)).

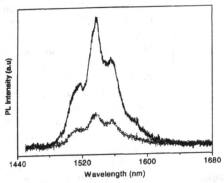

Figure 4. $^4I_{13/2} \rightarrow {}^4I_{15/2}$ emission spectrum recorded at 300 K with excitation in the $^2H_{11/2}$ multiplet (full line :Er^{3+} 0.6 at.% ; dotted line : Er^{3+} 0.17 at.%).

The $^4I_{13/2}$ state was excited through the $^2H_{11/2}$ multiplet. The emission spectrum recorded at room temperature is shown on Figure 4 for two different concentrations. No significant changes, in energies and linewidths, can be observed with increasing the Er^{3+} concentration. Under this excitation, the spectrum remains the same for each investigated sample; the luminescence intensity follows the same behavior with the Er^{3+} concentration.

Selective excitation spectral analysis monitoring the $^4I_{13/2} \rightarrow {}^4I_{15/2}$ main emission lines was performed at low temperature (T = 10K). The lines located at 1531 and 1537 nm are excited in the $^2H_{11/2}$ state although excitation spectra of the two others lines observed at 1545 and 1552nm show a weak excitation in the $^2H_{11/2}$ state and other efficient excitation peaks at 515.7 nm and 518 nm respectively (Fig.5a). These lines are off-resonance with absorption of 4f -intra shell transitions.

Emission spectra selectively excited in the different excitation lines are represented on Fig.5b. Three different spectra are recorded and they are ascribed to three different centers labeled in the following as C1, C2 and C3 for simplicity. Selective excitation in both lines observed around 510 nm provides emission intensity one order of magnitude centers labeled in the following as C1, C2 and C3 for simplicity. excitation wavelength (Fig.5b). As shown on Fig.5b, excitation in the intra -

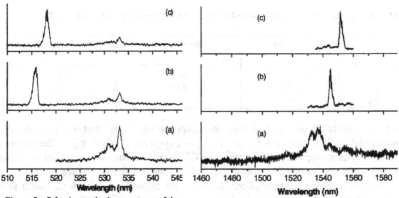

Figure 5a. Selective excitation spectrum of the $^4I_{13/2} \rightarrow {}^4I_{15/2}$ transition recorded at 10K for GaN:Er^{3+} (0.6 at.%)(a : $\lambda_{analysis}$ = 1537 nm ; b : $\lambda_{analysis}$ = 1545nm ; c : $\lambda_{analysis}$= 1552 nm)

Figure 5b. $^4I_{13/2} \rightarrow {}^4I_{15/2}$ Fluorescence spectrum recorded at 10K for GaN :Er^{3+} (0.6 at.%) with selective excitation (a : $\lambda_{excitation}$ = 533.7 nm ; b : $\lambda_{excitation}$ = 515.7 nm ; c: $\lambda_{excitation}$ = 518 nm)

4f- configuration gives rise to emission lines at 1532, 1537, 1545 and 1552 nm, the latter intensity being much weaker than that of emission recorded under excitation in non resonant lines. With increasing the temperature, C2 and C3 fluorescence decreases greatly and cannot be recorded.

The infrared emission was recorded at low temperature with the argon laser line at 496.5 nm as excitation. This wavelength does not resonantly excite Er^{3+} ions, but in this case, the excitation falls within the phonon sideband of the $^4F_{7/2}$ state and near the other excitation lines (at 515.7 and 518 nm). In the emission spectrum represented on Figure 6, four lines are then observed and their positions correspond to more intense ones of the three centers: 1531, 1537, 1545 and 1552 nm. This suggests that under this excitation, the fluorescence results from the contribution of all the centers together.

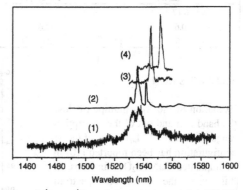

Figure 6. $^4I_{13/2} \rightarrow {}^4I_{15/2}$ emission spectrum recorded at 10 K for different excitation wavelengths (1 : 533.7 nm corresponding to excitation in the $^2H_{11/2}$ mulitplet ; 2 : 496 nm ; 3 : 515.7 nm corresponding to the main C2 excitation line ; 4 : 518 nm corresponding to the C3 main excitation line).

The temporal behavior of the $^4I_{13/2}$ state of the different centers is shown on Fig.7.The excitation pathway for center 1 is completely different from others. The rise and decay times deduced from the theoretical fit of the experimental curves are summarized in

TABLE 1. $^4I_{13/2}$ rise and decay times measured at 10K for the different centers (GaN: 0.6 at.% Er^{3+})

	$\lambda_{excitation}$ (nm)	Rise time (μs)	Decay time (ms)
C1	533.7	862	3.1
C2	515.7	64.5	1.718
C3	518	79	1.773

Tab.1. A simple model with a rise and exponential decay was used to fit the experimental decay curves: $\tau = C[\exp(-t/\tau_d) - \exp(-t/\tau_r)]$ where $\tau_{r,d}$ are the rise and the decay times respectively, C is a constant. C2 and C3 exhibit quite a similar dynamic behavior which means that they correspond to Er^{3+} ions in a similar environment.

The fluorescence transient of the $^4I_{13/2}$ state for center C1 has been measured as a function of the Er^{3+} concentration. The values of the decay and the rise times are summarized in Tabl.2. All time constants strongly decrease with increasing the Er^{3+} content in the samples as a consequence of Er^{3+} - Er^{3+} interactions. The thermal quenching of the $^4I_{13/2}$ decay rate as already been reported [13].

TABLE 2. $^4I_{13/2}$ rise and decay times measured at room temperature as a function of Er^{3+} concentration ($\lambda_{excitation}$ = 533.7 nm, analysis for center C1 at $\lambda_{analysis}$ = 1537 nm).

[Er^{3+}] (at.%)	Rise time (μs)	Decay time (ms)
0.17	98	2.43
0.6	33.6	1.41
1.9	21	0.310

3.3.DISCUSSION

Excitation of the $^4I_{13/2}$ multiplet below the band gap and out of the absorption range of the Er$^{3+}(^{2S+1}L_J)$ states of the 4f configuration has been reported [10] in Er-implanted GaN/Al$_2$O$_3$ thin films. In this case, the excitation spectrum consists of a broad band located between 460 and 530 nm. The luminescence intensity provided by such excitation wavelengths is one order of magnitude more intense than that recorded with a resonant 4f-4f excitation. These results are not

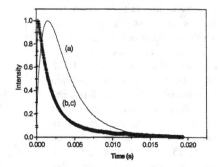

Figure 7. $^4I_{13/2}$ decay profile as a function of excitation wavelength recorded at 10K (a : $\lambda_{excitation}$ = 533.7 nm and $\lambda_{analysis}$ = 1537 nm corresponding to C1 ; b : $\lambda_{excitation}$ = 515.7 nm and : $\lambda_{analysis}$ = 1545nm corresponding to C2 ; c : $\lambda_{excitation}$ = 518 nm and $\lambda_{analysis}$= 1552 nm corresponding to C3) (full line : theoretical fit ; cross : experimental curves).

completely similar to ours since we observe a $^4I_{13/2} \rightarrow {}^4I_{15/2}$ emission spectrum which depends on the excitation wavelength (Fig.5b) and the excitation spectrum consists in narrow lines compared to ref. [10]. However, the energy of these extra excitation lines is well in the range of the broad excitation band and furthermore, the shape of the emission spectra of C2 and C3 are very similar to that observed in [10]. Due to the strong efficiency of these extra excitations, and to the fact that the intra-4f transitions have very low absorption cross sections compared to that of exciton suggests that excitation of C2 and C3 centers does not proceed via Er^{3+} 4f-4f intra-shell transitions but rather by an Er^{3+}-trap-related absorption line. We conclude that the C2 and C3 concentration are much mower than that of C1.

Despite a lower intensity in the infrared, the concentration of C1 center seems to be much higher than the other two. First, because it can be directly excited in the 4f-intrashell transitions which oscillator strengths are very weak. Also the complex behavior of the visible decay profile suggests an important contribution of interactions between Er^{3+} in center C1 in the de-excitation process. From the static and the dynamic spectroscopic properties, and the XRD spectra as a function of the Er^{3+} content, the C1 center corresponds to Er^{3+} ions in regular group III positions in the lattice. Moreover, from RBS channeling analysis [14] it has been shown that in the studied samples a great majority (≈ 95 %) of the Er^{3+} ions occupy substituted sites on the Ga sublattice even at relatively high Er concentrations. Most of the dopant ions substitute on Ga sites. The SSMBE growth method allows one to introduce RE ions from low to high concentration. The concentration quenching of the lifetime as shown on Fig.3 indicates that cluster formation is present even at low concentrations. For very high Er^{3+} concentration, a pure ErN phase is obtained as shown by XRD demonstrating the clustering of Er^{3+} in the GaN host [12]. The C2 and C3 centers can be ascribed to Er^{3+} ions in interstitial positions near defects produced by introduction in the lattice, as it is observed in other RE doped semiconductors [4, 5]. Another explanation could be the location for C2 and C3 such as vacancies on the N sub-lattice. The difference of the shape (broad band versus narrow lines) recorded for Er–trap related defects can be explained by considering the doping process used in [10] and in the present work. The implantation doping is a simple method and the doping level can be controlled independently of the growth conditions. However, this method provides introduction of damage in the GaN crystal which cannot be completely removed by annealing. So a higher concentration of defects or the presence of extended defects are expected in the implanted samples. This may explain a broad band that is observed compared to our samples obtained by MBE in which the defects should be more localized, giving rise to a discrete spectrum.

Moreover, RE induced gap levels and their role in the excitation process have already been demonstrated in $InP:Yb^{3+}$ [6], $GaAs:Er^{3+}$ [15] and $Er^{3+}:Si$ [16]. The main difference between RE in insulators and semiconductors, is that in the latter case the RE ions should be considered with the gap level they generate in the host. These RE-generated levels originate from local strain or from native defects states generated in the process of RE doping. However, these results apply for low gap semiconductors in which only one excited state ($^4I_{13/2}$ for Er^{3+} or $^2F_{5/2}$ for Yb^{3+}) is available. Further experiments are needed to extend these considerations to generated levels in the visible.

From the complex decay behavior observed for the visible emission, the quasi-absence of emission from intermediate states between the excited level and the emitting one, we can suggest that the process involved for the population of the $^4I_{13/2}$ state of C1 center is due to a cross relaxation process $(^4S_{3/2},^4I_{15/2}) \rightarrow (^4I_{11/2},^4I_{13/2})$ after migration of the excitation within the $^4S_{3/2}$ state, as it will be demonstrated in Section IV. The $^4I_{13/2}$ state is then populated through mainly the $^4I_{11/2} \rightarrow {}^4I_{13/2}$ radiative transition. The rise time observed is easily ascribed to the $^4I_{11/2}$ lifetime since at 10K multiphonon process probability is very low and 6 phonons are required to bridge the energy gap between those two levels, assuming a phonon cut-off frequency of 600 cm^{-1} as deduced by vibrational spectroscopy [17]. The value then obtained for low Er^{3+} concentrations is quite in agreement with values measured for this level in other lattices [18-20].

A lifetime of 3.1ms is obtained at 10K for the $^4I_{13/2}$ decay of C1 center for the sample doped with Er^{3+} (0.17 at. %). The occurrence of a cross relaxation process is strongly supported by the fact that emission from intermediate states such as $^4F_{9/2}$ and $^4I_{9/2}$ is much weaker than it is usually observed. This means that the multiphonon relaxation to those lower states is not efficient, then $^4I_{13/2}$, lying at lower energy, can only be populated by another process which overcomes the multiphonon relaxation one.

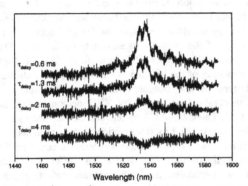

In the case of C2 and C3 centers, the process populating the emitting level is faster. We propose a a similar cross relaxation process between the related trap excited level and the Er^{3+} $^4I_{13/2}$ level. This process can explain why no emission from the $^4S_{3/2}$ level can be observed for these centers.

Figure 8. $^4I_{13/2} \rightarrow {}^4I_{15/2}$ Time resolved emission spectrum recorded at 10K with excitation in the $^2H_{11/2}$ (C1) multiplet.

Time-resolved emission spectra (TRS) from $^4I_{13/2}$ to the ground state for center C1 recorded at 10K with excitation in the $^2H_{11/2}$ multiplet are shown in Figure 8. These spectra exhibit several features which behave in the same way as the delay increases. This confirms that all of them belong to the same center i.e. C1, the spectrum recorded at the longest delay reveals a re-absorption of the resonant fluorescence observed at 1532 and 1537 nm. This is often observed for the first excited state of Er^{3+} and it is ascribed to a radiative energy transfer (radiative trapping) which gives an additional contribution to migration within the $^4I_{13/2}$ state.

This suggests also that the measured value of the decay is probably slightly longer that the real lifetime due to the radiative energy transfer which can give an apparent longer decay time.

4. Dynamic of C1 Center

We will consider C1 as the main center in the following, assuming that the concentrations of C2 and C3 are negligible compared to C1. The time development of the $^4S_{3/2} \rightarrow {}^4I_{15/2}$ fluorescence intensity deviates from a simple exponential dependence whatever the Er^{3+} concentration from 0.025 to 11.2 at.% (Fig.3). For the lower concentrations (0.17 and 0.6 at.%), the decay profile can be fitted to two exponential functions. This could be explained by the presence of two sites for the Er atoms. However, at higher concentrations the decay profile cannot be fitted by only two exponentials and therefore indicates that other processes such as energy transfer, diffusion are involved. First, the initial part of the decay profile deviates from a simple exponential dependence and the deviation from the classical law increases with increasing the Er^{3+} concentration. Furthermore, the long part of the decay follows an exponential behavior $\propto exp(-t/\tau)$, the decay rate $W=1/\tau$ remains nearly the same for low Er^{3+} concentrations (0.025 and 0.17 at.%) and gradually increases with the Er^{3+} content. Finally, for the higher Er^{3+} concentration (11.2 at.%), the overall decay of the Er^{3+} fluorescence becomes faster and approaches a single exponential. This is illustrated on Figure 3 where semi-logarithmic plots of the Er^{3+} decay for the different concentrations are shown.

4.1 THEORY

Several competing processes have to be considered for the de-excitation of the $^4S_{3/2}$ multiplet. The optical excitation will relax radiatively mainly to the ground state even if a radiative relaxation to the $^4I_{13/2}$ multiplet occurs. De-excitation by multiphonon relaxation to the next lower level will compete with the first process. This last process is operating in the studied samples since fluorescence transitions are observed from lower levels (i.e. $^4F_{9/2}$, $^4I_{9/2}$ and $^4I_{13/2}$). However, the quite efficient emission from the $^4I_{13/2}$ level compared to that of the intermediate states ($^4F_{9/2}$ and $^4I_{11/2}$) indicates that the multiphonon relaxation is not the main process involved in the filling of the $^4I_{13/2}$ multiplet. These other processes such as ion-ion interactions should be taken into account to explain the strong reduction of the observed decay as discussed in Section III.3. The complex decay profile and its concentration dependence indicate that the optical relaxation involves energy transfer processes among Er^{3+} ions.

Now to discuss about the nature of the non-radiative energy transfer we will consider a set of Er^{3+} ions acting as "sensitizers" and the other set being the "acceptors" which trap the energy and produce the quenching of the luminescence of the sensitizers; the acceptors could be other Er^{3+} ions or sinks such as "poisonous centers", the last kind is very often evoked but until now no nobody could give a proof of their real nature. So, in the following we will consider that Er^{3+} ions as sensitizers and activators which is a more realistic hypothesis. In this case, it is obvious that the concentration of acceptors will increase with the Er^{3+} concentration. The complex time development of the intensity of the sensitizer results from direct energy transfer and energy migration to acceptor. The characteristics of the sensitizer system relaxation can, however, be discussed conveniently in terms of three limiting cases: (A) direct relaxation – no diffusion, (B) fast diffusion, and (C) diffusion-limited relaxation. A careful examination of the sensitizer fluorescence transient as a function of the acceptor concentration will allow to characterize the sensitizer-sensitizer and sensitizer-acceptor dynamics and to identify the dominant mechanism. Among the three possible cases, only cases A and C predict a non-exponential behavior with increasing the sensitizer concentration. In case B fast migration of the optical excitation by resonant energy transfer occurs within the subset of sensitizers leading to a spatial equilibrium within the sensitizer system, the limiting step of the diffusion process being the sensitizer-acceptor transfer rate. Since the diffusion is fast, variations in the transfer time for different sensitizer-acceptor pairs are effectively averaged out and the sensitizer system exhibits a simple exponential. In case A, acceptors lying nearby sensitizer ions interact via exchange or multipolar forces and a direct energy transfer occurs. To take into account for the dispersion of the distances within sensitizer-acceptors pairs averaging over the sample volume is necessary to relate the microscopic transfer rates to the observable macroscopic signal $\phi(t)$ for multipolar coupling [21] :

$$\Phi(t) = \Phi(0)\exp\left[-\frac{t}{\tau_0} - \frac{4\pi}{3}\Gamma\left(1 - \frac{3}{s}\right)N_a R_0^3\left(\frac{t}{\tau_0}\right)^{3/s}\right],\tag{1}$$

where $\phi(0)$ is the initial excitation and R_0 is the critical transfer distance $(\alpha_{SA}\tau)^{1/s}$ defined as that separation at which the probability for energy transfer between a sensitizer-acceptor pair equals the intrinsic decay probability τ_0^{-1}, s =6, 8, 10 for dipole-dipole, dipole-quadrupole and 10 quadrupole-quadrupole coupling respectively. The overall decay is therefore characterized by an initial non-exponential part followed by an exponential decay at a rate governed by the intrinsic decay rate. Case C is more general and includes energy transfer within the sensitizer subsystem. Case C applies if the rate of the sensitizer-sensitizer is low and of the same order as the intrinsic decay rate. With this assumption, the decay of the total sensitizer system results from the contribution of excited sensitizers lying nearby acceptors which predominantly relax by direct ion-pair energy transfer and the contribution of other more distant sensitizers which must first diffuse into the vicinity of an acceptor before relaxation occurs. In this case, the sensitizer excitation density $\phi(t)$ as a function of time is found by solving a diffusion equation. For low acceptor concentrations, only a small fraction of the total number of excited sensitizers are within the critical distance of an acceptor and the sensitizer decay will be governed principally by intrinsic relaxation and by diffusion-limited relaxation to acceptors. The decay at long time after the excitation pulse follows an exponential function of time with a characteristic time :

$$\frac{1}{\tau} = \frac{1}{\tau_0} + \frac{1}{\tau_D},\tag{2}$$

where $1/\tau_D$ is the decay rate due to diffusion.

As the acceptor concentration increases, a larger fraction of the sensitizers are then within the critical interaction range of acceptors and energy migration becomes negligible. Yokota and Tanimoto [22] have obtained a general solution for the sensitizer decay function including diffusion within the sensitizer system and sensitizer-acceptor energy transfer via a dipole-dipole coupling :

$$\Phi(t) = \Phi(0)\exp\left[-\frac{t}{\tau_0} - \frac{4\pi^{3/2}}{3} N_A \left(\alpha_{SA}t\right)^{1/2} \left(\frac{1+10.87x+15.50x^2}{1+8.743x}\right)^{3/4} \right], \quad (3)$$

where $x = D \, \alpha_{SA}^{-1/3} \, t^{2/3}$, with D is the diffusion constant and α_{SA} is related to the sensitizer-acceptor interaction ($W_{SA}=\alpha_{SA}/R_{SA}^{6}$). At earlier times in the decay, diffusion is negligible and $\phi(t)$ reduces to Eq(1) with s=6. When t→∞, Eq (3) reduces to an exponential function of time with a lifetime approximately equal to that given by Eq (2). In case of a dipole-dipole interaction:

$$\frac{1}{\tau_D} = 8.6 \, N_A \, \alpha_{SA}^{1/4} \, D^{3/4}. \quad (4)$$

4.2. DETERMINATION OF THE MICROSCOPIC PARAMETERS OF THE ENERGY TRANSFER

From the above theoretical models, an non-exponential dependence with time is only predicted for cases A and C and an important point is to determine is the multipolar character of the interaction. This has been achieved by looking at the dependence of the earlier times of the decay. $Ln(\phi(t))+t/\tau_0$ is a linear function of $t^{1/2}$ which allows to deduce for a dipole-dipole interaction. The next step is to determine the nature of the energy transfer process i.e. direct relaxation or diffusion-limited, the concentration dependence of the long part of the decay which is exponential in both cases can help to choose between both models. In the diffusion-limited case, when t→∞ , the quenching rate $1/\tau_D$ as given by Eq.(2)

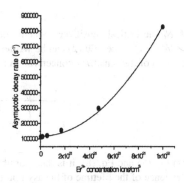

Figure 9. Asymptotic decay rate of the $^4S_{3/2}$ temporal behavior as a function of Er^{3+} concentration (T=300K) (full line : theoretical fit using Eq (6) ; dots : experimental data).

is concentration dependent and can be written as :

$$\frac{1}{\tau_D} = \frac{1}{\tau_{exp}} - \frac{1}{\tau_0} = KN_S N_A = KN^2, \quad (5)$$

where τ_{exp} is the experimental value of the lifetime in the long part of the decay, τ_0 is the intrinsic lifetime of the sensitizer without acceptor, N_S and N_A are the sensitizer and activator concentrations (here since Er^{3+} will act as sensitizer and acceptor, we can approximate to KN^2, K being a constant which depends on matrix elements and overlap integrals. As shown on Fig.9, the long part of the decay follows a quadratic dependence with Er^{3+} concentration which allows us to conclude to a diffusion-limited process. The best fit of the experimental values is obtain with $K = 7.2 \times 10^{-39} \, cm^6 \, s^{-1}$.

4.2.1. Determination of the intrinsic decay time (τ_0), the critical distance (R_0) and the sensitizer-activator coupling constant (α_{SA}).

From Eq. (5), K can be written as :

$$K = 8\pi \, \alpha_{SA}^{1/4} \, \alpha_{SS}^{3/4} , \qquad (6)$$

where α_{SA} is the sensitizer-acceptor coupling constant and α_{SS}, the sensitizer-sensitizer coupling constant. As we have already previously supposed, Er^{3+} ions will be considered as sensitizer and activator, then it is possible to simplify Eq.(6), the following expression is then obtained [23]:

$$K = \frac{9}{2\pi} \left(\frac{1}{N_0} \right)^2 , \qquad (7)$$

with N_0 the critical sensitizer concentration related to the critical transfer distance by $R_0 = (3/4\pi N_0)^{1/3}$. Then Eq.(2) can be written in a simple form for the asymptotic decay as a function of the sensitizer concentration :

$$\tau = \frac{\tau_0}{\left[1 + \dfrac{9}{2\pi} \left(\dfrac{N}{N_0} \right)^2 \right]} . \qquad (8)$$

From Eq. (5) and Eq. (8), it is then possible to extract τ_0 and N_0 from the concentration dependence of the lifetime of the asymptotic part of the decay.

The microscopic parameters N_0 and τ_0 deduced independently from Eq.(5) and Eq.(8) are in good agreement. The results are the following:
$N_0 = 4.31 \times 10^{21}$ ions/cm^3 (4.84 at.%); $\tau_0 = 8.55$ μs ; $R_0 = 3.8$Å
The sensitizer-acceptor transfer constant α_{SA} is related to R_0 and τ_0 through the relation $\alpha_{SA} = R_0^6 / \tau_0$, a value of 3.52×10^{-40} cm^6/sec is obtained for α_{SA}.

4.2.2. Determination of the acceptor concentration (N_A), the diffusion constant (D), the sensitizer-sensitizer transfer constant (α_{SS}), the sensitizer-sensitizer transfer rate (W_{SS}) and the sensitizer-activator transfer rate (W_{SA}).

In the diffusion-limited model, Eq.(3), at the earlier times of the decay, reduces to Eq.(1) which can be written in a more simple form for a dipole-dipole interaction as :

$$\Phi(t) = \Phi(0)\exp\left[-\frac{t}{\tau_0} - \gamma\, t^{1/2}\right],$$

(9)

with

$$\gamma = \frac{4\pi}{3}\Gamma\left(1-\frac{3}{s}\right)N_A\, R_0^3\, \tau_0^{-1/2}.$$

(10)

So, the initial part of the fluorescence transients depicted on Fig.3 were fitted using Eq. (9) with N_A as the only adjustable parameter. The results are summarized in Tabl.3.

TABLE 3. Sensitizer and acceptor concentrations

[Er^{3+}] at.%	[Er^{3+}] (10^{21} ions/cm^3)	N_A (10^{19} ions/cm^3)
0.025	0.022	-
0.17	0.151	0.34
0.6	0.534	0.69
1.9	1.691	1.13
5.3	4.717	1.76
11.2	9.968	2.77

Fitting $\phi(t)$ using the complete expression (Eq.(3)) as derived by Yokota and Tanimoto is rather difficult. However, the diffusion constant can be calculated using Eq.(4). The concentration dependence of the diffusion constant is depicted on Figure 10. Measured values of D for rare earth ions in insulators range from 10^{-11} cm^2 s^{-1} to 10^{-9}cm^2 s^{-1} [24], many orders of magnitude smaller than for excitons in molecular crystals [25]. Up to now very few studies on the spectroscopy of rare earth in semiconductors and especially on their interaction with free carriers have been reported to explain the spectroscopic properties observed in the visible region. Further experiments are in progress to explain such high values for the diffusion constant. The diffusion constant exhibits a ¾ dependence on the sensitizer concentration as predicted in the diffusion-limited model. The sensitizer concentration has been evaluated from the total Er^{3+} concentration minus the N_A which represents the part of Er^{3+} ions acting as activator centers. Looking at the expression of D as a function of N_A it is then possible to derive the sensitizer-sensitizer constant α_{SS}. A value of 8.77 x 10^{-37} cm^6.s^{-1} has been obtained for the sensitizer-sensitizer coupling constant.

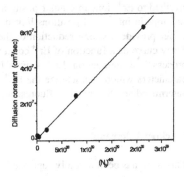

Figure 10. Diffusion constant as a function of sensitizer concentration. (dots: calculated values, full line: linear fit)

122

In the diffusion limited model, at the earlier times, the decay is mainly due to the direct sensitizer-acceptor transfer, then Eq. (10) can be written in the form :

$$\gamma = \frac{4\pi}{3} N_A R_0^3 W_{SA}^{1/2}.$$

(11)

From the fit of the earlier part of the decay and values of N_A and R_0 we deduce W_{SA}. A constant value equal to $3.68 \times 10^5 \ s^{-1}$ was obtained for the sensitizer-acceptor transfer rate in all the samples.

The asymptotic decay time $1/\tau_D$ plotted as a function of the sensitizer-sensitizer transfer rate expressed in units of the trapping rate in Log-Log scale (Fig.11) exhibits a linear dependence between 0.6 and 11.2 at.%, with a slope (equal to 0.6) less than unity which confirms the validity of using the diffusion limited model .

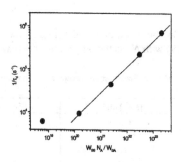

Figure 11 Asymptotic decay rate as a function of the trapping rate in a log-log plot ($W_{SS}N_A/W_{SA}$) (full line : linear fit ; dots : experimental data).

All the microscopic parameters which characterize the Er^{3+}-Er^{3+} interaction in the different GaN samples being determined, it is possible to simulate the transient fluorescence using the complete expression (Eq.(3)) for a diffusion-limited process. In Fig.3, the calculated curves are represented in comparison with the experimental data. As shown on Figure 3, this model explains the time dependence of the fluorescence intensity.

5. Conclusion

Three kinds of centers have been identified for Er^{3+} in GaN films. The main center is ascribed to Er^{3+} ions in regular positions in the lattice while two others (C2 and C3) to Er^{3+} ions in interstitial positions linked to defects. Interactions between Er^{3+} ions in C1 center provide a strong reduction of the $^4S_{3/2}$ lifetime. The profile of the experimental decay curve as a function of Er^{3+} concentration is well described by a diffusion-limited process, the interaction Er^{3+} - Er^{3+} being of dipolar nature. All the microscopic parameters which characterize the Er^{3+} - Er^{3+} interaction in the different samples were determined and the transient fluorescence has been simulated.

6. Acknowledgments

This work has been partially supported by the US Army Research Office.

7. References

1. Fang, X.M., Li, Y. and Langer, D.W. (1993) Radiative and nonradiative transitions in GaAs:Er, *J. Appl. Phys.*, **74**, 6990 - 6992.

2. Takahei, K. and Taguchi, A. (1995) Photoluminescence-excitation anlysis of Er- doped GaAs grown by metalorganic vapor phase deposition, *J. Appl. Phys.*, **77**, 1735 - 1740.

3. Takahei, K. and Taguchi, A. (1995) Photoluminescence analysis of Er- doped GaAs under host photoexcitation and direct intra-4f-shell photoexcitation, *J. Appl. Phys.*, **78**, 5614 - 5618.

4. Cedeberg, J.G., Culp, T.D., Bieg, B., Pfeiffer, D., Winter, C.H., Bray, K.L. and Kuech, T.F. (1999) Incorporation of optically active erbium into GaAs using the novel precursor Tris(3,5-di-tert-butylpyrozolato)bis(4-tert-butyle pyridine) erbium, *J. Appl. Phys.*, **85**, 1825 - 1831.

5. Culp, T.D.,. Cedeberg, J.G, Bieg, B., Kuech, T.F., Bray, K.L., Pfeiffer, D. and Winter, C.H. (1998) Photoluminescence and free carrier interactions in Er –doped GaAs, *J. Appl. Phys.*, **83**, 4918 - 4927.

6. Takahei, K., Taguchi, A., Nakagome, H., Uwai, K. and Whitney, P.S. (1989) Intra-4f-shell luminescence excitation and quenching mechanism of Yb in InP, *J. Appl. Phys.*, **66**, 4941 – 4945.

7. Thonke, K., Pressel, K., Bohnert, G., Stapor, A., Weber, J., Moser, M., Molassioti, A., Hangleiter, A. and Schoz, F. (1990) On excitation and decay mechanisms of Yb^{3+} luminescence in InP, *Semicond. Sci. Technol.*, **5**, 1124 - 1131.

8. Favennec, P.N., L'Haridon, H., Salvi, M., Moutonnet, D. and LeGuillou, Y. (1989) Luminescence of erbium implanted in various semiconductors IV,III-V and II-VI materials, *Electon. Lett.*, **25**, 718 -719.

9. Hömmerich, U., Seo, J.T., Abernathy, C.R., Steckl, A.J. and Zavada, J.M. (2001) Spectroscopic studies of the visible and infrared luminescence from Er doped GaN, *Mat. Sci. Eng.*, **B 81**, 116 - 120.

10. Przybylińska, H., Kozanecki, A., Głukhanyuk, V., Jantsch, W., As, D.J. and Lischka, K. (2001) Photoluminescence properties of Er doped GaN, *Physica B*, **308**, 34 - 37.

11. Lee, D.S., Heikenfeld, J., Birkhahn, R., Garter, M., Lee, B.K. and Steckl, A.J. (2000) Voltage-controlled yellow or orange emission from GaN codoped with Er and Eu, *Appl. Phys. Lett.*, **76**, 1525 - 1527.

12. Lee, D.S., Heikenfeld, J., Steckl, A.J., Hömmerich, U., Seo, J.T., Braud, A. and Zavada, J. (2001) Optimum Er concentration for in situ doped GaN visible and infrared luminescence, *Appl. Phys. Lett.*, **79**, 719 - 721.

13. Wilson, R.G., Schwartz, R.N., Abernathy, C.R., Pearton, S.J., Newman, N., Rubin, R., Fu, T. and Zavada, J.M. (1994) 1.54μm Photoluminescence from Er-implanted GaN and AlN, *Appl. Phys. Lett.*, **65**, 992 - 994.

14. Lorenz, K., Vianden, R., Birkhahn, R.H., Steckl, A.J., Da Silva, M.F., Soares, J.C. and Alves, E. (2000) RBS/Channeling study of Er –doped GaN fims grown by MBE on (111) Si substrates, *Nucl. Instr. Meth. B*, **161/163**, 946 - 951.

15. Hogg, R.A., Takahei, K. and Taguchi, A. (1997) Er-related trap levels in GaAs:Er,O studied by optical spectroscopy under hydrostatic pressure, *Phys. Rev. B*, **56**, 10255 - 10263.

16. Widdershoven, F.P. and Naus, J.P.M. (1989) BBBB, Mat. Sci. Eng. B, 4, 71 - 74.

17. Lee, D.S. and Steckl, A.J. (2001) Room-temperature grown rare-earth-doped GaN luminescence thin films, *Appl. Phys. Lett.*, **79**, 1962 - 1964.

18. Prokhorov, A.M., Kaminskii, A.A., Osiko, V.V., Timoshechkin, M.I., Zharikov, E.V., Butaeva, T.I., Sarkisov, S.E., Petrosyan, A.G. and Fedorov, V.A. (1977) Investigations of 3-μ stimulated emission from Er^{3+} ions in aluminium garnets at room temperature, *Phys. Stat. Sol. (a)*, **40**, K69 – K72.

19. Payne, S. A., Smith, L.K. and Krupke, W.F. (1995) Cross sections and quantum yields of the 3 μ emission for Er^{3+} and Ho^{3+} dopants in crystals, *J. Appl. Phys.*, **77**, 4274 - 4275.

20. Pellé, F., Gardant, N. and Auzel, F. (1998) Effect of excited state population density on nonradiative multiphonon multiphonon relaxation rates of rare earth ions, *J. Opt. Soc. Am.*, **15**, 667 - 679.

21. Inokuti, M. and Hirayama, F. (1965) Influence of energy transfer on exchange mechanism on donor luminescence, *J. Chem. Phys.*, **43**, 1978 - 1989.

22. Yokota, M. and Tanimoto, O. (1967) Effects of diffusion on energy transfer by resonance, *J. Phys. Soc. Japan*, **22**, 779 - 784.

23. Auzel, F., Bonfigli, F., Gagliari, S. and Baldacchini, G. (2001) The interplay of self-trapping and self-quenching for resonant transitions in solids; role of a cavity, *J. Lumin.*, **94-95**, 293 - 297.

24. Krasutsky, N. and Moos, H.W. (1973) Energy Transfer between the low lying energy levels of Pr^{3+} and Nd^{3+} in $LaCl_3$, *Phys.Rev. B*, **8**, 1010 - 1020.

25. Powell, R.C. and Soos, Z.G. (1975) Singlet exciton energy transfer in organic solids, *J. Lumin.*, **11**, 1 - 45.

HETEROGENEOUS DISPERSED SYSTEMS:
The New Materials Science Approach

V. KOSHKIN, YU. DOLZHENKO
National Technical University "Kharkov Polytechnic Institute",
21 Frunze St., Kharkov, 61002, Ukraine.
E-mail: Koshkin@kpi.kharkov.ua

Abstract. The peculiarities of thermodynamics of small particles with sizes of order of hundred Angstroms and less are discussed. It is shown that if the solubility of small particles in a solvent or in a melt cannot be long living, the impurity solubility in small particles appears to be for some orders of value greater compared to bulk samples and can preserve for a long time. This provides a new approach for materials preparation for different purposes.

Keywords: Small particles, thermodynamics, solubility.

1. Introduction

The article presents the investigation of thermodynamic peculiarities of condensed particles at a scale of 10 - 100 Angstroms. This very popular new field of solid state physics and engineering appears to be of special interest for general discussion of thermodynamics of heterophase systems. We will discuss the "equilibrium solubility" definition and determine the theoretical limit of a solubility in a bulk liquid, being really a limit of supersaturation determined by a virtual nucleation of condensed particles. We will discuss also general rules of solubility in small crystals appearing to be different comparatively to massive amounts of substances. Solubility in two contacting phases are self-consisting indeed. But we will analyze the two sides of this problem separately to clarify physical mechanisms. We will formulate also some new possibilities in materials science as the result of phenomena discussed.

2. Thermodynamics of Small Crystals Solubility in Liquids

2.1. WHAT IS AN EQUILIBRIUM SOLUBILITY ?

What is the way for equilibrium solubility measuring? Lumps of salt are thrown into a liquid and observations are made as to when the next portion of the solid phase being added into the resulting solution will at last remain insoluble, i.e. when the resulting solution and the solid phase will be in equilibrium. Just this value of N_∞ is taken as the limit solubility of the given solid substance in the given liquid. But Ostwald has shown

J.-C. Krupa and N.A. Kulagin (eds.), Physics of Laser Crystals, 125–134.
© 2003 *Kluwer Academic Publishers. Printed in the Netherlands.*

[1] that the limit concentration of a dissolved substance depends upon particles dispersion of a solid phase that a saturated solution is in the equilibrium with. What is solubility then? And how this difference can be put in correspondence with general thermodynamics: because if the breakage of one of the ingredients is incorporated into the process of solution preparation, the equilibrium result of the process changes, isn't it?

And finally, let us imagine the following experiment. There is a saturated solution obtained by the procedure described above, and therefore equilibrium relative to a solid phase it is in contact with. Let us draw off this solution, removing a solid phase. The concentration was limited and equilibrium as far as the solution was in an interaction with a crystalline phase. What is the limit concentration in a free solution without a solid phase, so to say, "absolute" solubility? Thus, what do we find reference books looking for the column "solubility"?

2.2. SOLUBILITY OF MASSIVE CRYSTAL IN A LIQUID

All calculations given below are related to ideal solutions which allow to clear up general phenomena pattern. Of course, it is possible to change "concentrations" for "activity" but corresponding coefficients will not add anything for the physical meaning.

Concentration of a liquid solution being in equilibrium with solid phase is determined by the equality of their chemical potentials

$$\mu_s = \mu_l, \tag{1}$$

where μ_s and μ_l are corresponding chemical potentials of the substance dissolved in both phases. In ideal solution approximation it is known to result in Schreder equation [2]:

$$d \ln N_\infty / dT = \Delta H_{m\infty} / RT^2 , \tag{2}$$

the integration of which with the limits of T_m to T and of I to N_∞ gives the expression:

$$\ln N_\infty = -\frac{\Delta H_{m\infty}}{R} \cdot \left(\frac{1}{T} - \frac{1}{T_m} \right), \tag{3}$$

which determines the limit concentration N_∞ (in mole fractions) of the dissolved substance being in equilibrium with crystal dissolved. In (3) T_m and $\Delta H_{m\infty}$ correspond to the temperature and heat of melting consequently. The main Schreder's assumption is: a crystal, transfering into a liquid phase "dissolves in itself", at the melting point the solubility appears to be equal to one.

Using Gibbs equation including contribution of phases interfaces into the thermodynamic potentials, Ostwald showed a solid phase dispersion defines a higher concentration of the dissolved substance (N_d) above the surface of the solid blob with the radius r as compared with that above the flat surface of a massive crystal. Ostwald's equation :

$$ln\left(N_d/N_\infty\right) = \frac{2\sigma V_m}{RT} \cdot \frac{1}{r} \quad , \tag{4}$$

gives the relation of this concentrations (σ is the specific surface energy, V_m – molar volume).

From Eq.(4) it becomes clear that the solubility N_d in a system, where a solid phase presents not an unlimited flat surface, but a set of small blobs with a strictly defined radius r, must be higher than N_∞ .

Meanwhile, the Exp.(4) is obtained like the Laplas-Thomson's relation and describes the pressure applied to the blob surface and directed to its centre but not the concentration of the solid substance dissolved in liquid. The latter, as we will see, is actually defined by the pressure, but indirectly, through its influence on thermodynamic parameters of condensed phase.

2.3. SOLUBILITY OF MONODISPERSED SOLID PARTICILES IN LIQUID

The mechanism of influence of dispersity on the solubility is easy to understand using the same Schreder's idea. Really, the Laplas pressure applied to the particle with the radius r results in the change of its melting point which can be easily calculated using the classical Gibbs expression of free energy for dispersed systems [1]:

$$T_d = T_m \cdot \left(1 - \frac{2\sigma V_m}{\Delta H_m r}\right), \tag{5}$$

where T_m and T_d are melting points of the massive and dispersed crystals accordingly. Substituting (5) to (3) and regarding ΔH_m as a first approximation to be independent of r, and the parameter $2\sigma V_m / rRT_m \ll 1$ at any values of r down to atomic sizes, one obtains:

$$ln N_d = \left[-\frac{\Delta H_m}{R} \cdot \left(\frac{1}{T} - \frac{1}{T_m}\right)\right] + \frac{2\sigma V_m}{RT_m} \cdot \frac{1}{r} \tag{5a}$$

or

$$ln\left(N_d/N_\infty\right) = 2\sigma V_m/RT_m r . \tag{5b},$$

The bracketed term in (5a) is sure to coincide with the expression for the equilibrium solubility above flat surface (3).

The Exp.(5b) differs from Ostwald's equation. The temperature dependence N_d in Exp.(5b) is determined only by the temperature dependence N_∞, and the ratio N_d/N_∞ is independent of temperature, i.e. at any temperature this ratio is constant.

But in fact, ΔH_m depends upon the particle's radius. Let us take into consideration this dependence. It is easy to show that the molar enthalpy of a monodispersed solid is:

$$H_s = H_{s\infty} - \frac{3V_m\sigma}{r},$$ (6)

where $H_{s\infty}$ is the enthalpy of a massive crystal mole. We use here the Hill's like approach [2]. The melted substance enthalpy does not surely depend upon the dispersity. Therefore the difference between solid and liquid phase enthalpies at melting point is:

$$\Delta H_{md} = H_s - H_l = \Delta H_m - \frac{3V_m\sigma}{r}.$$ (7)

Substituting (7) in (5a) one receives:

$$\ln N_d = -\frac{\Delta H_{m\infty}}{RT} + \frac{\Delta H_{m\infty}}{RT_m} + \frac{\sigma V_m}{R}\left(\frac{3}{T} - \frac{1}{T_m}\right)\cdot\frac{1}{r}.$$ (8)

The temperature range is naturally limited by the melting point T_m, and $T < T_m$, therefore $3/T > 1/T_m$, and (8) obtains the form accurate to the numerical factor (3 instead of 2), considering with the Ostwald's equation (4):

$$\ln\left(N_d/N_\infty\right) = \frac{3\sigma V_m}{RT}\cdot\frac{1}{r}.$$ (9)

Exps.(5) and (9) are obtained (as well as (4)), from the equilibrium condition (1) but in contrast to (4) with the consistent account of the Laplace pressure effect exactly upon the solid phase.

The mechanism of influence of solid particle sizes upon their concentration in solution is determined exactly through changing solid phase properties by its dispersion.

Finely dispersed particles of solid phase that have a noticeable influence upon solubility have sizes of an order of hundred of angstroms and less. Their shape is rather crystalline grains than spheres. If, for instance, these are cubes with the rib a, the numerical factors in all equations (4-7) should be changed.

Thus, (8) and (9) have the following forms:

$$\ln N_d = -\frac{\Delta H_{m\infty}}{R}\left(\frac{1}{T} - \frac{1}{T_m}\right) + \frac{2\sigma V_m}{R}\left(\frac{3}{T} - \frac{1}{T_m}\right)\cdot\frac{1}{a},$$ (10)

$$\ln\left(N_d/N_\infty\right) = \frac{6\sigma V_m}{aRT}.$$ (11)

Sometimes the so-called "shape factor" is introduced with the aim of taking into account the whole set of particles with different surface curvature including fragments with concave surface, i.e. with negative curvature [1]. This can be easily made, as well as more precise dependence of surface energy upon temperature and so on. But, as we will see later, these corrections hardly make any new sense, if the point of discussion is the solution being in *equilibrium* with the dispersed crystalline phase.

2.4. EQUILIBRIUM SOLUTION ABOVE A DISPERSED SOLID: IS IT POSSIBLE?

The equilibrium state of any system presupposes that there exist no time changes of its macroscopic characteristics. If there are solid particles of different sizes in the system then, according to Thompson's equation, the concentration of the liquid close by smaller particles is higher than that near the bigger ones. It predetermines the flow of atoms (either ions or molecules) from fine particles to larger-sizes ones, so as the former decrease in size down to disappearance and the larger ones grow. Finally, the whole of the solid phase will gather in a single monocrystalline particle with a shape of a regular polyhedron in conformity with Wulf's relationships, so far as any surface "roughness" is smoothed out owing to the same processes of dissolving protruding fragments (with positive curvature) and filling in cavities (with negative curvature). Therefore the "solution - dispersed solid phase" system, if a solid phase is not strictly monodispersed, when coming to equilibrium becomes "solution - massive crystal" system with perfectly smooth surface. The larger particles become, the slower goes the process of coming the equilibrium, still equilibrium solubility by "solid phase – solution" contact is only one: it is defined by classical Schreder's equation: this is N_∞ – the limit concentration of liquid above the smooth surface of a massive crystal.

The equilibrium solubility N_d exceeding N_∞ could exist in a sole case: if all solid particles had exactly the same size and shape. However, it is both unachievable experimentally and impossible theoretically since such a system is unstable relative to smallest size fluctuations. Actually, even a minor size change of a single particle brings immediately to considerable changes in local concentration so far as N_d depends exponentially upon r. Diffusive flows are proportional to the differences (gradients) of concentrations which results in very speedy loss (with rather small particle size) of monodispersity and consequently to the scenario described above.

Therefore, equilibrium N_d is nonexistent. At the same time, it results in the conclusion that the dispersed phase is principally unstable in the medium where atomic particles change is possible. However, it could be made not a conclusion but a precondition of the whole consideration: as a shattered crystal is not an equilibrium system because minimum of its free energy correspond to monocrystalline form. Thus the general principles of thermodynamics are quite all right here: the final product of the solution reaction does not depend upon the prehistory of its original ingredients.

Eq.(4) as well as its modifications Eqs.(5-9) could describe the equilibrium system only if there were a SOLE solid particle in it. It is outside possibilities of a typical chemical experiment, indeed.

However, the effect of enormous concentration increase of substances dissolved in melts that has been revealed in [3] by dissolving finely dispersed particles are, no doubt, authentic. They are not related to equilibrium but allow to obtain supersaturated solutions for long time. Expressions given in sect. 2.3 can be used for estimation their concentration. It should be mentioned though, that calculations of concentration using "average" values of particle radii hardly have any sense. As a matter of fact, the contribution to the solution concentration is made only by the smallest particles and this contribution grows exponentially fast with their radius reduction. The larger particles only lower the concentration of the solution so far as till there are smaller particles available, larger ones serve as nucleus of crystallization.

Measuring concentration in the solution above the polydispersed solid phase seems to give a method for sizing the most finely divided crystalline grains seems to give a method for sizing the most finely divided crystalline grains of polydispersed phase. The investigation of the kinetics of the solution concentration changes makes may be possible to reconstruct the size distribution of solid particles. It certainly needs a separate consideration. But the calculation of concentrations at a base of particle sizes determined in independent experiments hardly has any relation to reality, so far as neither optical nor electronic microscopy, nor gas adsorbtion – desorbtion methods can not, for different reasons, determine the presense and portion of smaller size particles.

2.5. VIRTUAL SOLID NUCLEI, "ABSOLUTE" SOLUBILITY, AND THE LIMIT OF SUPERSATURATION

Meanwhile, it is the dependence of local concentration upon particles radius that defines the "absolute" solubility just when a solution does not contact a solid phase.

As is well known, a solid phase growth from a solution or melt is preceded by a nucleus formation the radius of which r_c is determined by the complete free energy accounting its volume and surface components. If spontaneously arisen solid phase formation has the radius r smaller than the critical one, then such formation is unstable, and it is disappearing, if $r > r_c$, then nuclei begin to grow. Here is the expression for r_c (see, for example, [1]):

$$r_c = 2\sigma V_m \Big/ \Delta H_{m\infty}.$$ (12)

If the solution "infused with" dispersed particles and, consequently, non-equilibrium will clear of solid phase, the precipitated nucleus with a very small radius r (somewhat greater than r_c) begins at once to dissolve accordingly to the Laplace-Thompson-Ostwald law. While dissolving it diminishes down to the value of $r < r_c$ and break down. "Absolute" solubility N_a is determined by the equilibrium between a solution and dispersed nuclei with critical radius. The latter break down in a moment of birth. These are virtual solid particles. Having substituted r_c from (12) into (5a) and (9) we obtain respectively:

$$ln(N_a/N_\infty) = \frac{\Delta H_{m\infty}}{RT_m} = \frac{\Delta S_m}{R}$$ (13)

(ignoring the dependence of melting enthalpy of particle sizes) and

$$ln(N_a/N_\infty) = \frac{3}{2}\frac{\Delta H_{m\infty}}{RT}$$ (14)

taking it into account.

Both equations show that absolute solubility N_a exceeds the value of equilibrium solubility N_d and N_∞. In the first case the ratio N_a/N_d is independent of temperature, by accounting the relation $\Delta H_m(r)$ the ratio N_a / N_d is dependent on temperature: with the growth of the latter the value of N_a approaches (from above) N_d.

It is obvious that N_a is experimentally unachievable but the procedure of its approach could be as following. A highly concentrated solution is prepared, it is cleared of the solid phase (which, as we have seen, promotes its phase stability). Then, gradually, very small quantities of dispersed solid phase are added, particle sizes successively diminishing. One may use the temperature factor as well.

Thus, maximum solubility of solid substance in liquid is its concentration in free solution that is not in contact with the solid phase. This is the limit supersaturation in the Van der Vaals sense. One may use this value for estimation of supersaturation.

3. Thermodynamics of Media Substance Solubility in Small Crystals

3.1. TWO WAYS FOR SURFACE ENERGY DECREASING

As it was shown in the Part 2, the solubility of dispersed solid can not be equilibrium in general. But for isolated condensed particles in a definite media, a long term of non equilibrium state may be expected. The main idea which will be taken into account is that a decreasing of surface as a motive force for decreasing of free energy of two phases system (accounting its surface part) is not a single way for the free energy minimization.

Another way is the formation of solution of media substance in a small condensed (solid or liquid) particle, as well as its substance in the media if the formation of such solutions decreases an energy per unit surface at phases boundaries. Thermodynamics uses both ways for general free energy minimization. The formation of such solutions may lead to a growth of solid phase internal energy. In the case of small solid solubility of components, the energy change per single atom introduced in a solid $U > 0$, this is disadvantageous for a system. A competition between condensed particle internal energy increasing and changes of specific surface energy accompanying a dissolution of a media substance in condense phase (and vise versa), determines the final result. This idea is discussed in more details in the separate communication.

3.2. THE SOLUBILITY OF MEDIA SUBSTANCE IN SMALL CRYSTALS

Let us account contributions of different parts of Helmholz free energy ΔF of a lone B crystal (with N atoms per unit volume) immersed in a melt of a substance A. The equilibrium concentration c of A in B is determined by the minimum of whole free energy F, including its volume and surface parts

$$\Delta F = \left(UcN + NkTc\ln c\right)V + \sigma(c)S + kT\varphi S \ , \tag{15}$$

where S and V are surface square and volume of a crystal, $\sigma(c)$ is the specific surface energy of crystal B in the A melt, a dependence of $\sigma(c)$ on a concentration of A in solid B is presumed; k is the Boltzman constant, T - temperature. φ is a surface specific entropy. Let us suppose that A - melt presents an infinite media. We supposed also that concentration of A in solid B is small ($c \ll 1$), this is accounted in the term in brackets

Let σ_{AB} is the surface energy at the boundary of pure B crystal and pure A melt. The main point: what is a dependence of specific surface energy of the concentration c of A in B crystals? It is naturally to suppose that $\sigma(c)$ is proportional to the concentration of solid solution of A in B:

$$\sigma(c) = \sigma_{AB} + \beta c . \tag{16}$$

β is the coefficient which can be easily determined, taking into account, that at $c = 1$, $\sigma(c)$ is coinciding with the specific surface energy σ_{0A} at the boundary of pure A crystal and its melt. Thus $\beta = \sigma_{0A} - \sigma_{AB}$.

The equation:

$$c = exp\left[-\left(U - \beta\frac{V_m}{N_A} \cdot \frac{S}{V} + 1\right)\middle/kT\right], \tag{17}$$

gives the equilibrium concentration of A in B. The specific surface entropy is almost independent of concentration [1], thus derivative is small and omitted in (17). V_m is the mole volume of B, N_A is the Avogadro number.

The most important is the sign of β. It is determined by chemical relations between A and B atoms. If the bond energy A-A is significantly greater than A-B one, the surface energy σ at the boundary A-crystal - A-melt is less than σ_{0AB} and so $\beta < 0$. The expression (17) for equilibrium concentration of impurities in small particle looks like for a bulk sample but with the effective U contribution decreased for a factor, determined by surface to volume relation.

Such relation of bonds energies determines very small mutual solubility of A and B. Vice versa, if the A-B bonds are significantly more strong than A-A ones, so coefficient $\beta > 0$. This leads to a great bulk solid solubility and even to ordered phases or solid compounds formation (in physical and chemical terms respectively).

The analysis of minimized Helmholz free energy of a phase with accounting boundaries contributions gives the expression for the relation of equilibrium concentration c of A in B in the small particle to its concentration c_0 in massive samples, being classical expression which obtained easily if the surface contribution is negligible:

$$c/c_0 = exp\left(\frac{\beta V_m}{kTN_A} \cdot \frac{S}{V}\right) \tag{18}$$

Using the lattice parameter a and number η of atoms in an elementary cell ($V_m = \left(N_A \cdot a^3 \eta^{-1}\right)$) one obtains for little cube with a rib of ma

$$c/c_0 = exp\left(\frac{6\beta a^2}{\eta kT} \cdot \frac{1}{m}\right), \tag{19}$$

$S/V = 3/r$ for spheres, or $6/a$ for cubs with r radius or ma rib respectively, m is the number of elementary cells in a solid particle.

3.3. ESTIMATION FOR REAL SYSTEMS

What is a scale of effects? For $T = 1000$ K, $\sigma = 500$ erg/cm^2, $m = 20$ and $\eta = 4$ the *equilibrium A solubility* in B-particles appears for a factor of 1.5 more than in massive; by $m = 10$, a relation is 16, at $\sigma = 1000$ erg/cm^2 the c/c_0 relation is 200.

Similar relations are valid for thin films and layers. So we obtained the paradoxical result: the less the usual (macroscopic) solubility the more solubility in nanoform. This aspect of a problem will be discussed in the separate communication. Thus we obtain the possibility to provide the solids doped by impurities which cannot be introduced into the macroscopic species.

In contrary, if $\beta > 0$, a solubility in nanoparticles decreases exponentially with their size. Thus one have to expect the deepest purification of such small particles in an appropriate dissolvent. The self consisted problem of solid – liquid equilibrium in a restricted liquid volume with small condensed particles distributed will be discussed in a separate communication.

4. Conclusions

We had seen the equilibrium state with dispersed solid particles participation is impossible. At the same time we had understood that not liquid only but solid solution as well leads to the decreasing of general free energy of a system. The last gives a possibility to create new doped solids with a long existing time taking into account a low rate of diffusion in solids at not too high temperatures.

Numerical estimations delivered explains experimental observations [4, 5], showing, that in thin films with small metal particles in a media of another metal (being almost indissoluble in massive), the mutual solubility appears to be great. This is a way for a preparation of new materials laser materials in particular with desirable impurities in any desirable matrices.

Let us mention that an elaboration of gamma- and Ro-lasers which needs for radiation stability in addition to special luminescent parameters have to be adopted to small particles peculiarities. The general approach one may find in [6]. Two parameters: defects instability zones radii and sizes of small solid particles have to be account for a theory of dispersed particles peculiarities under the ionizing irradiation. This will be a subject of a next communication.

Acknowledgments for The Center for Fundamental Researches of Ukraine for a support this research in part. Thanks to Prof. V. Slyozov for numerous fruitful discussions.

134

5. References

1. Frolov, Yu.G. (1982) *Kurs kolloidnoy himii. Poverhnostnye javlenija i dispersnye sistemy*, Himija, Moskva
2. Hill, T.L. (1963) *Thermodynamics of Small Systems*, J.Wiley, N.Y. – Amsterdam.
3. Cherginets, V.L., Deineka, T.G., Demirskaya, O.V., Rebrova, T.P. (2002) Potentialmetric investigation of oxide solubilities in molten KCl–NaCl eutectic. The effect of surface area of solid particles on the solubilities, *J. Electroanal. Chem.* **531**, 171-178.
4. Zubkov, A.I., Il'insky, A.I., Podgornaya, O.A., Sevruk, V.I., Sokol-Prusskii, Ya. G. (1990) On the aging possibility of fast quenched Cu-Mo alloys, *Fizika Metallov i Metallovedenie* **10**, 197-198.
5. Gleiter, H., Weissmuller, J., Wollershtein, O. and Wurschum, R. (2001) Nanocrystalline materials: a way to solids with tunable electronic structure and properties?, *Acta materialia* **49**, 737-745.
6. Koshkin, V.M. (2002) Instability zones and short living defects in physics of crystals, *Low Temperature Physics* **28**, 695-712.

ELECTRONIC STRUCTURE OF DOPED AND IRRADIATED WIDE BAND-GAP CRYSTALS

N.A. KULAGIN
Kharkov National University for Radioelectronics,
av. Shakespeare 6-48, 61045 Kharkov, Ukraine
E-mail: nkulagin@bestnet.kharkov.ua

Abstract. This work is devoted to the development of ab initio calculation of the electronic structure of either clusters or doped and unperfected crystals. In addition, analysis of different experimental data for γ-irradiated pure and doped wide-band gap compounds is presented. Change in the spectroscopic properties of sapphire, perovskite, garnets and other doped compounds as well as ion oxidation state in pure and doped γ-irradiated crystals are discussed in detail using the X ray spectroscopy experimental results.

1. Introduction

The electronic states of component and impurity ions entering in the composition of solids are responsible of the solid ground properties. In particular, the oxidation state of the ions and the symmetry of the crystalline lattice determine the structure of the energy bands, optical, dielectrical and other properties of the crystals. The electronic configuration of ions with unfilled d- or f- shell gives rise to the symmetry of the cluster formed by the nl – ion and its close surrounding [1-3].

Irradiation in most of the wide-band-gap crystals doped with nl- ions creates different color and point defects that induce additional optical absorption bands in solids [1-3]. The types of color center (charge trapped in anion or cation vacancy, change in oxidation state, interstitial ions and more complex defects) and the intensity of the optical absorption (emission) bands depend on the crystallographic structure of the compounds as well as on the crystal growth conditions along with the types and concentration of impurities. Spectroscopic methods were widely used to study the radiation induced defects in crystals but not so far successfully.

It is well known that the ions of iron, lanthanum and actinium group elements (i.e. ions with unfilled 3d-, 4f-, 5f- electronic shell, nl- ions, shortly) show oxidation states from "+2" to "+6" and even "+7" in different compounds. Thermal treatment, irradiation, variation of the crystal growth conditions, etc may induce changes in oxidation state of the ions (components and impurities). For example, $Sr^{2+}Ti^{4+}O_3$ single crystals can contain more than 20 % of Ti^{3+} ions upon a thermal treatment [3-5]; K^+ ion in KCl single crystals change its oxidation state to K^o under the influence of UV-, X ray, γ- or electron beam irradiation [5-7].

J.-C. Krupa and N.A. Kulagin (eds.), Physics of Laser Crystals, 135–162.

It is well known that the doped nl- ions change their oxidation state more efficiently upon irradiation or thermal treatment than the regular crystal component ones [2-3, 6-8]. Defect accumulation during irradiation induces also changes in the doped ion oxidation state. For lanthanum or iron group ions, this effect may be studied by using ESR, optical absorption (or emission), thermo-stimulated processes such as thermo-stimulated conductivity (TSC) and thermo-stimulated luminescence (TSL) and other methods[9-11]. It should be noticed that the experimental determination of the ion oxidation state for most of the doped and component ions in the solid is a complex problem. For this purpose, we propose an effective method based on the valence shift of X Ray lines called VSXRL.

The radiation induced defects in ruby: α-Al_2O_3:Cr, perovskite; $YAlO_3$:Cr, garnets: $Y_3Al_5O_{12}$:Cr/Nd, $Gd_3Sc_2A_3O_{12}$:Cr:Ca [12-14] for doses higher than 10^2 Gy give rise to additional optical absorption bands at 217, 260, 360 and 460 nm for ruby, [9], 295. 370, 425 and 500 nm for the perovskite [13], and 253, 282, 417 and 488 nm for $Y_3Al_5O_{12}$:Cr garnet [15]. The concentration of the Cr^{3+} ions can be determined via the $\frac{1}{2} \leftrightarrow \frac{1}{2}$ transition intensity measurement and the concentration of the colour and radiation induced defects can be estimated by using the half-width value of the ESR line attributed to the $1/2 \leftrightarrow 3/2$ transition [10]. Upon temperature treatment of the irradiated samples at 600 to 1200 K during 600 s, the extra optical absorption bands can be removed.

To understand the physical processes occurring in the irradiated crystals, we have developed an ab initio energy calculation for clusters. The Hartee-Fock-Pauli method and Heitler-London approximation were used as basis in this approach [16-25]. The self-consistent field theory for clusters and crystals permits to calculate the wave functions and to simulate the physical properties of clusters and solids. Optical, and X-ray etc. spectra of nl- ions in crystals were studied with high accuracy [3, 4, 12, 14].

The cluster Me^{n+}:$[L]_k$, consisting in the Me or RE ions with its surrounding ligands, L, is the elementary grouping for the study of the energy level scheme of the doped solids, in particular, laser crystals. The number of ions in the cluster is not defined. Similar clusters play an important role in the study of liquids, gas phases and plasma.

In this work, we consider the foundations of an ab initio theory for the Me^{n+}:$[L]_k$ cluster electronic structure calculation in different solids. For example, we have considered the electronic structure of Cr^{n+} ions in ruby and garnet, Cu and Ni or Zn ions in supraconductors and different RE^{n+} ions in oxide and fluoride compounds. Me^{n+}:$[L]_k$ clusters for different Me ions and ligands, with different oxidation states ,"+n", coordination number, k, and inter-atomic distance, R, were studied. This approach was also used to investigate the theoretical and experimental dependence of the X ray energy ($K_{\alpha, \beta}$-lines, for 3d- and 4f- ions and $L_{\alpha, \beta}$-lines for RE or actinides (An)) on the ion oxidation state. We will discuss the main paths leading to a change in the oxidation state of nl- ions and in the electronic structure of the doped crystals upon irradiation.

2. Method of the Self-consistent Field for Clusters and Solids.

A symbiosis of the Hartree-Fock-Pauli approach and the Heitler-London model has been used to study the electronic structure of clusters and solids [19-25]. The theoretical values are compared to experimental ones for optical and X-ray spectra of Me and RE /An ions in wide band gap crystals [3, 16, 27-32].

The foundation of the theory is the well known Born-Oppenheimer approximation for unrestricted Me^{n+}:$[L]_k$ cluster. For the electronic part of the total wave function of the crystal, we may write the following equation:

$$\left[-\frac{1}{2}\sum_i \Delta_i - \sum_i V(r_i, R) + \sum_{i<j}\frac{1}{r_{ij}}\right]\Psi_l(r, R) = W_l(R)\Psi_l(r, R).$$ (1)

For nl- ions Eq.(1) may be written as:

$$\left[H_{free}(r) + H_{int}(r, R)\right]\Psi_l(r, R) = E_l(R)\Psi_l(r, R),$$ (2)

where, H_{free} is the Hamiltonian for the free ion and H_{int} describes electron–ion and electron-electron interactions:

$$H_{int}(r, R) = \sum_i \frac{Z_i}{|r - R_i|} + \sum_{i>j}\frac{1}{|r_i - r_j|},$$ (3)

For the first step, we may consider the electronic structure of Me^{n+}:$[L]_k$ cluster, consisting of the central Me ion with nl^N- ground configuration surrounded by k ligands, L, at a distance R. The electronic configuration of the ligand contains the occupied electron shells, $n'l'^{4l'+2}$, only. The symmetry of the cluster corresponds to the symmetry of the crystallographic structure of the compound.

In the framework of the Heitler–London approximation, the wave function of the cluster can be written as:

$$\Phi(n_1 l_1^{N_1} n_2 l_2^{N_2}, G) = \overset{\triangleleft}{A}\Psi_1(n_1 l_1^{N_1}, G_1)\Psi_2(n_2 l_2^{N_2}, G_2)$$
$$A = \frac{1}{\left[(N_1 + N_2)!\right]^{1/2}}\sum_{i<j}\varepsilon_{ij}P_{ij}$$ (4)

In Eq.(4), the term A describes the anti-symmetrization of the wave functions of the central ion, $\Psi_1(nl^N)$, and ligand, $\Psi_2(n'l'^N)$, relatively to the rearrangement of the electrons for the ion in the crystal.

Using the one-electron approximation for the wave functions of the central field approximation, the energy of the cluster may be written as [3, 12]:

$$E(Me^{+n}:[L]_k) = E_0 + kE_1 + k'(E_z + E_c + E_{ex}),$$ (5)

where E_0 (E_1) is the energy of the central ion and ligand respectively, in the free state.

The terms E_z, E_c and E_{ex} in Eq. (5) correspond to the energy of the interaction of electrons with the strange nucleus, the Coulomb interaction and the exchange interaction for all electrons. k', is a numerical coefficient depending on the symmetry of the cluster. Expressions for the terms E_z, E_c and E_{ex} in Eq.(5) may be written as:

$$E_z = \sum_m O_{m_1 m_1'} O_{m_2 m_2'} [Z_2 (m_1 \mid m_2)(m_1 \mid r_{z_1}^{-1} \mid m_2') + Z_2 (m_2' \mid m_1)(m_2 \mid r_{z_1}^{-1} \mid m_1')] +$$

$$+ Z_2 \sum_m O_{m_1 m_1'}^{\overline{m}_1 \overline{m}_1'} O_{m_2 m_2'} (m_1 \mid m_2)(m_1' \mid m_2)(\overline{m}_1 \mid r_{z_1}^{-1} \mid m_2') +$$

$$+ Z_1 \sum_m O_{m_1 m_1'} O_{m_2 m_2'}^{\overline{m}_2 \overline{m}_2'} (m_2 \mid m_1)(m_2' \mid m_1)(\overline{m}_2 \mid r_{z_1}^{-1} \mid m_1').$$

$$E_C = \sum_m O_{m_1 m_1'} O_{m_2 m_2'} \left(m_1 m_2 \mid r_{12}^{-1} \mid m_1' m_2' \right) \tag{6}$$

$$E_{ex} = -A(SS')[\sum_m O_{m_1 m_1'} O_{m_2 m_2'} \left(m_1 m_2 \mid r_{12}^{-1} \mid m_1' m_2' \right)] +$$

$$+ \sum_m O_{m_1 m_1'}^{\overline{m}_1 \overline{m}_1'} O_{m_2 m_2'} (m_1 \mid m_2)(\overline{m}_1 m_2 \mid r_{12}^{-1} \mid \overline{m}_1 m_1') +$$

$$+ \sum_m O_{m_1 m_1'} O_{m_2 m_2'}^{\overline{m}_2 \overline{m}_2'} (m_2 \mid m_1)(\overline{m}_2 m_1 \mid r_{12}^{-1} \mid \overline{m}_2 m_2') +$$

$$+ \sum_m O_{m_1 m_1'}^{\overline{m}_1 \overline{m}_1'} O_{m_2 m_2'}^{\overline{m}_2 \overline{m}_2'} (m_1 \mid m_2)(m_2 \mid m_1')(\overline{m}_1 \overline{m}_2 \mid r_{12}^{-1} \mid \overline{m}_1 \overline{m}_2')],$$

where: $\left(m_1 \mid m_2' \right)$ is the two electrons two-centers penetrating integral of the wave functions of nl-ion and n'l'-ligand;

$\left(m_1 \mid r_{z_2}^{-1} \mid m_2' \right)$ is the one electron two-centers matrix elements of operator describing the interaction of the electron with the strange nucleus;

$\left(m_1 m_2 \mid r_{12}^{-1} \mid m_1' m_2' \right)$ is the two electrons two-centers matrix element of the Coulomb or exchange interaction. Indexes Z_1, m_1 and Z_2, m_2 in Eq.(6) correspond to nl- ion and ligands.

Spin dependent coefficient, $A(S_1 S_2)$, was determined in [3, 12] as:

$$A(SS') = -\frac{1}{2} + \frac{2(S \mid (S \cdot S') \mid S)}{[S, S']} \delta(S, S_1 \pm 1/2)[\delta(S', S_2' \pm 1/2) - \delta(S', S_2' \mp 1/2)]. \tag{7}$$

In Eq.(7), $[S] = 2S + 1$ and coefficients $O_{mm'}$, $O_{mm'}^{\overline{mm'}}$ were calculated in [3] by the following way:

$$O_{mm'} = N \sum_{\alpha_1 L_1 M} (l^N, \alpha L S \| l^{N-1}(\alpha_1 L_1 S_1), l)(l^{N-1}(\alpha_1 L_1 S_1), l) \| l^N, \alpha' L' S') \times$$

$$\tag{8a}$$

$$\times \begin{bmatrix} L_1 & l & L' \\ M_1 & m & M' \end{bmatrix} \begin{bmatrix} L_1 & l & L \\ M_1 & m & M \end{bmatrix}$$

$$O_{\overline{m}\overline{m}}^{\overline{m}\overline{m}} = N(N-1) \sum_{\alpha_1 L_1 \alpha_2 L_2 S_2 \alpha_3 L_3 Mm} (l^N, \alpha LS \| l^{N-1}(\alpha_2 L_2 S_1), l)(l^{N-1}(\alpha_1 L_1 S_1), l \| l^N, \alpha' L'S') \times$$

$$\times (l^{N-1}, \alpha_1 L_1 S_1 \| l^{N-2}(\alpha_3 L_3 S_3), l)(l^{N-2}(\alpha_3 L_3 S_3), l \| l^{N-1}, \alpha_2 L_2 S_1) \begin{bmatrix} L_1 & l & L \\ M_1 & m & M \end{bmatrix} \times \quad (8)$$

$$\times \begin{bmatrix} L_2 & l & L \\ M_2 & m' & M \end{bmatrix} \begin{bmatrix} L_3 & l & L_1 \\ M_3 & \overline{m} & M_1 \end{bmatrix} \begin{bmatrix} L_3 & l & L_2 \\ M_3 & \overline{m}' & M_2 \end{bmatrix}$$

The one-particle fractional parentage coefficients: $(l^N, \alpha LS \| l^{N-1}(\alpha_1 L_1 S_1), l)$ and

$(l^{N-1}, \alpha_1 L_1 S_1 \| l^{N-1}(\alpha_2 L_2 S_2), l)$ and 3j-symbol $\begin{bmatrix} L_3 & l & L_2 \\ M_3 & \overline{m}' & M_2 \end{bmatrix}$ which were introduced in

Eq. (8) are listed for d- and f- free ions in [33] and in [22] for ions in crystals.

We have used the central field approximation where the one-electron wave function is determined as: $\phi_{nlm}(r, \theta, \varphi) = P(nl | r) Y_{lm}(\theta, \varphi)$. The system of equations for the radial functions of the nl-ion, $P(nl | r)$ and the ligand, $P(n'l' | r)$ was taken as the result of minimization of Eq.(5) with respect to the radial wave functions of the nl- ion and ligands (see detail in [2-3, 13-18]).

The following system of equations may be written for each ion in the cluster [3, 16, 18]:

$$\left[\frac{d^2}{dr^2} + \frac{2}{r} Y'(nl | r) - \varepsilon_{nl} - \frac{l(l+1)}{r^2} \right] P(nl | r) = X'(nl | r) + \sum_{n' \neq n} \varepsilon_{n'l} P(n'l | r), \quad (9)$$

where ε_{nl} is one-electron energy. The Coulomb and exchange potentials - $Y'(nl | r)$ and $X'(nl | r)$ differ from the original Hartree-Fock potentials [25] by the following additional terms.

$$\Delta Y(nl | r) = r/2 \sum_{k, k_1, n'l'} [a_{ll'}^{kk_1} Y_{kk_1}(n'l', n'l' | r) + b_{ll'}^{kk_1} Y_{kk_1}(nl, n'l' | r)], \quad (10)$$

$$\Delta X(nl | r) = - \sum_{k, k_1, n'l'} [\alpha_{ll'}^{kk_1} Y_{kk_1}(nl, n'l' | r) + \beta_{ll'}^{k} r^{k_1}] P(n'l' | r),$$

Eq.(10) describe the self-consistent field of the cluster depending on all electrons of the system. Here, the tensor function - $Y_{kk_1}(nl, n'l'|r)$ and the coefficients: $a_{ll'}^{kk_1}$, $b_{ll'}^{kk_1}$, $\alpha_{ll'}^{kk_1}$, β^k are given by the following expressions:

$$Y_{kk_1}(nl, n'l' | r) = \delta(k, k_1 + k_2) \frac{r^{k_1}}{R^{k+1}} \int P(nl | r') P(n'l' | r') r'^{k_2} dr'.$$

$$a_{ll'}^{kk_1} = v \sum_m O_{m_1 m_1'} O_{m_2 m_2'} (m_1 m_2 \mid r_{12}^{-1} \mid m_1' m_2')_\varphi + A(SS') v' \sum_m O_{m_2 m_2'}^{\overline{m}_1 \overline{m}_1'} O_{m_2 m_2'}^{\overline{m}_1 \overline{m}_2'} (m_1 \mid m_2')^*$$

$$\cdot (m_2 \mid m_1')(\overline{m}_1 \overline{m}_2 \mid r_{12}^{-1} \mid \overline{m}' \overline{m}_2')_\varphi$$

$$b_{ll'}^{kk_1} = v . A(SS') \sum_m O_{m_1 m_1'}^{\overline{m}_1 \overline{m}_1'} O_{m_2 m_2'} (m_1 \mid m_2')(\overline{m}_1 m_2' \mid r_{12}^{-1} \mid \overline{m}_1' m')_\varphi \qquad (11)$$

$$\alpha_{ll'}^{kk_1} = v' A(SS') \sum_m O_{m_1 m_1'} O_{m_2 m_2'} (m_1 m_2 \mid r_{12}^{-1} \mid m_2' m_1')_\varphi,$$

$$\beta_{ll'}^{k} = \frac{Z_2 v'}{R^{k+1}} \sum_m O_{m_2 m_2'} (m_1 \mid m_2)[O_{m_1 m_1'} + O_{m_1 m_1'}^{\overline{m}_1 \overline{m}_1'} (m_1 \mid m_2)(m_1 \mid r_{Z_2}^{-1} \mid m_2')_\varphi$$

Eq. (11) are including the angular part of the matrix elements of different operators, only. v and v' are numerical coefficients. It is easy to see that solutions of Eq. (9) for the cluster depend on the wave functions of all ions in the cluster.

The energy of Stark-levels of LSJ-state in crystalline field for nl - ion may be written as:

$$E(nl^N \mid \alpha\alpha' LSJ\Gamma) = E_0 + \sum_k f_k (l^N, \alpha\alpha' LS) F_k (nl, nl) +$$
$$+ \chi(LSL'S', J)\eta(nl) + \sum_{k,q,i} B_{kq} Y_{kq} (\Theta_i \Phi_i), \qquad (12)$$

where E_0, is the center of gravity of nl^N- configuration, $F_k(nl,nl)$, theSlater integrals, $\eta(nl)$, the spin-orbit constant, and B_{kq}, the crystal field parameters (for d-ions in a cubic field, Dq). The radial integrals are determined as:

$$F_k(nl, nl) = D_k \int P^2 (nl \mid r) \frac{r_<^k}{r_>^{k+1}} P^2 (nl \mid r') dr dr',$$

$$F^k(nl, n'l') = \int P^2 (nl \mid r) \frac{r_<^k}{r_>^{k+1}} P^2 (n'l' \mid r') dr dr', \qquad (13)$$

$$G^k(nl, n'l') = \int P(nl \mid r) P(n'l' \mid r) \frac{r_<^k}{r_>^{k+1}} P(n'l' \mid r') P(n'l' \mid r') dr dr',$$

where D_k is a numerical coefficients depending on k. The Racah integrals B and C for d-ions and $E_i(nf,nf)$ for f-ones are the linear combinations of Slater integrals [23-25].

For preliminary calculation of the electronic level energy of ions in the cluster or solids, we may use semi-empirical values of the crystal field parameters: B_{kq} for RE, An ions or Dq for Me ones [3, 28]. For crystals with unknown crystal field parameters, we may use the following procedure proposed in [27, 31].

$$B_{kq} = \sigma_k B_{kq}^{ref}, \qquad (14)$$

where B_{kq}^{ref}, is the reference value corresponding to an ion in a given electronic configuration embedded in the crystal. Parameter σ_k describes the change of the radial distribution and inter-atomic distances for the ion investigated in comparison to the reference one:

$$\sigma_k = \frac{(nl \mid r^k \mid nl)}{(nl \mid r^k \mid nl)_{ref}} \cdot \frac{R_{ref}^{k+1}}{R^{k+1}}. \tag{15}$$

The radial integrals (average values of r^k for nl- shell), $(nl \mid r^k \mid nl)$ and $(nl \mid r^k \mid nl)_{ref}$, describe the electronic distribution:

$$(nl \mid r^k \mid nl) = \int P(nl \mid r) r^k P(nl \mid r) dr. \tag{16}$$

In Eq. (15) R^{k+1} and R^{k+1}_{ref} are the inter-atomic distances for nl- ion in investigated and reference compounds respectively.

As shown in [3, 5] and [27-32], Eq. (5–15) are describing the radial distribution of the electronic density in the cluster. Using these equations, we can study the change of the electronic distribution during the transfer of the free nl-ion into the crystal for different compound. For all 3d- or 5f-ions and some 4f-ones, we have obtained results that closely correspond to the semi empirically determined data for Slater (or Racah) integrals and for level energies, in particular, for ions in doped oxide crystals.

It should be noted that the radial wave functions of all electrons in the ion change under the free ion-to-crystal transition. As shown in [30-32], the radial part of the electronic density for d- or f-electrons changes significantly under this transition. For Me^{2+}- Me^{4+} ions, we observed an expansion of the 3d-shell and a strong decrease of the Slater integrals accordingly to the experimental determination [34]. For RE^{2+} ions, a similar situation is observed: the average values of k in the radius of nl- shell, $(nl \mid r^k \mid nl)$, increases and Racah integrals decrease under free ion-to-crystal transition. For RE^{3+} and RE^{4+} the situation is more complex because there is a relative shift of the electronic density of the 4f-electrons on the one hand and external 5sp-ones on the other hand. For different clusters and some 4f-ions in the solid state, we observed a decrease of the average radius of the 4f-shell and a change in the energy and electronic density distribution of the initial electrons, for example, 1s- and 2p- ones.

When the relative energy of the core electrons changes, the energy of X ray lines is shifted. Results of the calculation of the energy of K-, L-lines ...etc., for all d- and f-ions closely correspond to the experimental values measured in free ions and for ions in compounds. Therefore, our approach is describing well the change of the electronic density distribution during a transition from one compound to another as it occurs during a structural phase transition. It should be noticed that the relative error for the X ray line energy measurement, for example $L_{\alpha 1}$- line of RE or An ions is less than 0.5% [36].

We have studied the nephelauxetic effect and the pressure dependence on the energy level scheme of nl- ions in different compounds using Eq.(9). The R dependence on the energy level scheme of nl^N- ions was investigated by using data obtained for the radial integrals $F_k(nl,nl)$ and $\eta(nl)$ for definite R and also the R- dependence on the crystal field parameters, Dq or B_{kq}.

It should be noted that the solutions of Eq. (9) depend on αLS -term of nl- ion. It was shown in [35-36] that the radial wave function, $P(nl \,|\, LS \,|\, r)$ and the radial integrals depend on αLS state of the ion. The difference in values of the radial integrals for ground and excited states may account for 300% (for example, $4f|r^6|4f$ integral for Pr^{+3} in $LaCl_3$ [3, 18]).

Eq. (9) were used successfully to calculate energies in a cluster with boundary conditions for either the free ion or the crystal with Wigner-Zeits boundary conditions [3, 5, 12, 18, 27-32]. The so-called Roothaan's radial wave functions were used for the ligands: O^{2-}, F^- and Cl^-. Additional potentials (10) for clusters were introduced in our calculation which is using the numerical programmes presented in [3]. The Fourier-transform ligand wave function at the centre of the cluster was used for calculation of the additional terms (10).

It should be noted that the influence of the relativistic effects on the ion energy with 3d-, 4f- and 5f-shell is comparable, at the first order, to the influence of the surroundings. We used the relativistic Hartree-Fock-Pauli (HFP) approximation which is more effective for these ions. HFP-method operates with the main part of Breit relativistic Hamiltonian: velocity-dependence on electron mass, contact, $C(nl)$, and spin-contact, $Q(nl)$ interactions which were also taken into account in our approach. The calculated results closely correspond to the experimental data.

3. Change of the Electronic Structure of the Ions under Free Ion-to-cluster Transition.

For the calculation of the change of the electronic structure of ions during free ion-to-crystal transition, we have considered the $Me^{n+}:[L]_k$ cluster where Me stands for ions of the beginning and the end of the iron, rare earths and actinide groups of elements. We have used analytical functions for the ligands and numerical functions of the Hartree-Fock-Pauli approximation for the free 3d-ion as an initial parameter set for the calculation.

For 3d-ions, the values of the radial integrals during the free ion-to-crystal transition and the radial distribution of the electronic density change significantly. Calculated results for Cr^{3+} ion in $Cr^{3+}:[O^{2-}]_6$ cluster are presented in Tab.1. The radial integrals, E the total energy, the average radius of nl- shell for ion in the cluster are increasing comparatively to the free ion values. In the contrary, the values of the Slater integrals, the spin-orbit constant and the radial integrals, $F_k(3d,n'l')$ or $G_k(3d,n'l')$, are decreasing. Changing in the radial integrals is similar for all 3d-ions and for all clusters. The radial integral values for the cluster depend on the oxidation state and configuration of the nl- ion as well as type and number of ligands. The results of calculation of different radial integrals are presented in Tab. 1-4 for the free ions with $3d^N$, $3d^{N-1}4s$ and $3d^{N-1}4p$-configurations and for ions in $Me^{n+}:[O^{2-}]_k$ clusters with different R.

TABLE 1. Theoretical values of radial integrals for Cr^{3+} ions in $Cr^{3+}:[O^{2-}]_6$ clusters for different R

Integral	Free ion / R(Å) = 2.1	1.96	1.9	1.8	1.5
Configuration 3d³					
$F^2(3d,3d)$, cm⁻¹	87080	72010	58644	50932	45863	44795
$F^4(3d,3d)$, cm⁻¹	54582	42380	35644	30881	27796	29599
$\eta(3d)$, cm⁻¹	290.9	245.1	220.2	194.8	167.5	74.9
$(3d|r|3d)$, a.u.	1.093	1.351	1.561	1.721	1.839	2.100
Configuration 3d²4p					
$F^0(3d,4p)$, cm⁻¹	91284	65490	69606	71629	74294	
$F^2(3d,4p)$, cm⁻¹	22295	9455	12485	14705	21010	
$G^1(3d,4p)$,cm⁻¹	7778	2513	4924	7046	14430	
$G^3(3d,4p)$,cm⁻¹	7193	2001	3838	5471	11135	
$\eta(3d)$, cm⁻¹	331.9	322.1	303.8	288.7	238.1	
$\eta(4p)$, cm⁻¹	642.0	97.6	129.6	153.3	198.8	
$(3d|r|3d)$ a.u.	1.018	1.064	1.148	1.219	1.474	
$(4p|r|4p)$ a..u.	2.734	3.538	3.314	3.210	3.045	
$(4p|r|4p)$ a..u.	2.734	3.538	3.314	3.210	3.045	
$\Delta E(3d^3–3d^24p)$, eV	17.8	21.8	16.1	9.9	11.6	

Tab.1 includes data for radial integrals for free Cr^{3+} ion or $Cr^{3+}:[O^{2-}]_6$ cluster (configuration $3d^3$ and $3d^24p$). In Tab. 2, the theoretical and semi-empirical energy levels for free Cr^{3+} ion or $Cr^{3+}:[O^{2-}]_6$ clusters are listed.

TABLE 2. Dependence of Cr^{3+} ions energy levels on R_{Cr-O} (cm⁻¹)

R_{Cr-O}(Å)\ Level	²E	²T₁	⁴T₂	²T₂	⁴T₁(t₂²e)	⁴T₁(t₂e²)
Theory						
2.0	14850	15652	16500	22171	24229	37661
1.96	14220	14969	18100	21538	25811	40324
1.9	12500	13113	20500	19392	27659	44176
Ruby						
Experiment	14433	15087	18133	21318	24767	39067
Semi-empirical	14354	14989	18108	21355	24843	39362

TABLE 3. Semi-empirical and theoretical data for B, C and Dq for Cr^{3+} ions in different crystals (cm⁻¹)

Integral	α-Al₂O₃	Y₃Al₅O₁₂	Gd₃Sc₅O₁₂	Gd₃Sc₂Ga₃O₁₂	$Cr^{3+}:[O^{2-}]_6$
B	682	725	740	740	789
C	3120	3373	3578	3578	2829
Dq	1787	1650	1500	1500	1750

Tab.3 shows data for B, C, Dq for Cr^{3+} ion in different crystals. The theoretical energy level scheme of Cr^{3+} ion in cluster $Cr^{3+}:[O^{2-}]_6$ calculated for R = 1.96 Å closely corresponds to the level scheme in ruby and for R = 1.90 Å in YAG garnet [12, 14, 17]. The numerical data of Tab. 1-3 make possible the calculation of the energy level scheme for any crystal doped by Cr^{3+} ions, in particular under pressure [28]. The power of this approach is shown through the study of the radiation induced defects in ruby. In the γ-or electron irradiated ruby (at doses higher than 10^2 Gy), the following additional optical bands appear in the absorption spectra at 217, 260, 360 and 460 nm (see Fig.7) [11]. Comparison of the theoretical data of Tab. 4 with experimental ones [2-3] permits to attribute the additional bands at 217, 360 and 460 nm to the Cr^{4+} ion in an octahedral environment. These optical bands are assigned as electronic transitions in $Cr^{4+}:[O^{2-}]_6$ cluster. Transition of Cr^{3+} to Cr^{4+} oxidation state upon irradiation of ruby or perovskite, $YAlO_3$:Cr, or garnet, $Y_3Al_5O_{12}$:Cr, and other oxide crystals doped with Cr was detected by ESR, VSXRL, and others methods [3-4]. In Tab. 4, the results of the theoretical energy level scheme calculation in Cr^{4+} ions in $[O^{2-}]_6$ cluster (octahedral environment) are presented.

TABLE 4. Theoretical level scheme for ion Cr^{4+} in ruby (Dq = 1990 cm^{-1}, B = 1050 cm^{-1} and C = 3873 cm^{-1})

$^{2S+1}\Gamma(t,e)$–level	Energy,cm^{-1}	$^{2S+1}\Gamma(t,e)$- level	Energy, cm^{-1}
$^3T_1(t^2_2)$	0	$^1T_2(t_2e)$	34909
$^1E(t^2_2)$	15002	$^3A_2(e^2)$	38391
$^1T_2(t^2_2)$	15618	$^1T_1(t_2e)$	38391
$^3T_2(t_2e)$	18045	$^1E(e^2)$	54429
$^1A_1(t^2_2)$	30962	$^1A_1(e^2)$	75683
$^3T_1(t_2e)$	31939		

As we can see, the energy of the triplet – triplet transitions (Tab. 4) corresponds to the experimental energy of the additional optical absorption bands in the irradiated ruby. It is quite simple to calculate the energy of triplet $\leftrightarrow$ triplet transitions for $Cr^{+4}:[O^{2-}]_6$ cluster for any crystal: perovskite or garnet doped with Cr ions, by changing the parameters of the $Cr^{+4}:[O^{2-}]_6$ cluster according to the respective crystallographic data. For the latter one, B and C integrals are the same as in ruby and Dq increases up to 2250 cm^{-1}.

Let us consider now, the electronic structure of the Me ions for high-temperature Cu containing superconductors (the so-called, 1-2-3 ceramics) where Cu^{n+} ions play an important role [30, 37-38]. We have studied the electronic state of the ground 3d^N- and excited 3d^{N-1}4s-configurations of Cu^{3+} and Cu^{2+} ions. The results for $Cu^{n+}:[O^{2-}]_k$ - clusters are presented in Tab. 5 - 7 [30]. According to these data, the minimum value of the total energy, E_{min} for $Cu^{n+}:[O^{2-}]_6$ cluster is observed for Cu^{3+} ion in the octahedral cluster, at R = 1.89 - 1.95 Å. E_{min} for Cu^{2+} ion is observed for tetrahedral cluster Cu^{2+}: $[O^{2-}]_4$ and R = 1.85 Å [30-31]. These data closely correspond to the experimental ones [37-38]. It is important to note that the energy of 3d $\rightarrow$ 4s transitions in Cr^{2+} ions in clusters $Cu^{2+}:[O^{2-}]_k$ strongly decreases for R = 2.0 Å, k = 6 and R = 1.85 Å, k = 4.

Similar calculations for Ni^{2+} embedded in the same environment [30-31] did not show the same effect as well as for Zn^{n+} ions.

The electronic structure of $Me^{n+}:[L]_k$ clusters for other ligands, in particular, F$^-$ and Cl$^-$ ions was also studied in detail. Results for Me^{n+} in a fluorine environment with different R and k values are reported in [3] and [27-30]. The calculated qualitative and quantitative changes of the Slater integrals for these clusters are close to the experimental ones. The $F_k(3d,3d)$ change for $Me^{n+}:[F]_k$ cluster is weaker than for the oxygen environment. It is well known that Slater integrals for $Me^{n+}:[Cl]_k$ cluster change

TABLE 5. Theoretical values of radial integrals for Cu^{3+} and Cu^{2+} ions in $Cu^{n+}:[O^{2-}]_k$- cluster

Integral	Free ion/R(Å)	2.5	2.0	1.95	1.9	1.8	Units
$E_0(Cu^{3+})$, k=6	-1636.7563	.7564	.7539	.7551	.7551	.7557	a.u.
ε_{3d}	2.2391	2.21756	2.1578	2.1436	2.1273	2.1042	a.u
$(3d\|r\|3d)$	0.8422	0.8414	0.8427	0.8434	0.9445	0.8463	a.u.
$F^2(3d,3d)$	129327	99498	98753	98711	98662	98618	cm^{-1}
$F^4(3d,3d)$	80874	62188	61719	61680	61660	61625	cm^{-1}
$\eta(3d)$	925.2	910.8	909.5	909.3	908.8	908.1	cm^{-1}
$E_0(Cu^{2+})$, k=6	-1638.1232	.0841	.0839	.0835	.0832	.0723	a.u.
ε_{3d}	1.4789	1.4477	1.3872	1.3735	1.3532	1.3356	a.u.
$(3d\|r\|3d)$	0.8837	0.8989	0.9013	0.9025	0.9043	0.9070	a.u.
$F^2(3d,3d)$	119110	93078	91987	91987	91796	91569	cm^{-1}
$F^4(3d,3d)$	74806	58120	57495	57495	57365	57251	cm^{-1}
$\eta(3d)$	845.9	825.0	820.8	820.8	818.9	815.1	cm^{-1}
$E(3d^9 \to 3d^8 4s)$	1.89	1.71	1.43	1.63	1.79	-	eV

under free ion-to-crystal transition more strongly than for fluorine but less than for oxygen environment (for R corresponding to the crystallographic structure). These data are presented in Tab. 6. Thus, the relative change of $F_k(3d,3d)$ is 22 % for $Cu^{2+}:[O^{2-}]_6$, 19 % for the fluorine cluster (R= 2.1 Å and k = 6) and 20 % for the chlorine one (R= 3.0 Å and k = 6). The radius of the 3d-shell increases by 6 %, 2 % and 4 %, accordingly. Similar results were obtained for the spin-orbit coupling constant of Me-ions in the respective clusters. Slater or Racah integrals for all 3d-ions decrease under free ion-to-crystal transition by 10 to30 % as well as the value of $\eta(3d)$, in close agreement with experiment. The average "size" of the ions, $(3d\|r\|3d)$ increases by 8 to22 % when R decreases.

TABLE 6: Values of radial integrals for Cu-ion with $3d^8 4s$-configuration in the cluster $Cu^{2+}:[Cl]_6$ and $Cu^{2+}:[F]_6$ (in a.u.).

Integral	R(Å), 2.0	1.9	1.8	1.7
$Cu^{2+}:[Cl]_6$				
ε_{3d}	3.2874	3.3559	3.4291	3.5070
ε_{4s}	2.4984	2.5494	2.5964	2.6355
$(3d\|r\|3d)$	0.8506	0.8498	0.8485	0.8463
$(4s\|r\|4s)$	2.0454	1.9956	1.9402	1.8796
$F^0(3d,4s)$	0.5405	0.5520	0.5657	0.5819
$G^2(3d,4s)$	0.0585	0.0616	0.0653	0.0698
$\Delta E(3d^9 \to 3d^8 4s)$, eV	8.509	8.810	9.227	9.814
$Cu^{2+}:[F]_6$				
ε_{3d}	1.7362	1.7278	1.6924	1.6569
ε_{4s}	1.0026	1.0012	0.9959	0.9924
$(3d\|r\|3d)$	0.8474	0.8471	0.8465	0.8467
$(4s\|r\|4s)$	2.3290	2.3326	2.3459	2.3556
$F^0(3d,4s)$	0.4883	0.4875	0.4844	0.4821
$G^2(3d,4s)$	0.0448	0.0446	0.0436	0.0431
$E(3d^9 \to 3d^8 4s)$, eV	7.848	7.848	7.906	7.927

Let us consider the electronic density in nf- ions in the clusters. The $4f^3 5d$-configuration is the ground configuration of Nd^{2+} ion and the $4f^4$ is the excited one. But in solids the opposite situation is observed: the ground configuration of Nd^{2+} in solids is the $4f^4$-one. The total energy of the $4f^3 5d$- and $4f^4$- configuration of the Nd^{2+} ions in free state are the following: $E_0(4f^3 5d) = -9590.5881$ a.u., and $E_0(4f^4) = -9590.5358$ a.u. For the ion in the cluster $Nd^{2+}[O^{2-}]_8$, the corresponding energies are $E_0(4f^3 5d) = -9588.4364$ a.u. and $E_0(4f^4) = -9589.3669$ a.u. The value of $(4f|r|4f)$ goes from 0.9559 to 1.1356 a.u. for $4f^4$ state and from 0.9000 to 0.8971 a.u. for the $4f^3 5d$-one for ion in the free state and in the cluster respectively. The average value of $(5d|r|5d)$ changes from 2.4457 to 4.0262 a.u. during the free-state-to cluster transition of Nd^{2+} ion with $4f^3 5d$-configuration.

For RE ions, the best results were obtained for $RE^{2+}:[L]_k$ clusters, only. For RE^{3+} in the oxygen or chloride cluster, a decrease of the radial integral values by 5 to 10 % (less than the experimental shift magnitude) was observed. As expected, fluorine cluster, leads to a small change in radial integrals. In [3] and [27-29], data for $RE^{n+}:[L]_k$ clusters that are presented depend on the method of calculation of the free ion wave functions. For clusters with 5f-ions with oxidation states, An^{2+} -An^{4+}, results are similar to the ones obtained for 3d-ion clusters. Values of $F_k(5f,5f)$ and $\eta(5f)$ for actinides in the cluster decrease by 15 to 25 % accordingly to [31-32].

4. Change in the Parameters of the Electronic Structure of Ions in Cluster along with Inter-atomic Distance and Coordination Number.

After consideration of the different numerical results for clusters, we have studied the dependence of the radial integrals on the distance Me-L and on the number of ligands in the first coordination sphere. For example, we have studied the dependence of the level energies of Cr^{3+} ions in ruby under pressure and Cr^{4+} ions in octahedral and tetrahedral sites in garnets. In Tab. 1 ,2, the different theoretical and semi-empirical or experimental level energies of Cr^{3+} ions in cluster $Cr^{n+}:[O^{2-}]_k$ for different R are listed. The energy of $3d \rightarrow 4s$ excitation for different clusters with Cu^{2+} or Cu^{3+} ions are presented in Tab. 5 where R changes from 3.0 to 1.6 Å with an increment of 0.05 Å and the number of oxygen ions (or other ligands) varies from 4 to 6. Together with the calculated data published in [17, 21], the present data show a non-linear dependence of E_0, the total energy of $3d^N$-configuration, on R. On the contrary, the dependence of the radial integrals $F_k(3d,3d)$ or $\eta(3d)$ on k and R is linear at $R > R_{cr}$ where R_{cr} is the critical value of the inter-atomic distance when the interpenetration of the wave functions of nl-ion and ligands starts to strongly increase.

Our calculation shows an important influence of the ligand type and R-magnitude on the cluster electronic structure. Thus, values of $(3d|r|3d)$ or $(4s|r|4s)$ for oxygen surrounding increase when R decreases (below R_{cr}). For chlorine and fluorine environment, the values of $(nl|r|nl)$ and R change similarly.

We have obtained a quite good set of data for the energy of the Stark levels of Cr^{3+} ions in ruby and under pressure. It should be noted once more, that we have observed increasing values for $(3d|r|3d)$ in $Me^{n+}:[L]_k$ clusters below a critical value R_{cr}. We have noticed that the collapse of nl-shell may start for $R < R_{cr}$. The R_{cr} value depends on the type of Me-ion, ion oxidation state, type and number of ligands. For Cu^{3+} ion in oxygen surrounding, R_{cr} is equal to 1.6 Å while for Cu^{2+} ion, $R_{cr} = 1.7$ Å. For fluorine and chlorine ligands, and for nf- ions, the value of R_{cr} is higher by 15 to 20 %.

The value of R_{cr} is determined by the overlap integral $(nl|2p)$ for Me-O (or Me-F) pair and by $(nl|3p)$-integrals for Me-Cl pair. At $R < R_{cr}$, the shell interpenetration is impossible according to the Pauli law. As a result, we observed for example, a decrease of the values of $(3d|r|3d)$ at $R < 2.3$ Å for $Cu^{2+}:[Cl^-]_6$ cluster (see Tab. 6).

The collapse of the 4f-shell starts at values of R_{cr} higher than those for Me-ions. When R decreases, the $(5p|2p)$-overlap integrals increase more rapidly than $(4f|2p)$-integrals. This rapid increase of the $(5p|2p)$-value leads to the expansion of the ion size and a shift of the maximum of the electronic density of 5sp-shells. The latter process leads to a decrease of the screening constant for 4f-electrons and $(4f|r|4f)$-values start collapsing.

The changing in parameters for Me-ions under free ion-to-$Me^{n+}:[F]_k$ cluster transition is not as strong as for the oxygen neighbouring. Different distribution of the external shells of Me ion and ligands is the cause of different R-dependences of radial integrals and energy of 3d-4s interaction for 3d-ions (see Tab. 6).

Our investigation of the electronic density of different nl- ions in the free state and in clusters as well as the calculation of radial integrals for nl- ions in the $Me^{n+}:[L]_k$-clusters for O^{2-}, F^- and Cl^- ligands with various inter-atomic distances lead to the following conclusions:

i) the results of our calculation for Me (An) ions in the clusters closely correspond to the experimental ones regarding the energy level schemes of the doped ions in oxides, fluorides and chlorides. The adopted procedure of calculation of the electronic structure of the clusters seems correct for the description of the dependence of the energy and radial-integrals on the inter-atomic distance, ligand types and in particular, for pressure dependence in spectra of Me-ions in solids.

ii) the minimum values of the energy of $3d \rightarrow 4s$ electronic transition for $Cu^{2+}:[O^{2-}]_4$ and $Cu^{3+}:[O^{2-}]_6$ clusters at $R = 1.85$ Å and 1.95 Å were calculated in agreement with the experimental ones for R in $YBa_2Cu_3O_7$ superconductors. For $Cu^{2+}:[O^{-2}]_4$ cluster at $R = 1.8$ Å, we observed a strong decrease (about 40 %) of the energy of $3d \rightarrow 4s$ excitation and an increase of the energy of 3d-4s interaction. Similar changes in the energy of $3d \rightarrow 4s$ transition and 3d-4s interaction were not observed for other ions and clusters [38, 39].

5. Theoretical Data for Energy of K- and L- X Ray Lines

For most of the compounds containing impurities, in particular laser crystals, the determination of the electronic state of component and doped ions is a hard task as well as the determination of the concentration of the mixed valence ions.

Standard experimental data for nl- ions, especially 3d-ones, in different compounds are well known [3, 43]. In reverse, data for the "Valence Shift" in the nl- ions are very scarce [44-45]. Different experimental results for energy of X-ray lines in actinides are given in [45] and were used as reference for the estimation of the accuracy of our theoretical approach.

Effects of irradiation in 4f-ions in solid state have been studied in detail in different compounds by O. Sumbaev et al. by using high resolution spectrometer [37]. According to this work, the experimental dependence of X-ray lines energy of RE^{n+} ions with the number of 4f-electrons is linear: energy of $K_{\alpha\beta}$ lines increases when the number of 4f-electrons increases (oxidation state of RE ion decreases).

In [3] and [36], theoretical values for energy of K- and L-lines for Me and RE ions in free state are reported. We have published similar data for An ions [42] and performed measurements of X ray line energy for nl-ions in $Me^{+n}:[L]_k$ clusters.

The procedure of the theoretical study of the X ray spectra for nl-ions is similar to the one already presented in Section 2-4: the energy of the transition of a core electron from an initial state, $n'l'$, to a final one, $n''l''$ is the difference between the energy of the configurations $n'l'^{-1}_j nl^N$ and $n''l''^{-1}_j nl^N$. For example, X-ray lines are determined by the following way: $1s_{1/2}nl^N \rightarrow 2p^5_{1/2,3/2}nl^N$ ($K_{\alpha1,2}$ lines,); $2p^5_{3/2}nl^N \rightarrow 3d^9_{3/2,5/2}nl^N$ ($L_{\alpha1,2}$ lines) etc. Therefore, the energy of X-ray line, E_x, may be defined as:

$$E_x = E^N_{n'l'} - E^{N\pm1}_{n''l''}. \qquad (17)$$

In Eq.(17), $E^N_{n'l'}$ is the energy of nl^N- configuration with vacancy in $n'l'$- shell. It should be noticed that $n''l''^{-1}_j nl^N$- configurations are highly excited ones and calculation of their energy is a complex problem for the free ion and for clusters. We performed calculation for all ions of Me, RE and An groups in oxidation state from «+2 to +4». Data are reported in [3] and [35-36] and experimental data for nf- ions are presented in [37]. The energy of X-ray lines for Me ions are well known [3] and for example, in Tab. 7, we present these energies for different RE and An ions. For investigation of the stability of the electronic state of nl- ions in crystals, we have used the so-called valence shift of X-ray line procedure, VSXRL. The shift of X-ray lines is due to change in ion oxidation state (change of the number of nl- electrons, N). VSXRL is determined as:

$$\Delta E_x = E^N_x - E^{N\pm1}_x, \qquad (18)$$

where E_x^N ($E_x^{N\pm1}$) is the energy of X-ray lines of ions with nl^N and $(nl^{N\pm1})$-configuration.

It is well known that energy of X-ray lines depends on the number of nl- electrons [3, 5]. For example, the value of VSXRL for the ME^{n+} to ME^{n+1} transition in 3d-, 4f- and 5f-ions is about 100 meV [35-37] accordingly to Eq.(18). The highest value of VSXRL was obtained for K_α-lines in An ions while the smallest one was observed for L_α-lines.

Different calculated values of K- and L X-Ray line energy for nf- ions are listed in Tab. 7. It should be noted that the X-ray line energies of all d- or f- ions increase with the number of nl- electrons. The negative shift of X-ray line energy occurs when the energy, E_x, decreases along with an increase of the ion oxidation state. For RE ions, the shift of X-ray lines are greater than for the actinides by 10 to 20 %.

Comparing our results with the experimental data [16], we may conclude that for K_α- and K_β-lines of nf- ions, the mismatch is typically of 1 to 3 %. For L_α-lines in An ions, the agreement is surprisingly good (misfit of 0.5 %) but for others lines the relative deviation is worse and may reach about 6 %.

The change in energy of X-ray lines in nf- ions with ionization of nf- electrons depends on the relative shift between $n'l'$- and $n''l''$- shells.

TABLE 7. Theoretical values of energy of K_α- and L_α- lines
in RE and An ions (in eV.)

RE	E ($K_{\alpha 1}$)	E($K_{\beta 1}$)	E($L_{\alpha 1}$)
Nd $^{2+}$	37337,290	42250,942	42246,471
Nd $^{3+}$	37336,610	42248,875	5369,708
Nd $^{4+}$	37335,741	42246,471	5368,449
Eu^{2+}	40071,174	45371,500	5852,357
Eu^{3+}	40070,420	45369,567	5848,793
Eu^{4+}	40069,405	45367,192	5847,622
Gd^{2+}	42921,661	48620,131	6062,622
Gd^{2+}	42920,521	48618,235	6063,971
Gd^{2+}	42920,118	48615,998	6060,614
Yb^{2+}	52156,939	58580,850	7421,563
Yb^{3+}	52155,897	58578,564	7426,668
Yb^{4+}	52155,007	58576,031	7429,100
U^{2+}	95912,345	108809,007	13639,793
U^{3+}	95912,037	108808,394	13639,528
U^{4+}	95911,663	108807,659	13639,200
Np^{2+}	98307,960	111517,588	13970,953
Np^{3+}	98307,642	111516,963	13970,674
Np^{4+}	98307,262	111516,221	13970,346
Bk^{2+}	108277,894	122779,219	15339,303
Bk^{3+}	108277,543	122778,545	15339,005
Bk^{4+}	108277,142	122777,775	15338,668

The electronic density of external $(n+1)s$- and $(n+1)p$- electrons in nl- ions participating into the X-ray transition is described by a function with nodal points and maxima in the inner region of the atom. In relation with the shift of these shells, the value of the screening constant for n'l'- and n"l"- electrons participating in X-ray transition are also changing. The screening constant value for 1s-shell, σ_{is}, (for example, for $K_{\alpha 1}$-line) changes more strongly than σ_{2p}. As a result of our calculation, the energy of $K_{\alpha 1}$-line decreases when the oxidation state of nf- ion increases. Data in Tab. 7 permit to draw a simple conclusion: ionization of nf- electron (and nd-, too) leads to a change in the electronic density distribution of the 1s-shell. Changing in the density distribution of 2p (3p) and 3d-shells is smaller.

We have determined the relative concentration of the ions with different oxidation states in the crystal with an uncertainty of 7to10 % by using a X-ray micro-analyser or a two-crystal X-ray spectrometer [3, 4, 12]). The study of the VSXRL for $ME^{n+} \leftrightarrow ME^{n\pm 1}$ transition is very useful for consideration of mixed valence nl- ions in compounds.

6. N- and Z - Dependence on E(nl^N, $\alpha\alpha'$LS), E_x and ΔE_x

The analytical dependence of the energy levels E(nl^N, $\alpha\alpha'$LS) of nl^N- ions on number of active electrons and nucleus charge is a well known method in solid state physics [5, 19, 35]. This approach was founded by G. Racah in his famous papers [40-41].

The analytical Z- and N-dependences for different ion parameters were used successfully for prediction of spectroscopic properties of ions in the free state and solids [5, 42]. Theoretical N-dependence of the electrostatic parameters for nl^N- configurations was calculated in [12, 39, 42] where Racah's method was used. In [35-36], the analytical dependence of the energy of X-ray lines and optical transitions was studied for 3d- and 4f-ions. Considering the electronic structure of nl^N- and $n'l'^{-1}nl^N$-configurations, the energy of the different $K_{\alpha,\beta}$- and $L_{\alpha,\beta}$-lines for all Me, RE and An ions were calculated [3, 35-36]. The energy of the X-ray lines (see Eq.(13)) was taken equal to the difference between the average energy of $n'l'^{-1}nl^N$- and $n''l''^{-1}nl^N$-configurations with the full self-consistent field calculation for each configuration.

The experimental dependence of the energy of X-ray lines of RE^{n+} ions on the number of 4f- electrons was found linear [37]. The energy of $K_{\alpha,\beta}$-lines increases when the number of 4f-electrons increases (oxidation state of RE ion decreases).

Theoretical values for the energy of K- and L-lines for Me and RE ions in free state were reported in [3, 36]. In [42], we have published similar data for An ions. Experimental results for Me and RE ions are well known but for actinides, data are rather scanty [43-45]. Some experimental results for energy of X-ray lines of An are given in [45]. They were used as a reference for the estimation of the accuracy of the theoretical approach used.

Let us consider the dependence of the energy of nl^N- configuration, energy of X-ray lines on the ion oxidation state.

Based on expression (12), the level energies of nl^N- ions in free state or clusters can be rewritten as follow [19] and [41]:

$$E(nl^N \mid \alpha\alpha' LSJ) = E_0 + \sum_k e_i(l^N, \alpha\alpha' LS)E_i(nl,nl) + \chi(LSL'S')\eta(nl), \quad (19)$$

$$E_0 = \sum_{n_i l_i} N(n_i l_i)I(n_i l_i) + \frac{1}{2}\sum_{n_i l_i} N(n_i l_i)[N(n_i l_i) - 1] \times$$

$$\times \left\{ F_o(n_i l_i, n_i l_i) - \frac{1}{4l_i + 1}(l_i \parallel C^2 \parallel l_i)^2 F_2(n_i l_i, n_i l_i) + \frac{1}{4l_i + 1}(l_i \parallel C^4 \parallel l_i)^2 F_4(n_i l_i, n_i l_i) \right\} +$$

$$+ \sum_{n_i l_i \neq n'l'} N(n_i l_i)N(n'l') \left\{ F_o(n_i l_i, n'l') - \frac{1}{(4l_i + 1)^{1/2}(4l' + 1)^{1/2}}\sum_k (l_i \parallel C^2 \parallel l')^2 G_k(n_i l_i, n'l') \right\}$$

In Eq. (19), $(l \parallel C^k \parallel l)$ is a sub-matrix element of operator of spherical harmonics, C^k [20-21], and $N(n_i l_i)$ is the number of electrons in $n_i l_i$-shell. It easy to see that the dependence of E_0 is a square function of N. The angular coefficients, $e_i(l^N, \alpha\alpha' LS)$, are the linear combinations of $f_k(l^N, \alpha\alpha' LS)$ and according to G. Racah [40, 41] and [42], they can be presented as matrix elements of the irreducible unit tensor for group R_{2l+1}. A simple mathematical procedure allows to follow the square dependence of the angular coefficients on N. The N-dependence of the radial integrals was studied by approximation of the numerical data for different N. Similar theoretical N-dependence for the electrostatic, $F_k(nl,nl)$ and $\eta(nl)$, integrals and angular coefficients for the energy levels of nl^N- configurations of Me, RE or An ions were studied in detail in [35-36, 42]. Linear N-dependence of $F_k(nl,nl)$ is only valid for these ions as a first approximation. Significant deviation of the N-dependence from the linear one for $F_k(nl,nl)$ and $\eta(nl)$ integrals were observed for ME^n - to- ME^{n+1} transition. For nf- ions, the deviation results from the well known lanthanide (actinide) contraction [28]. In final form, the N-dependence of the level energy, $E(nl^N \mid \alpha\alpha' LSJ)$, is a function of the forth degree of N

[3,5]. Following J. Slater [19], we can consider N as a parameter and develop the effective occupation number approach. By this way, the expression for $E(nl^N|\alpha\alpha'LSJ)$ we can be rewritten as:

$$E(nl^N \mid \alpha\alpha'LSJ) = A(\alpha\alpha'LSJ) + B(\alpha\alpha'LSJ)\cdot N_{eff} + C(\alpha\alpha'LSJ)\cdot N_{eff}^2 +$$
$$+ D(\alpha\alpha'LSJ)\cdot N_{eff}^3 + G(\alpha\alpha'LSJ)\cdot N_{eff}^4, \tag{20}$$

where A, B, C, D and G are numerical constants depending on $\alpha\alpha'LSJ$-state of nl- ion. This approach is very useful for the consideration of correlation effects and estimation of the energy of the cluster and energy of the nl- ion - ligand interaction [42]. The correlation energy can easily be estimated using Eq.(20). For example, the dependence of the relative energy of αLSJ -term for N_{eff} for Z = 23 and of Z = 26 is shown in Fig.1,2. Using the figures 1,2 and related experimental data, we can estimate N_{eff} and the main contributions of the excited configurations to the energy levels for V and Mn ions. For V^{3+} ion, $N_{eff} = 2.2$ and the excitation of the core electrons to 3d-shell play an important role. For Mn ions (see Fig.2) an opposite picture is observed: $N_{eff} = 4.7$ and excitation of 3d-electrons to external shells is more effective.

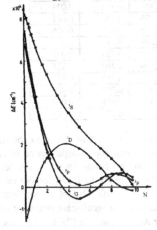

Figure 1. N- dependence of LS-energy terms for Z= 23.

According to theoretical data, the energy of the X-ray lines of nl- ions is related, at first approximation, to the second-degree function of N:

$$E_x = a + b\cdot N + c\cdot N^2, \tag{21}$$

where the constants a, b and c depend on the nucleus charge and the type of transition.

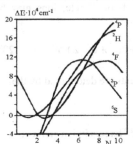

Figure 2. N- dependence of LS- energy terms for Z=25

For X-ray transitions with significant interaction of nl"-electrons with nl-active shell (these are 4p- or 4d-electrons for lanthanides and 5p-, 5d-ones for actinides), we should calculate the energy of interaction of the active nl-electrons with n'l'- and n"l"-core electrons that depends on N^2. In final form, the N-dependence of E_x may be represented by the third-degree function of N [3, 36]:

$$E_x = E_0 + a\cdot N + b\cdot N^2 + c\cdot N^3, \tag{22}$$

For K- and L-lines of Me, RE and An ions, Eq. (21) describes the N-dependence on E_x with a relatively high accuracy. The values of E_0, a and b, for Me and RE ions are presented in [3, 5] and [35]. For example, data for nf- ions collected in Tab. 8 can be used to calculate the energy of X ray lines. It should be noticed that the constant, a, is positive and constant b negative and smaller than a for all X-ray lines. In addition, for all nl-ions (the ratio $|b|/a$ is equal to 0.12 to 0.15 for Ac and 0.03 to 0.02 for No).

TABLE 8. Values of the coefficients a and b for the N-dependence of the energy of different X ray lines in actinides

Ion	$K_{\alpha 1}$- line			$L_{\alpha 1}$- line		
	E_0	a	-b	E_0	a	-b
Ac	88951.080	0.431	0.070	12671,789	0,424	0.052
Th	91233.519	0.467	0.034	12989,542	0,326	0.032
Pa	93553.072	0.540	0.034	13311,565	0,479	0.030
U	95910.145	0.605	0.033	13637,868	0,539	0.029
Np	98305.122	0.659	0.032	13968,481	0,580	0.028
Pu	100738,542	0,691	0,031	14303,356	0,640	0.027
Am	103210,648	0,724	0,030	14641,226	0,682	0.026
Cm	105721,613	0,767	0,028	14986,249	0,711	0.025
Bk	108272,134	0,826	0,026	15334,323	0,728	0.023
Cf	110862,572	0,882	0,025	15686,806	0,740	0.021
Es	113493,397	0,920	0,024	16043,560	0,784	0.021
Fm	116165,234	0,935	0,023	16405,048	0,790	0.020
Md	118877,847	0,977	0,022	16770,396	0,851	0.020
No	124428,762	1,035	0,021	17140,694	0,875	0.019

The Z-dependence of the energy of $\alpha\alpha'LSJ\Gamma$–state of nl- ion in free state or in solids may be studied by a similar procedure. For X-ray lines, the Z-dependence of a and b constants is linear according to Eq.(19 - 21). For d- and f-ions, in particular for actinides, the constant a increases along with Z from 0.431 for Ac to 1.035 for No and at the same time, the values of $|b|$ decreases from 0.070 to 0.020. The $|b|$ value, upon Ac-to-Th nuclear transition, changes from 0.070 to 0.034, i.e. a significant decrease of the $|b|$ values is observed for $K_{\alpha 1}$-line. Change in $|b|$ for $AcK_{\beta 1}$-line is 0.128 for Ac I and 0.071 for Ac IV. For $L_{\alpha 1}$-line, the $|b|$ parameter varies from 0.003 to 0.036 during Ac - Th nuclear transition.

The Z-dependence of the E_0 values for all X-ray lines may be described by a square-law function with a high accuracy according to Eq.(19). The final form of the equation describing data of Tab.8 or similar ones for 3d- or 4f- etc. ions may be expressed by the formula:

$$E_x = E_0(c + \alpha \cdot Z + \beta \cdot Z^2) + a(c' + \alpha' \cdot Z) \cdot N + b(c'' + \alpha'' \cdot Z) \cdot N, \qquad (23)$$

where values Z, N and $E_0, \alpha, \alpha', \alpha'', \beta, \beta', \beta'', c, c', c''$ constants are determined by the oxidation state of the nl- ion.

7. Analysis of Experimental Data for Radiation Induced Defects in Un-perfect Crystals

First of all, we will discuss the experimental data related to a change of nl- ion oxidation state in clusters. We will consider the experimental data resulting from investigation of the radiation induced defects into pure and doped with Me- or/and RE-ions in oxide single crystals, such as sapphire (α-Al_2O_3); garnets ($Y_3Al_5O_{12}$, $Gd_3Ga_5O_{12}$;

$Gd_3Ga_2Sc_3O_{12}$, $Gd_3Sc_2Al_3O_{12}$) and perovskites ($YAlO_3$, $SrTiO_3$). Other wide band-gap compounds, KCl and CsI:Tl single crystals were also considered to draw the main conditions for creation of irradiation induced color centers and point defects by using the theoretical approach above presented.

7.1. IRRADIATION, EXPERIMENTAL METHODS AND CRYSTALS

The defects and colour centres were created by the following irradiations:
- X-rays, E = 50 - 200 KeV, dose - 10^{-2} - 10 Gy, T = 77, 300 K;
- γ-irradiation, E = 0.6 and 1,25 MeV, dose - 10 - 10^4 Gy, T = 77, 300 K;
- electron beam, E = 2 - 10 MeV, dose 10^8 - 10^{15} cm^{-2}, T = 77, 300 K etc.

It is well known that upon irradiation, additional bands may appear in most of the wide band gap crystal optical absorption spectra. For the study of the oxidation state stability of the compound component and doped nl- ions, the valence shift of X-ray lines method (VSXRL) was used while for the colour centre investigation in irradiated crystals, optical luminescence, thermo-luminescence and thermo-conductivity methods described in detail in [10-16, 45] were used.

Colour centres before and after irradiation of solids were thoroughly studied in sapphire single crystal, α-Al_2O_3. We have made an attempt to investigate the influence of the crystals growth parameters on types and concentration of point defects in sapphire crystals grown by different methods such as Verneuil (V), Czochralski (Cz), Stepanov (S), Cyropulos (C), horizontal directed crystallization (DC). Different compounds: super pure (sp), pure (p), recrystallized (r), and standard ones were used for the crystal production under different atmospheres in the growing chamber: hydrogen, oxygen (V), vacuum (DC, C), argon (Cz, S). The temperature gradients in the growth zone are maximum for the Verneuil technique and minimum for the Cyropulos one. The maximum growth rate is possible with the Stepanov method.

Ruby, α -Al_2O_3:Cr, with Cr ion concentration up 10^{-3} to $5 \cdot 10^{-1}$ wt % were grown by the Verneuil technique while YAlO3 perovskite crystals doped with 10^{-3} to 10^{-1} wt % of Cr ions and co-doped with 10^{-2} to 1 wt % Nd ions were grown by the Czochralsky method or directed crystallization one. Garnet crystals were grown by the vertical directed method, horizontal directed method or the Czochralsky one. The concentration of Me or/and RE-ions was varied from 10^{-2} to $5 \cdot 10^{-1}$ wt % but it should be noticed that most of the samples grown with the standard compounds contain different Me-ions such as Cr and Fe, as un-controlled impurities.

7.2. RESULTS FOR CRYSTALS DOPED WITH ME IONS

We have studied the oxidation state of the following doped ions: V, Cr, Mn, Fe, Co, Ni etc and the related colour centres in oxides, before and after γ- and e-irradiation or/and thermal treatment. The concentration of the doped ions are $5 \cdot 10^{-3}$ to 10^{-1} wt %. The different ways of changing the impurity oxidation state upon irradiation or thermal treatment were discussed in [3-5].

7.2.1. Optical Properties of Oxide Single Crystals.
Pure Sapphire. Optical absorptions in the pure sapphire single crystals grown by different methods are shown in figure 1 and data of ESR, optical, TSC and TSL

154

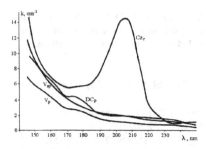

Figure 3. Optical absorption spectra of
the sapphire grown by different methods

investigation are reported in Tab.9 together with the impurity concentration. It should be noticed that no absorption bands were recorded upon irradiation in the super pure sapphire crystal. The main colour centres in the crystals containing impurities are due to trapped charges in oxygen and cation vacancies [19-21].

Irradiation of sapphire changes the band intensity in the optical spectra.

TABLE 9. Spectral parameters of the sapphire grown by different techniques

Sample	Abs.edge nm	Optical Bands, nm	AA bands nm	TL intensity 320-420 nm	TSC peaks T, K	TL bands nm
V	195	206*, 225*, 260*, 400* 570	206, 225, 280, 475	-	388, 578	690
V_p	142	175, 206, 230, 400*	206, 225, 280, 475*	4	385, 560 507	320, 690 690
V_{sp}	142	185*,206*, 230*	No AA	0.1	430, 507* 560	420, - 690
DC	145	175, 206* 235*	206, 230	no TL	398*, 507*	-
DC_r	142	175, 206, 235	206, 230 280*, 470*	8	373, 506, 565	320, 420 690
Cz_r	143	180*, 206	206, 475	2	430, 580	320, 420 420, 690
C_r	142	198, 225*	No AA	no TL	387*, 426* 485*	-
S_r	142	175, 206 230	206, 230	1	390,418, 430, 506*	420 -

According to data [9], absorption bands are related to one- or two-electrons trapped in oxygen vacancies (the so-called F-centres [1]) and electrons trapped in cation sites. The types and concentration of the point defects depend on the crystal growth parameters.

Pure Garnet. Similar experiments were carried out for pure $Y_3Al_5O_{12}$ garnet [15, 46], to study the change of the intensity of the optical bands of structural point defects upon irradiation. Optical absorption spectra of $Y_3Al_5O_{12}$ in VUV-UV region were reported in [46].

Doped Sapphire. The concentration of the doped ions and irradiation induced defects in sapphire was determined by using ESR method. The concentration of Cr^{3+} may be evaluated via the intensity of $\frac{1}{2} \leftrightarrow -\frac{1}{2}$ ESR line while the total concentration of defects may be estimated via the half-width of $3/2 \leftrightarrow 1/2$ transition in Cr^{3+} ion [10].

In ruby, it is well known that upon irradiation with any type of radiations, the intensity of $\frac{1}{2} \leftrightarrow -\frac{1}{2}$ line of Cr^{3+} decreases by 15 to 20 % and the half-width of 3/2

<-> 1/2 line of Cr^{3+} increases by 20-25 % [3, 9, 22] when Cr^{3+} concentration is ranging from 10^{-3} to 10^{-1} wt %.

Doped Garnet. Figs. 4, 5 present data on the dependence of the optical absorption spectra on the concentration of Cr^{3+} ions (curves 1 and 2), upon irradiation (curves 1' and 2') and the dependence of the of ½ ↔ -½ ESR line of Cr^{3+} ions upon γ-irradiation of $Y_3Al_5O_{12}$:Cr. The latter dependence shows the existence of many steps in the process of creation and dissociation of the irradiation induced defects in garnet crystals [15, 46]. According to these data, the stability of the Cr^{3+} ions in garnet using the optical spectra only, is unpredictable.

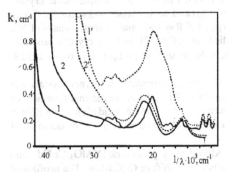
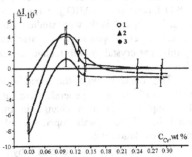

Figure 4. Optical absorption spectra of $Y_3Al_5O_{12}$:Cr before (1) and after (2) irradiation

Figure 5. Intensity of ½ ↔ - ½ Cr^{3+} ions in $Y_3Al_5O_{12}$:Cr with C_{Cr} and dose of γ-irradiation 10^3 (1), 10^4 (2) 10^5 (3) Gy

7.2.2. X-ray Spectroscopy of 3d- and 4f-Ions in crystals

We have studied the change in ion oxidation state of different nl- ions in wide band-gap crystals upon irradiation or thermal treatment by using X-ray spectroscopy. Experiments were performed on a modified Cameca microanalyser equipped with a X-ray spectrometer (Fig. 6). Experimental procedure was described in detail in [3, 5] and [4, 7, 12, 14].

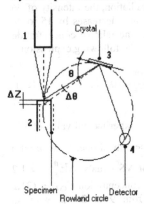

Figure 6. Simplified scheme of microanalysor

Our approach was tested by recording the chemical shift of L_α-lines for RE ions (Eu, Tm, Yb etc.) in fluoride compounds (REF_2 - REF_3) and by comparing the data to the energy for K_α-lines obtained by O. Sumbaev et al. [37] with a high resolution X-Ray spectrometer.

The VSXRL method is based on the dependence of K-, L- etc. X ray lines on the oxidation states of the ions. The scheme of our set-up is shown in Fig. 6. An electron gun (1) with current of 10 nA excites a sample spot of 1 μm in diameter (2) and the X-rays emitted are analysed by a crystal analyser (quartz 1010) (3) and detected with a proportional counter (4). The X-ray line intensity is measured for every step of the rotating crystal-analyzer. The profiles were calculated by Lorentz functions. The results obtained for various impacts on the sample surface differed by 1 to 3 %. The accuracy of the energy determination of $REL_{\alpha 1}$-line varied from 10 to 100 meV and for $MeK_{\alpha 1}$-line from 8 to 60 meV. The energy of X-ray line can be calculated by using Eq.(15) which provides, for.example, the theoretical energy value for $TiK_{\alpha 1}$-line at $E(K_{\alpha 1}) = 4508.8$ eV and the theoretical value for the valence shift, ΔE $(Ti^{3+} \rightarrow Ti^{4+}) = 0.86$ eV [3]. The valence shift for $L_{\alpha 1,2}$-lines in lanthanides is similar. The valence shift was corrected according to the electron energy losses in the sample compounds.

We have studied the oxidation states and valence shifts in ions of the I - VII group elements on the one hand and Me and RE ions, on the other hand in different crystals before and upon irradiation or thermal treatment. Firstly, Yb in simple oxide crystal, (Yb_2O_3), perovskites (ABO_3), garnets $(Y_3Al_5O_{12})$ and some glasses and secondly, Tm in TmS, Tm_2O_3 and TmF_x were studied. Energy of X Ray lines and oxidation states of Nd, Eu, Gd, Tb etc. doped ions were also studied carefully [47]. In these crystals produced by different crystal growth techniques, the valence shift of NdL_{α}-line is (0.86 ± 0.28) eV.

The concentration of the doped 3d-ions in ruby, perovskite and garnet crystals was $5*10^{-3}$ to 10^{-1} wt %. In general, γ-irradiation $(^{60}Co, 10 - 10^4$ Gy$)$ or thermal treatment of the samples leads to a change in impurity or component oxidation state. Upon irradiation of the crystals doped with nl- ions, we observed an increase or decrease of the ions oxidation state.

For example, γ-irradiation of ruby $(\alpha$-Al_2O_3:Cr$)$, or perovskite (YAlO$_3$:Cr:Nd), and garnets $(Y_3Al_5O_{12}$:Cr$)$ is accompanied by a negative shift of $CrK_{\alpha 1}$-line. The profiles of the $CrK_{\alpha 1}$-lines for ruby before (1) and after (2) irradiation are shown in Fig. 4. Mathematical processing of the X- ray line profile by using Lorentz function permits to evaluate the valence shift of $CrK_{\alpha 1}$-line at $\Delta E(CrK_{\alpha 1}) = (0.36 \pm 0.08)$ eV. As already mentioned, the VS value depends on the impurity concentration and the initial crystal growth parameters as well as the thermal treatment. A negative shift of $CrK_{\alpha 1}$-line is related to the transition $Cr^{3+} \rightarrow Cr^{4+}$. It is well known that upon irradiation, the intensity of $\frac{1}{2} \leftrightarrow -\frac{1}{2}$ ESR line in ruby decreases by 15 to 20 %. We can estimate the relative concentration of Cr^{4+} ions by using the following expression:

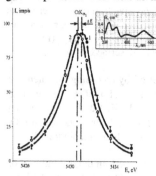

Figure 7. Profile of $CrK_{\alpha 1}$ line before (1) and after (2) γ-irradiation of ruby.
Insertion: AA of γ-irradiated ruby

$$C(Cr^{+4}) = \frac{\Delta E_x^{exp}}{\Delta E_x^{theor}} \cdot 100 \%, \qquad (24)$$

where ΔE_x^{exp}, is the experimental value of VS, and ΔE_x^{theor}, the theoretical one. If we used a theoretical value for VS equals $\Delta E_x^{theor} = 1.2$ eV [3], the relative concentration $C_{Cr4+} = (31.2 \pm 6.7)$ % varies from 9 to 25 % with an uncertainty of $\pm 7\%$. This magnitude closely corresponds to the data obtained via ESR experiments. Upon thermal treatment of the γ-irradiated ruby at 900 K during 600 s, the absorption bands are removed and the intensity of the ESR line returns to its initial value. It was not possible to determine VS for $CrK_{\alpha 1}$-line in ruby before irradiation and upon thermal treatment [5]. The optical absorption bands of the γ-irradiated ruby are presented in Fig. 7.

Similar experiments were carried out to estimate VS for $CrK_{\alpha 1}$-line in perovskite and garnet crystal doped with Cr ions [13, 15]. For γ-irradiated $YAlO_3$:Cr:Nd, VS is equal to $\Delta E = (0.31 \pm 0.08)$ eV while for $Y_3Al_5O_{12}$:Cr:Nd, the value $\Delta E = (0.22 \pm 0.08)$ eV was obtained with a strong dependence of VS on the concentration of Cr ions [49].

Similar experiment were performed for NdL_{α}-line in $Y_3Al_5O_{12}$:Nd or $YAlO_3$:Cr:Nd crystals. The energy of NdL_{α}-line does not depend on irradiation but depends on crystal growth conditions [49]. For example, a positive shift of NdL_{α}-line was obtained for samples grown by MDC. The value of VS NdL_{α}-line is equal to $\Delta E = (0.18 \pm 0.08)$ eV and the estimated concentration of Nd^{2+} ions is near $C_{Nd^{+2}} = (15 \pm 7)\%$.

We used VSXRL method to study other garnets for example, $Gd_3Sc_2Al_3O_{12}$ doped by Cr and Ca. Upon a thermal treatment in vacuum, the optical absorption spectra of the crystal changed significantly [14] with very intense additional optical absorption bands (Figs. 8, 9). We may assume that the optical bands may be attributed to Cr^{4+} ions located in tetrahedral positions of GSAG crystals. Moreover, the valence shift of $CrK_{\alpha 1}$-line for $Gd_3Sc_2Al_3O_{12}$:Cr:Ca (Mg) crystals upon thermal treatment is similar to one obtained for the irradiated ruby. It should be noticed that the theoretical energy for the triplet-triplet transitions in $[Cr^{4+}]_6$- clusters do not correspond to the experimental one for GSAG crystals. In Figs. 8, 9, experimental optical spectra recorded in GSGG:Cr:Mg and GSAG:Cr:Mg before (1) and after (2) the thermal treatment in oxygen atmosphere are shown with the labels of the $d \leftrightarrow d$ transitions of Cr^{4+} in tetrahedral sites (Tab. 10). A quite good fit between theoretical and experimental data is observed for this crystal.

TABLE 10. Energy levels of Cr^{4+}:$[O^{2-}]_4$ cluster

Γ	$Y_3Al_5O_{12}$		$Gd_3Sc_2Ga_3O_{12}$	
	λ_{theor}, nm	λ_{exp}, nm	λ_{theor}, nm	λ_{exp}, nm
$^3A_2$	-	-	-	-
^1E	10950	11000	847	-
$^3T_2$	964	964	1052	1052
$^3T_1$	640	640	661	600
$^1A_1$	627	-	507	504
$^1T_2$	517	-	475	504
$^1T_1$	453	-	407	410
$^3T_1$	410	-	410	410

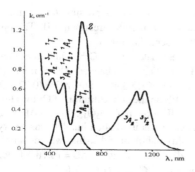

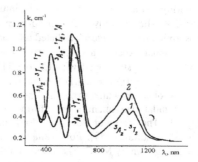

Figure 8. Optical absorption spectra of
GSAG:Cr:Mg before (1) and after(2) treatment

Figure 9. Optical absorption spectra of
GSGG:Cr:Mg before (1) and after(2) treatment

7.3. VALENCE STABILITY OF THE MATRIX COMPONENT AND DOPED ns- AND np- IONS IN WIDE BAND-GAP CRYSTALS UPON IRRADIATION

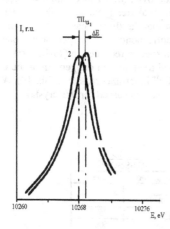

This part is devoted to the change in oxidation state of s- or p-ions upon irradiation. It is well known that KCl single crystals become blue after X rays or intense UV irradiations. VSXRL for K- and Cl-ions before and upon γ-irradiation showed a negative shift of $K_{\alpha1}$-X-ray line with a value of $\Delta E_x = (0.46 \pm 0.06)$ eV [50]. According to the theoretical and experimental data, we can conclude that a part of K^+ ions changes their oxidation state to K^0. The irradiation induced procedure in KCl may be described by the following transition:

$$K^+ + e- => K^o, \tag{25}$$

Figure 10. Profile of TlL$_{\alpha1}$- line before (1)
and after (2) after γ-irradiation

where K^+ ion captures an electron from Cl⁻ ion.

Creation of the Cl^--Cl^0 pair was explained in [6] by using the excitonic way.

Similar experiments were carried out for CsI single crystals doped with Tl ions. The concentration of the doped ions was varied from 10^{-3} to $2 \cdot 10^{-1}$ wt %. Additional optical bands in CsI:Tl crystals upon γ-irradiation by doses of 10^2 Gy were observed. The spectra consist in the following optical absorption bands: 365, 395, 440, 470, 485, 530, 555 and 670 nm. Regarding X-ray lines, VS for CsL$_{\alpha1}$-, IL$_{\alpha1}$–lines was impossible to estimate but for TlL$_{\alpha1}$-lines, VS is equal to $\Delta E_x = (0.42 \pm 0.18)$ eV (Fig.10). We can conclude that the oxidation state of Cs^+ and I⁻ ions did not change upon irradiation of the crystal but some Tl^+ ions change to Tl^0. Additional optical bands of CsI:Tl are due to different clusters.

8. Discussion.

We may conclude that upon irradiation of KCl and CsI:Tl, the cations: K^+ and Tl^+ change their oxidation state to Me^0. In the same way, the 3d-ions doped into oxide single crystals change also their oxidation state. The phenomena can be explained by the following irradiation induced processes:

$$
\begin{aligned}
Cr^{3+} &\rightarrow Cr^{4+} + e & V^{3+} &\rightarrow V^{2+} + h \\
Fe^{3+} &\rightarrow Fe^{4+} + e & Co^{3+} &\rightarrow Co^{2+} + h \\
Mn^{3+} &\rightarrow Mn^{4+} + e & Ni^{3+} &\rightarrow Ni^{2+} + h \\
Mn^{3+} &\rightarrow Mn^{2+} + h & &
\end{aligned}
\tag{26}
$$

It is well known that the stability of Me ion oxidation state can be estimated by comparing the ionization energy of Me ion, I_{Me}^{n+} with the value of the Madelung's constant for the cation site in crystals:

$$
\alpha_M = -\sum_i Z_i / R_i. \tag{27}
$$

For example, for sapphire $\alpha_M = -35.2$ eV and therefore, the oxidation state of Me^{n+} ions with $I_{Me}^{n+} > \alpha_M$ is stable. It is easy to see that for Me ions in the oxide crystals investigated, the main oxidation state is Me^{3+} and I_{Me}^{n+} value increases along with the ion charge. It should be noted that Me^{2+} oxidation state for different Me ions into different oxides may be observed, too. Me^{2+} state results from Eq.(27) and appearance of oxygen vacancies during the crystal growth. Upon irradiation of the doped crystals, the value of α_M changes with the creation of defects in the nearest environment of Me ion [51]. Change of α_M resulting from the $Me^{3+} \rightarrow Al^{3+}$ substitution (change of inter-atomic distances) leads to the appearance of unstable oxidation states. Appearance in the first coordination sphere of either holes trapped by anions or electrons trapped in anion vacancies changes α_M significantly (around 2 eV). The change of Madelung's constant of cation site upon irradiation or thermal treatment (for example, Cr^{4+} ions in tetra-sites of complex garnets) is causing the oxidation state change in Me ion. If the main types of defects are known, $\Delta\alpha_M$ can be calculated and the ion oxidation state upon irradiation estimated.

The change in ion oxidation state on the left side of Exp. (26) is determined by the oxygen vacancy, V_O^{2-} in its nearest environment. Irradiation of crystals leads to electronic excitation of Me^{n+} ion and the electron capture by V_O^{2-}. Therefore, the intensity of Cr^{4+} optical bands in octahedral sites in ruby, $Y_3Al_5O_{12}$:Cr or YAlO₃:Cr etc., upon irradiation depends on the concentration of structural defects. Additional optical bands in these crystals may be observed at 217, 360 and 460 nm (ruby), 253, 417 and 488 nm ($Y_3Al_5O_{12}$:Cr), and 295, 425 and 500 nm (YAlO₃:Cr). They are related to Cr^{4+}:[O^{2-}]₆ cluster It should be noted that Cr^{4+} ions were observed in tetra-site of complex garnets, too.

The electronic states of all ions, Me and RE ions, in solid state may be discussed within theframework of the discrete (HFP approach for clusters) or effective occupation number approximations. For the first case, level energy schemes for nl↔nl transitions can be determined as well as the concentration of ions with different oxidation states. For the estimation of the amount of Me^{n+} ions changed into $Me^{n\pm1}$ upon irradiation or thermal treatment, the simple Eq.(24) can be used or Eq.(20, 23) for the determination of the effective occupation number (or effective charge). It will provide useful information about the correlation effects and estimation of the energy of the optical bands.

9. Conclusion

The *ab initio* method used for the investigation of the electronic structure of $Me^{n+}:[L]_k$ clusters for ions with unfilled nl-shell and the calculations performed for clusters at different values of R and different types of ligands allow to draw several conclusions. Firstly, the available experimental data on both optical and X ray spectra for the impurity ions of the iron and actinium groups can be completely described within our approach. The energy level schemes calculated for $Cr^{3+}:[O^{2-}]_6$ cluster at $R = 1.96$ Å are close to the experimental spectra recorded in ruby. The theoretical R-dependence of the energy of Cr^{3+} ions describes correctly the pressure dependence on spectra of ruby, too. The *ab initio* calculated level scheme of Cr^{4+} ions in octahedral and tetrahedral environment corresponds to the optical spectra of ruby, $YAlO_3$ perovskite and garnet. The results of the study of the valence shift of $CrK_{\alpha 1}$-line make possible to assume that the additional optical absorption bands in these crystals are due to change in oxidation state of impurity. In oxygen - containing clusters, the relative energy of the excited $3d^{N-1}4s$- and $3d^{N-1}4p$-configurations decreases as R varies from 2.0 to 1.9–1.8 Å. The most interesting results were obtained for copper ions where a sharp decrease in the excitation energy and an increase in the d–s-interaction energy are observed for Cu^{2+} ion surrounded by four oxygen ions at distances corresponding to the structure of high-temperature ceramics. This result has been obtained only for Cu^{2+} ions and is not observed for Cu^{3+} ions or the other $Me^{n+}:[L]_k$ clusters considered. It should be noted that such a phenomenon is not observed for Ni and Zn ions surrounded by oxygen atoms as well as for the Cu^{2+} and Cu^{3+} ions with fluorine and chlorine ligands. In these cases, the dependences of the excitation energy and the $d \leftrightarrow s$ interaction energy is practically linear. The energy of the X-ray spectral characteristic lines of the nl- ions and the valence shift of these lines calculated within our approach are close to the experimental values. This makes possible the use of X-ray spectral lines for the investigation of changes in the valence of the host and impurity nl- ions in solids.

The study of the electronic structure of $Me^{n+}:[L]_k$, energy of X-ray lines and optical spectra of nl-ions on the one hand and investigation of the influence of irradiation or thermal treatment in doped crystals on the other hand allow to draw the simple conclusion that stability of the oxidation state of ions in crystals is determined by the consideration of the ionization energy of nl- electron and the Madelung's constant for the cation site [5].

The method of the effective occupation number was presented as an original approach of the correlation effects, energy of the nl-ion $\leftrightarrow$ ligand interaction and study of X-ray spectra and valence shift of X-ray lines in solids upon irradiation or thermal treatment.

10. References

1. Stoneham, A. (1968) *Theory of Defects in Solid State*, McGrow Hill, London.
2. Abragam, A., Bleaney, B. (1970) *Electron Paramagnetic Resonance of Transition Ions*, Claredon Press, Oxford.
3. Kulagin, N., Sviridov, D. (1986) *Methods of Calculation of Electronic Structure of Free and Impurity Ions*, Nauka Press, Moscow.
4. Kulagin, N., Ozerov, M. (1993) Electronic states of Ti^{+n}- ions in SrTiOx and TiOy single crystals, *Phys .Sol. State* **35**, 2463-2469.
5. Kulagin, N., Sviridov, D. (1990) *Introduction to Doped Crystals Physics*, High School Publ., Kharkov.
6. Luschik, Ch.B., Vitol, I.K. (1977) Desintegration of electronic excitations to the radiation defects into ionic crystals, *Usp. Fiz. Nauk* **122**, 223-251.
7. Kulagin, N., Podus, L., Zaitseva, J., Vo Chiong, Ki (1984) Change of electronic state of K-ions under γ- irradiation of CsJ single crystals, *Phys. Sol. State* **26**, 234-236.
8. Vinetsky, V.L., Kalnin', Ju.Ch., Kotomin, E.A., Ovchinikov, A.A. (1990) Radiation stimulation agregation of the Frenkel's-defects in solid states, *Usp. Fiz. Nauk* **160**, 1-33.
9. Bessonova., T.S. (1982) Defects in Sapphire, *Problems of Nuclear Physics* **16**, 3-20, High School Publ., Kharkov.
10. Apanasenko, A., Kulagin, N. (1981) On the valency transitions of impurities in γ-irradiated corrundum", *J. Appl. Spectroscopy* **35**, 135-137.
11. Kulagin, N., Litvinov, L. (1985) The defects and spectral properties of sapphire grown by melting methods, *Cryst, Res. Technology* **20**, 1667-2672.
12. Kulagin, N., Sviridov, D. (1984) The Spectra of Chrome Ions and Irradiation Influence on Ruby *J. Physics (London)* **C17**, 4539-4546.
13. Kulagin, N., Trojan-Golovjan, G. (1993) Stabilisation of electronic state of chromium ions in ABO₃ crystals, *Sov. Opt. Spectroscopy* **74**, 141-147.
14. Krutova, L., Kulagin, N., Sandulenko, V., Sandulenko, A.(1989) Electronic state and positions of chromium ions if garnet crystals, *Phys. Sol. State* **31**, 170 -175.
15. Kulagin, N., Ozerov, M., Rokhmanova, V (1987) Influence of γ- irradiation to the electronic state of chromium ions in Y₃Al₅O₁₂ single crystals, *J. Appl. Spectroscopy.* **46**, 612-615.
16. Kulagin, N.(1985) Hartree-Fock equations for doped ions in crystals, *Phys. Sol. State* **27**, 2039-2044.
17. Kulagin, N., Sandulenko, V. (1989) Ab initio theory of electronic spectra of doped crystals. Chromium ions in oxide compounds, *Phys. Sol. State* **31**, 243-249.
18. Kulagin, N.(2002) Ab initio calculation of the electronic structure and spectra of nl ions, *Phys. Sol. State* **44**, 1484-1490
19. Slater, J. (1968) *Consistent Field Methods for Molecules and Solid States*, J. Wiley, N.Y.
20. Morrison, J. (1987) *Many-Body Calculation*, Springer Verlag, New York.
21. Cowan, R.D.(1981) *The Theory of Atom Structure and Spectra,* University California Press, Berkely
22. Sviridov, D., Smirnov, Ju. (1977) *Theory of the Optical Spectra of the Ions of the Transition Metals*, Nauka, Moscow.
23. Wybourne, B.G. (1965) *Specroscopic Properties of Rare Earth*, J. Wiley, N.Y.
24. Froese-Fisher, C. (1977) *The Hartree-Fock Methods for Atoms*, J. Wiley, N.Y.
25. Hartree,.D. (1957) *Atomic Structure Calculation*, J. Wiley, London.
26. Carnall, W.T. Goodman, G.L., Rajnak, K., Rana, R.S. (1987) A systematic analysis of the spectra of the lanthanides doped into single crystals LaF₃, *J. Chem. Phys.* **90**, 3443-3457.
27. Kulagin, N. (1987) Hartree-Fok equations for doped ions, *Phys. Sol. State* **27**, 2039-2044.
28. Kulagin, N., Sviridov, D. (1982) Theoretical dependence of levels scheme of chromium ions in ruby under pressure, *Dokl. AS USSR* **266**, 616-620.
29. Kulagin, N., Kuliev, Sh. (1991) Electronic structure of Cu-clusters in super-conductors, *Phys. Sol. State* **31**, 3382-3385.'
30. Kulagin, N. (1996) d-s interaction in Me ions in superconductors, *Physica* **B222**, 173-179.
31. Dojcilovich, J., Kulagin, N. (1996) Anomalies of the temperature dependence of the dielectric properties of SrTiO₃, *Phys. Sol. State* **38**, 2941-2949.
32. Kulagin,.N (1998) Theory of electronic structure of rare earths and actinides in Rn^{+n}:[L]ₖ clusters, *Physica* **B245**, 52-60.

162

33. Jucys, A.P., Savukinas, A.J.(1973) *Mathematical Foundations of the Atomic Theory*, Mintis, Vilnius

34. Sviridov, D., Sviridova, R. Smirnov, Ju (1976) *Optical Spectra of the Ions of the Transition Metals in Crystals*, Nauka, Moscow.

35. Kulagin, N. (1983) Dependence of X ray energy on number of on optical electron in nl-ion, *J. Physics (London)* **B16**, 1695 -1702.

36. Kulagin, N. (1995) Theory of actinides: $5f^N$-configurations and X ray spectra. *J. Physics (London)* **B28**, 373-381.

37. Sumbaev, O. (1978) Shift X ray K lines under valency change and isomorphic phase transitions in rare earths, *Usp. Fiz. Nauk.* **124**, 281-306.

38. Muller, K.A. (2000) Recent experimental insights into HTSC materials, *Physica* **C341**, 11-18 .

39. Matheiss, L.F. (1988) Electronic band properties and superconductivity in $La_{2-y}X_yCuO_4$, *Phys.Rev.Lett.* **58**, 1028-1032.

40. Racah, G. (1943) Theory of complex spectra .III, *Phys. Review* **63**, 367-383

41. Racah, G. (1949) Theory of complex spectra .IV, *Phys. Review* **76**, 1352-1365

42. Kulagin, N.A., Zalubovsky, I.I. (1981) Effective occupation number in the theory of the spectra of $3d^N$ and $4f^N$ ions, *J.Physics (London)* **B14**, 1537-1547.

43. Barinskyy, L.B., Nefedov, V.I. (1966) *X-Ray Spectroscopy Investigation of the Charge of atoms and Molecules*, Nauka, Moscow

44. Sandstrom, A.E. (1957) X ray spectra of actinides, in S.Flugge (ed.) *Encyclopedia of Physics*, part XXX, J.Wiley, N.Y., 182-196.

45. Kulagin, N., Apanasenko, A., Kazakov, N. (1983) Conditions of stability of different states of magneze's impurity in α -Al_2O_3, *J.Appl.Spectr.* **38**, 988-993

46. Kulagin, N., Ovechkin, A., Antonov, E. (1985) Colour centers of γ- irradiation $Y_3Al_5O_{12}$, *J. Appl . Spectroscopy.* **43**, 478-484.

47. Vo, Chiong Ki, Zaitseva, Ju., Kulagin, N. et al. (1984) Valency shift of L- lines of Eu and Yb ions in separate compounds, *Phys. Sol. State* **26**, 3521-3525

48. Zaljubovsky, I.I., Kulagin, N.A., Litvinov, A.L., Podus, L.P. (1981) Change of valency of chromium ions under irradiation of ruby. *Phys.Sol.State* **23**, 846-849

49. Antonov, E., Bagdasarov, H., Kazakov, N. et al. (1984) Influence of crystals growth conditions to radiation spectral properties of $Y_3Al_5O_{12}$:Nd, *Crystallography Reps.* **29**, 175-176.

50. Kulagin, N., Podus, L., Kovaleva, L. et. al. (1984) Influence γ- irradiation to valency change of ions in CsJ single crystals doped with tallium, *Izv. AS USSR. Neorg. Math.* **17**, 700-702

51. Kulagin, N. (2000) Mixed valency of the rare earth and actinium ions in solid states. *J.Alloys and Comps.* **300-301**, 348-352.

OVERLAP POLARIZABILITY AND COVALENCY IN DIATOMIC MOLECULES AND EUROPIUM COMPLEXES

R.Q. ALBUQUERQUE, O.L. MALTA[*]
Departamento de Química Fundamental
CCEN - UFPE, 50590-470, Recife, PE, Brazil.

Abstract. The concepts of overlap polarizability and ionic specific valence have been initially applied to the series of alkali halides RX, where R = Li, Na and K, and X = F, Cl, Br and I. The values of overlap polarizability for this diatomic series have been calculated and a new covalency scale based on this quantity has been proposed, showing a good correlation with Pauling's covalency scale. The charge factors, g, appearing in the simple overlap model for the ligand field in lanthanide compounds have been associated to the ionic specific valences calculated for a series of trivalent europium compounds. The Stark levels for the complexes $Eu(TTA)_32H_2O$, $Eu(btfa)_32H_2O$ and $Eu(btfa)_3phenNO$ have been calculated using this association, and the root-mean-squares deviations obtained were 1.97%, 2.94% and 5.02%, respectively, in comparison with experimental values. The overlap polarizability values calculated for the Eu^{3+} compounds have also been compared with the energy of the $^5D_0 \rightarrow {}^7F_0$ transition in order to study the nephelauxetic effect and a good linear correlation has been obtained, providing a support to the fact that covalency is of importance in this effect.

Keywords: *covalency, overlap polarizability, alkali halides, lanthanides*

1. Introduction

The understanding of covalency in chemical systems such as coordination compounds or doped crystals is of great importance in the description of ligand field interactions [1]. Covalency is usually defined as the degree in which the electrons in a chemical bond are shared between atomic species. From this concept, Pauling was the first to propose a scale of covalency for diatomic molecules based on the difference between the electronegativities of the two atomic species A and B [2].

The polarizability of a molecular system, α_{MOL}, can be interpreted as a measure of the tendency of deformation of its electronic cloud when an external electric field is applied. The quantity α_{MOL}, to a good approximation, can be expressed as a sum of chemical bond polarizabilities, α_{CB}. This partitioning scheme is better justified when the interaction between bonds is not very strong. Considering a single directional AB bond, α_{CB} can be further decomposed as a sum of three parts: α_A, α_B and α_{OP}^* [3]. The major

*Corresponding author. E-mail: oscar@renami.com.br

J.-C. Krupa and N.A. Kulagin (eds.), Physics of Laser Crystals, 163–170.
© 2003 *Kluwer Academic Publishers. Printed in the Netherlands.*

contribution to α_{CB} comes from the polarizabilities of the atoms A e B (α_A and α_B) and a minor contribution comes from the overlap polarizability (OP), α_{OP}^*, associated with the overlap region. The charge q occupying this region is shared by A and B. If this charge is highly polarizable, then it is shared more effectively by A and B, giving as a result a highly covalent bond and, therefore, α_{OP}^* can be directly related to the covalency in a chemical bond.

It is well known that in a classical treatment of an harmonic oscillating point charge, the polarizability of this oscillator is proportional to the value of charge squared, the proportionality constant being the force constant of the oscillator [4]. An analogous relation has been postulated between α_{OP}^* and the overlap charge q, involving the bond force constant, giving rise to the concept of ionic specific valence (ISV) [3]. The ISV is interpreted as the capacity of the atomic species in a single bond to donate charge to the formation of the chemical bond. This concept can be associated to the charge factors, g, appearing in the simple overlap model (SOM) for the ligand field in lanthanide compounds [5, 6]. These charge factors have been determined semi-empirically in the literature [7]. The association between ISV and g provides a theoretical way to calculate them for all the ligating atoms in a given lanthanide compound.

The aim of the present work is to apply these concepts initially to a series of alkali halides RX (R = Li, Na, K and X = F, Cl, Br, I) and to propose a scale of covalency based on the OP. Then, the case of more complex molecular systems (europium compounds) is treated. Stark levels in some Eu^{3+} complexes are calculated from the association of the charge factors with the ISV's of the ligating atoms and the nephelauxetic effect is successfully rationalized in terms of the OP, suggesting that covalency is of importance in this effect.

2. Theoretical

The polarizability of a molecular system can be calculated using the well-known quantum-mechanical expression in the absence of an external electric field:

$$\alpha = 2e^2 \sum_n \frac{\langle a|\hat{r}|n\rangle\langle n|\hat{r}|a\rangle}{(E_a - E_n)} . \tag{1}$$

In Eq.(1), e is the electronic charge, E_a and E_n are the energies of the fundamental and excited levels, respectively, $\hat{r}$ is the position operator of the electron and the summation runs over the n excited states of the molecule. This expression is applied to diatomic molecules containing only single bonds, and the molecular orbitals $|a\rangle$ and $|n\rangle$ are treated using the monoelectronic approximation. The summation in Eqn.(1) can be truncated by considering only the first excited state of the molecule (the LUMO), once it gives the major contribution to α. The final expression for α_{OP}^*, considering these approximations is [3]:

$$\alpha_{OP}^{*} = \frac{e^2 R^2 \rho^2}{2\Delta E}. \tag{2}$$

In Eq.(2), e is the electronic charge, R is the inter-nuclear distance, ΔE is the energy difference between HOMO and LUMO and ρ is the overlap integral involving the valence orbitals of the atomic species in the diatomic molecule. The relation between α_{OP}^{*} and q^2 leads to the following expression for the ISV [3]:

$$ISV = v_a + v_c = R\sqrt{\frac{k}{2\Delta E}}, \tag{3}$$

where k is the bond force constant. The ISV can be written as a sum of v_a and v_c, which are the capacities of the anion and cation, respectively, to donate charge for the formation of the chemical bond. It is assumed that the formation of the chemical bond occurs through a Lewis acid-base reaction $(A^+ + B^- \rightarrow AB)$. Thus, the condition $v_a > v_c$ should be satisfied.

In the case of the diatomic species, the quantities R, k and ΔE have been determined through an *ab-initio* calculation using the basis STO-6G* in the Gaussian 98 program. The overlap integral ρ has been calculated using the Rico-STO program [8]. The expression proposed for the new covalency scale based on α_{OP}^{*} for the diatomic molecules has the form:

$$C_{OP} = 1 - e^{-A\alpha_{OP}^{*}}, \tag{4}$$

where A is an adjustable parameter.

The ISV and α_{OP}^{*} have been also calculated for the following europium compounds: $Eu(btfa)_3 2H_2O$ and $Eu(btfa)_3 PhenNO$ [9], $Eu(TTA)_3 2H_2O$ and $Eu(TTA)_3 2DBSO$ [10], $Eu(PicNO)_3 terpy$ [11], $Eu(TTA)_3 2TPPO$ [12], $Eu(TTA)_3 2DMSO$ [13], $YOCl : Eu^{3+}$ [14], $LaF_3 : Eu^{3+}$ [15], $Eu^{3+}_{(gaseous)}$ and $Eu^{3+}_{(aqueous)}$ [16]. In the last compound it has been considered 9 water molecules coordinated to the Eu^{3+} ion. For the crystals, it has been considered only the first coordination sphere around the Eu^{3+} ion.

The bonds between the Eu^{3+} ion and the ligating atoms have been treated individually, the parameters ΔE, k, R and ρ being calculated for each bond. Because of the high positive charge in the europium ion, $v_a \gg v_c$, which justifies the approximation $ISV \approx v_a$. The ΔE parameter has been calculated as the energy difference between the 4f orbitals (considered as the LUMO) and the valence orbitals of the ligating atoms (HOMO). The overlap integral, ρ, has been calculated using the expression given in the SOM [5, 6]. The geometries and force constants have been calculated using the SPARKLE II model [17] implemented in the MOPAC93R2 program.

The charge factors g appearing in the SOM have been associated to the ISV values calculated to each ligating atom ($g \equiv ISV \approx v_a$). From this approximation, the so-called ligand field parameters, B_q^k, and the energies of the Stark levels for the complexes

Eu(TTA)$_3$2H$_2$O, Eu(btfa)$_3$2H$_2$O and Eu(btfa)$_3$PhenNO have been calculated. The experimental energies of the Stark levels have been obtained from the emission spectra of these complexes. The root-mean-squares deviations for these predictions have been calculated from the expression:

$$\sigma = \left[\sum_i^n \frac{1}{n} \left(\frac{E_i^{EXP} - E_i^{CALC}}{E_i^{EXP}} \right)^2 \right]^{1/2} .100\% , \qquad (5)$$

where n is the number of experimental Stark levels, E_i^{EXP} and E_i^{CALC} refer to the experimental and calculated values of the Stark levels, respectively.

3. Results and Discussion

The values of the variables calculated for the diatomic species are shown in Tabl.1.

TABLE 1. Calculated values of inter-nuclear distances (R), overlap integrals (ρ), force constants (k), energy differences between HOMO and LUMO (ΔE), overlap polarizabilities (α_{OP}^*) and ionic specific valences (ISV) for the alkali halides.

Molecule	R (Å)	ρ	ΔE (a.u.)	k (mdyn/Å)	α_{OP}^* (Å^3)	ISV
LiF	1.4034	0.2791	12.4110	4.9370	0.0890	1.56
LiCl	1.9138	0.3663	10.6990	2.3350	0.3308	1.58
LiBr	2.0331	0.3922	10.2700	2.2030	0.4457	1.66
LiI	2.2630	0.4284	9.7180	1.7000	0.6963	1.67
NaF	1.7351	0.2366	10.6870	5.8820	0.1136	2.27
NaCl	2.2251	0.3192	9.6370	2.6390	0.3770	2.06
NaBr	2.3156	0.3468	9.4600	2.6690	0.4909	2.17
NaI	2.5689	0.3778	8.7540	1.9800	0.7745	2.16
KF	2.0130	0.1852	9.6010	2.3840	0.1042	1.77
KCl	2.5819	0.2594	9.7610	1.5680	0.3309	1.83
KBr	2.6926	0.2818	9.4490	1.5500	0.4386	1.93
KI	2.9709	0.3114	9.0990	1.2600	0.6775	1.95

Some trends can be noted on the OP values calculated for the series RX. The α_{OP}^* increases for a fixed alkaline metal when the size of the anion increases: α_{OP}^* (RF) < α_{OP}^* (RCl) < α_{OP}^* (RBr) < α_{OP}^* (RI). When the inter-nuclear distance increases, the attraction on the overlap charge by the two nuclei is weaker, increasing the mobility of this charge. As a result, the overlap charge becomes more polarizable, giving rise to a higher value of α_{OP}^*.

The parameter A in Eq.(4) has been optimized through a comparison between C$_{OP}$ and Pauling's covalency (calculated from reference 2), giving the value of 1.5. The comparison between Pauling's covalency and the proposed covalency scale is shown in figure 1, from which a good correlation between the two scales can be observed. Although it requires a greater computational effort, the advantage of this new scale is that it is based on a quantity that is directly related to the concept of covalency, and, once the electronic structure is known, it can be applied not only to neutral species but also to charged species.

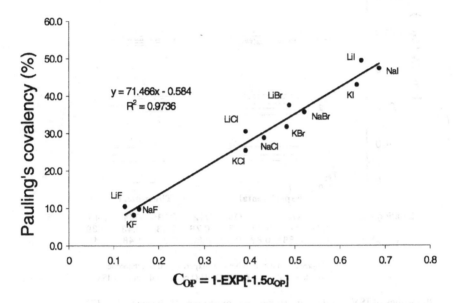

Figure 1. Comparison between Pauling´s covalency and the covalency based on α^*_{OP}.

The variables involved in the calculation of ISV and α^*_{OP} are shown in Tabl.2 for the complex $Eu(TTA)_3 2H_2O$. A comparison between the values of α^*_{OP} for the diatomic molecules (Tabl.1) and for the Eu^{3+} complexes (Tabl.2) show, as expected, that the degree of covalency in the complexes is very low. The mean value of the ISV for the oxygens of the β-diketonate ligands (= 0.63) is close to the value of the charge factors commonly found for these kind of ligands (≈ 0.8), showing that the proposed relation g ≡ ISV is quite satisfactory.

In the prediction of the Stark levels it has been obtained σ = 1.97%, 2.94% and 5.02% for the complexes $Eu(TTA)_3 2H_2O$, $Eu(btfa)_3 2H_2O$ and $Eu(btfa)_3 PhenNO$, respectively. The same prediction using the semi-empirical g's has given σ = 2.28%, 2.31% and 4.99% for the complexes, respectively. The comparison between the experimental and theoretical Stark levels for $Eu(TTA)_3 2H_2O$ is shown in Fig.2 as well as

168

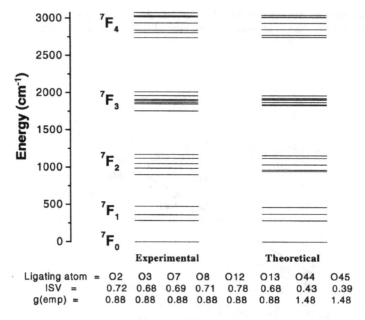

Figure 2. Comparison between experimental and theoretical
Stark levels for the Eu(TTA)₃2H₂O compound using g ≡ ISV.

the values of ISV and the corresponding semi-empirical charge factors [7].

TABLE 2. Calculated values of inter-nuclear distances (R), overlap integrals (ρ), force constants (k), energy differences between HOMO and LUMO (ΔE), overlap polarizabilities (α^*_{OP}) and ionic specific valences (ISV) for each bond between the ligating atoms and the Eu^{3+} ion in the complex Eu(TTA)₃2H₂O.

Complex	Bond	R (Å)	ρ	ΔE (a.u.)	k (mdyn/Å)	α^*_{OP} (Å³)	ISV
	Eu-O2	2.3070	0.0500	0.3871	0.3255	0.0091	0.72
	Eu-O3	2.3135	0.0495	0.4164	0.3165	0.0083	0.68
	Eu-O7	2.3566	0.0464	0.3685	0.2731	0.0086	0.69
Eu(TTA)₃2H₂O	Eu-O8	2.3113	0.0497	0.3736	0.3084	0.0093	0.71
	Eu-O12	2.3319	0.0482	0.2984	0.2902	0.0112	0.78
	Eu-O13	2.3380	0.0477	0.3866	0.2851	0.0085	0.68
	Eu-O44	2.4937	0.0381	0.7788	0.2008	0.0031	0.43
	Eu-O45	2.5333	0.0360	0.8324	0.1699	0.0026	0.39

With this new theoretical approach for the calculation of the charge factors it is possible to attribute a different g to each of the ligating atoms, while in the semi-empirical methodology a single value of g has been attributed to a given group of ligating atoms.

The nephelauxetic effect has been investigated for the Eu^{3+} compounds studied through the comparison between the energy of the $^5D_0 \rightarrow {}^7F_0$ transition and the mean value of α^*_{OP}. When the bonds are formed in the complex, the inter-electronic repulsion in the Eu^{3+} ion decreases, which leads to a smaller energy gap between the $^5D_0$ and $^7F_0$ levels. The red shift observed for the $^5D_0 \rightarrow {}^7F_0$ transition (see Fig.3) is accompanied by the increase of α^*_{OP}, suggesting that the nephelauxetic effect is mainly dominated by covalent effects.

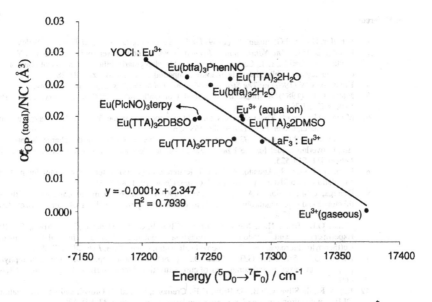

Figure 3. Correlation between the nephelauxetic effect and the mean value of α^*_{OP} for the europium compounds studied. NC is the coordination number.

4. Conclusions

A new covalency scale for diatomic species has been proposed through an exponential expression involving the overlap polarizability. It has been found a good agreement between this new scale and Pauling's covalency scale.

The very low values of the overlap polarizability calculated for the europium complexes show, as expected, that covalency in these systems is very low. The identification of the ISV with the charge factors appearing in the SOM has allowed to predict g in a completely theoretical way, allowing to attribute a different g to each of the ligating atoms in the complex. The prediction of Stark levels has shown a good agreement with the experimental ones, indicating that the present approach for the calculation of g is quite satisfactory.

The good correlation between the energy of the $^5D_0 \rightarrow {}^7F_0$ transition and α^*_{OP} suggests that the nephelauxetic effect can be rationalized in terms of covalent effects.

5. Acknowledgements

The authors acknowledge the CNPq, CAPES (brazilian agencies), RENAMI (Molecular and Interfaces Nanotechnology Network) and IMMC (Millenium Institute for Complex Materials) for financial support.

6. References

1. Schläfer, H.L. and Gliemann, G. (1969) *Basic Principles of Ligand Field Theory*, J. Wiley, L.
2. Pauling, L (1960) *The Nature of the Chemical Bond*, Cornell University Press, Ithaca, N.Y.
3. Malta, O.L., Batista, H.J., Carlos, L.D. (2002) Overlap polarizability of a chemical bond: a scale of covalency and application to lanthanide compounds, *Chem. Phys.* **282**, 21-28.
4. Kondratyev, V.(1967) in: *The Structure of Atoms and Molecules*, 2.ed., Mir, Moscow.
5. Malta, O.L. (1982) A simple overlap model in lanthanide crystal-field theory, *Chem. Phys. Letters* **87**, 27-29.
6. Malta, O.L. (1982) Theoretical crystal-field parameters for the YOCl :EU^{3+} system a simple overlap model, *Chem. Phys. Letters* **88**, 353-356.
7. Albuquerque, R.Q., Rocha, G.B., Malta, O.L., Porcher, P. (2000) On the charge factors of the simple overlap model for the ligand field in lanthanide coordination compounds, *Chem. Phys. Letters* **331**, 519-525.
8. Rico, JF, Lopez, R, Aguado, A, Ema, I, Ramirez, G (1998) Reference program for molecular calculations with Slater-type orbitals, *J. Comp. Chem.* **19**, 1284-1293.
9. Donega, C.M., Junior, S.A., de Sa, G.F. (1996) Europium(III) mixed complexes with beta-diketones and o-phenanthroline-N-oxide as promising light-conversion molecular devices, *Chem. Comm.* **10** 1199-1200.
10. Malta, O.L., Brito, H.F., Menezes, J.F.S., Silva, F.R.G.E., Donega, C.M., Alves, S. (1998) Experimental and theoretical emission quantum yield in the compound Eu(thenoyltrifluoroacetonate)3.2(dibenzyl sulfoxide), *Chem. Phys. Lett.* **282**, 233-238.
11. De Sa, G.F., Silva, F.R.G.E., Malta, O.L. (1994) Synthesis, spectroscopy andphotophysical properties of mixed-ligand complexes of europium III and terbium III, *J. Alloys Comp.* **207**, 457-460.
12. Reyes, R., da Silva, C.F.B., de Brito, H.F., Cremona, M. (2002) Growth and characterization of OLEDs with europium complex as emission layer, *Braz. J. Phys.* **32**, 535-539.
13. Brito, H.F., Malta, O.L., Menezes, J.F.S. (2000) Luminescent properties of diketonates of trivalent europium with dimethyl sulfoxide, *J. Alloys Comp.* **303**, 336-339.
14. Holsa, J., Porcher, P. (1981) Free ions and crystl-field parameters for REOCl : Eu^{3+}, *J. Chem. Phys.* **75**, 2108-2117.
15. Carnall, W.T., Crosswhite, H., Crosswhite, H.M. (1977) *Energy Levels Structure and transition Probabilities of the Trivalent Lanthanides in LaF₃*, Argonne National Laboratory Report, Argonne.
16. Frey, S.T., Horrocks W.D. (1995) On correlating the frequency of the $^7F_0 \Rightarrow {}^5D_0$ transition in EU(III) complexes with the sum of nephelauxetic parameters for all of the coordinating atoms, *Inorg. Chim. Acta* **229**, 383-390.
17. .de Andrade, A.V.M., da Costa, N.B., Simas, A.M., de Sa, G.F. (1994) Sparkle model for the quantum-chemical AM1 calculation of europium complexes, *Chem. Phys. Letters*, **227**, 349-353.

RED-INFRARED EMISSION OF Tm^{3+} IONS IN YVO_4 AND $LiNbO_3$ CRYSTALS BY MULTI-PHOTON EXCITATION WITH HIGH POWER 798 nm LASER DIODE

Taiju Tsuboi[a], Ruan Yongfeng[b] and Nocolai Kulagin[c]

[a]*Department of Information & Communication Science, Faculty of Engineering, Kyoto Sangyo University, Kamigamo, Kyoto 603-8555, Japan*
Email: tsuboi@cc.kyoto-su.ac.jp
[b]*Department of Physics, Tianjin University, Tianjin 300072, China*
[c]*Department of Physics, Kharkiv National University for Radioelectronics, Lenin av. 14, 61166 Kharkiv, Ukraine*

Abstract. Luminescence spectra of Tm^{3+} ion have been investigated for YVO_4 and $LiNbO_3$ crystals by excitation with 798 nm laser diode of high power in a range from 100 mW up to 10 W. In addition to the one-photon excited infrared emission bands, emission bands are observed at 700 and 1208 nm in YVO_4 by high power excitation and the similar emission bands are observed at 702 and 1216 nm in $LiNbO_3$. It is observed that each of the 700 and 1208 nm emission intensities in YVO_4 has quadratic pump-power dependence in a pump power range of 0.6-4 W and cubic dependence in 4-10 W range. Same result is observed for the 702 and 1216 nm emission in $LiNbO_3$. The observed 1208 nm emission band is attributed to the $^1G_4 \rightarrow {}^3H_4$ transition which is caused by two- and/or three-photon excitation processes. Discussion is made on the luminescence process for these up-conversion and up-conversion-pumped emission.

1. Introduction

Tm^{3+} ions doped in ionic crystals show photoluminescence in red-infrared region. Of several luminescence bands, infrared luminescence bands at about 2000 nm and 1400 nm are interested because they are possible to use for medical applications because liquid water has a strong absorption band in such a spectral region [1-4], amplifier in the S^+-band optical communications [5, 6], and ground or space remote sensing for LIDAR and metrology [7]. Especially the 2000 nm laser has been performed using YVO_4 crystal [1, 3, 8-9]. The infrared Tm^{3+} laser has been also achieved using $LiNbO_3$ crystal [4, 10-12].

Although the luminescence properties of Tm^{3+} ions in YVO_4 and $LiNbO_3$ have been studied by several investigators [13-21], most studies have

171

J.-C. Krupa and N.A. Kulagin (eds.), Physics of Laser Crystals, 171–185.
© *2003 Kluwer Academic Publishers. Printed in the Netherlands.*

been made on the down-conversion where luminescence at lower frequency is produced by excitation light with higher frequency. Up-conversion where luminescence at higher frequency is produced by excitation light with lower frequency has been observed in Tm^{3+}-doped crystals, but few works have been performed on YVO_4. Blue and Red up-conversion Tm^{3+} lasers have been performed by pumping $BaYb_2F_8$ crystal and pumping fluorozirconate and fluoride fibers with diode lasers or infrared lasers [11,22-25].

In this paper we study the up-conversion and its pump power dependence under excitation of Tm^{3+}-doped YVO_4 and $LiNbO_3$ with a 798 nm laser diode. So far, Tm^{3+} up-conversion has been studied using relatively low power lasers of less than about 200 mW. High-power laser diodes are now commercially available. Here we use high power laser of a range from 0.1 W up to 15 W and study what happens in the up-conversion process when high power pumping is undertaken.

Two infrared Tm^{3+} emission bands have been observed at 1700-1950 and 1400-1500 nm in the infrared region for YVO_4 [13-16]. The former and latter emission bands are attributed to radiative electronic transition from the first and third lowest excited states $^3F_4$ and $^3H_4$, respectively. However, emission band due to the transition from the second lowest excited state $^3H_5$ has not been observed in both YVO_4 and $LiNbO_3$. We investigate if such an emission appears by pumping with high power laser and if the other emission bands appear in red-infrared region.

2. Experimental Procedure and Results

Single crystals of YVO_4 doped with Tm^{3+} ions (0.47 mol%) were grown using the floating zone method. On the other hand, single crystals of $LiNbO_3$ with Tm^{3+} (1.0 mol%) were grown using the Czochralski method and they were Z-cut and polished. Absorption spectra of the crystal were measured with Cary-5E and UV-3100 spectrophotometers. The spectral resolution was 0.2 nm. Crystals were excited using two kinds of cw AlGaAs laser diodes. One laser (called LD1) emits 788 nm radiation with maximum power of 50 mW, while the other one (called LD2) is a Spectra Diode Labs. SDL-3460-P6 quantum-well high power, fiber-coupled linear array laser diode. The LD2 laser emits 798 nm with maximum power of 15 W. Emission spectra were measured in 400-1750 nm spectral region with an Advantest Q8381A optical spectral analyzer containing a Si-photodiode. The spectral resolution was set to be 2 nm. Luminescence from crystal was introduced to the analyzer through a 2-m long optical fiber. The spectral analyzer has a limit of the measurement at 1750 nm.

In Fig.1 absorption spectra of Tm^{3+} ions in YVO_4 and $LiNbO_3$ crystals at 296 K are shown. The spectra are similar to those observed in various Tm^{3+}-doped crystals and glasses. Therefore the bands observed in spectral regions of 1600-1900, 1150-1250, 770-820, 680-710, 650-680, 465-480 and 360-365 nm are due to the electronic transitions from the $^3H_6$ ground state to the $^3F_4$, $^3H_5$, $^3H_4$, $^3F_3$, $^3F_2$, $^1G_4$ and $^1D_2$ states of Tm^{3+}, respectively [13,18-20].

Of these bands, the $^3F_2$ band is the weakest in both crystals, while the $^3H_4$ band has the highest peak in YVO$_4$ and the $^1G_4$ band has the highest peak in LiNbO$_3$. It is found that all the absorption bands are broader in YVO$_4$ than in LiNbO$_3$, and, unlike the case of YVO$_4$, each of these bands consists of several sharp bands in LiNbO$_3$. Sharp and intense rise due to the absorption edge of host is observed at about 356 nm in both crystals.

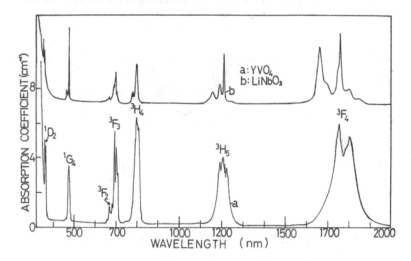

Figure1. Unpolarized absorption spectra of Tm^{3+}-doped YVO$_4$ crystal (a) and Tm^{3+}-doped LiNbO$_3$ crystal (b) at 296 K. The excited states of the absorption bands are indicated above each band. The baseline of the Tm^{3+}-doped LiNbO$_3$ spectrum is shifted to avoid the overlap with the Tm^{3+}-doped YVO$_4$ spectrum.

When YVO$_4$ crystal is excited with low power LD1 laser, three emission bands are observed in 600-1750 nm region at room temperature. The first band has a peak at 816 nm, the second band consists of several sharp lines which are located in the 1400-1510 nm region (this band is called 1450 nm band because its center is at about 1450 nm), and the third band has a peak at 1648 nm as shown in Fig.2. The latter band (called 1648 nm band) consists of two components with peak at 1648 and about 1715 nm. This band is the most intense of the three bands. These emission bands are quite similar to those observed in various Tm^{3+}-doped crystals. When crystal is excited with the LD2, four additional bands are observed at 435, 476, 700 and 1208 nm in YVO$_4$ (see Figs.2 and 3). The 435 nm emission band consists of two bands with peaks at about 432 and 438 nm, the 700 nm emission band consists of three bands with peaks at 693, 700 and 706 nm, and the1208 nm band consists of two bands with peaks at 1208 and 1224 nm. The 700 nm band is about 12 times bigger than the 1208 nm band.

When LiNbO$_3$ crystal is excited with the high power laser LD2, similar emission bands are observed in the red and near infrared regions as shown in

Fig.4; in addition to the emission bands observed by low power pumping, the emission bands corresponding to the 700 and 1227 nm bands of YVO_4 are observed in 680-725 and 1120-1250 nm regions, respectively. As the case of YVO_4, the latter band is much weaker than the former one. These bands consist of narrower bands than the case of YVO_4 as shown in inset of Fig.4: the former band (called 702 nm band, hereafter) consists of at least six bands with peaks at 685, 692, 697, 702, 710, and 718 nm, while the latter band (called 1216 nm band)

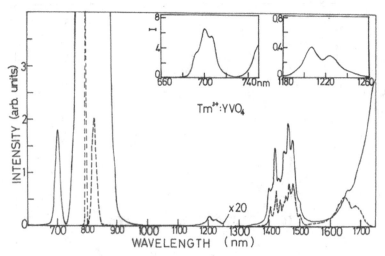

Figure2. Emission spectra of Tm^{3+}-doped YVO_4 crystal excited by 788 nm and 798 nm radiations of LD1 and LD2 lasers with power of 50 mW and 2W (broken and solid lines, respectively) at 296 K. The intense bands at 788 and 710-840 nm are the excitation laser radiations. In the spectrum obtained by the 798 nm excitation, the spectrum of 650-1250 nm is enlarged by 20 times. Inset shows the enlarged spectrum of the 700 and 1208 nm emission bands which were measured with a spectral resolution of 1 nm. I: emission intensity in arbitrary units

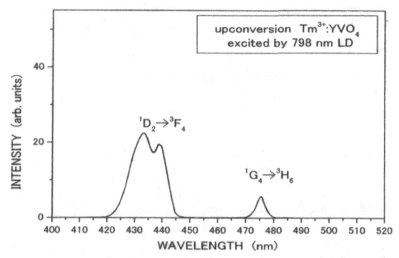

<figure>
Figure 3. Emission spectrum, in the 400-520 nm region, of Tm^{3+}doped
YVO$_4$ crystal excited with the 798 nm laser diode (power: 8 W) at 296 K.
</figure>

consists of three intense bands with sharp peaks at 1177, 1216 and 1234 nm
and three weak bands at 1135, 1162 and 1198 nm. This is consistent with the
result of absorption spectra where the absorption bands of LiNbO$_3$ are sharper
than those of YVO$_4$. Same 1216 nm emission band was obtained by the
excitation of 478 nm laser emitted from an optical parametric oscillator which
was pumped with the 355 nm third harmonic of Nd:YAG laser (see Fig.5). The
478 nm laser was used to excite the $^1G_4$ state directly. The 1216 nm emission
band is located close to the absorption band due to the $^3H_6 \rightarrow ^3H_5$ transition.

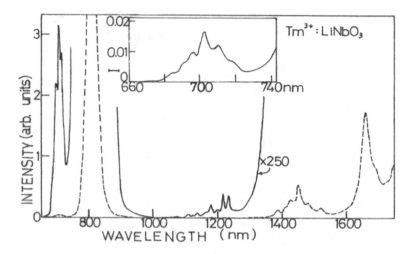

Figure 4. Emission spectra of Tm^{3+}-doped LiNbO$_3$ crystal observed by excitation with 798 nm radiations at 296 K. A 250-times enlarged spectrum is shown by solid curve. Inset shows the enlarged spectrum in 660-745 nm region. I: emission intensity in arbitrary units

We investigated the pump power dependence of these Tm^{3+} emission bands. The intensities of the 816, 1450 and 1648 nm bands of YVO$_4$ are linearly proportional to the pump power when pump power of the 798 nm laser is below about 100 mW. The 700 and 1208 nm bands, which are observed by excitation with high power laser, show different power dependence. Fig.6 shows the log-log plotted pump power dependence for YVO$_4$. Each of the two emission bands has two straight lines: one straight line has a slope of about 2 at lower pump powers of 0.6- 4 W and the other has a slope of about 3 at higher pump powers of 4-9 W. The change of slope appears at almost the same power (about 4 W) for both emission bands. Fig.7 shows the log-log plotted pump power dependence for LiNbO$_3$. For the 702 and 1216 nm emission bands, the intensities of the 702 and 1177.9 nm peak were selected, respectively. It is observed that, like the case of the 1208 nm emission band of YVO$_4$, both of the 702 and 1216 nm bands consist of two straight lines, one has a slope of about 2 at lower pump powers of 1-5 W and the other has a slope of about 3 at higher pump powers of 5-9 W. The change of

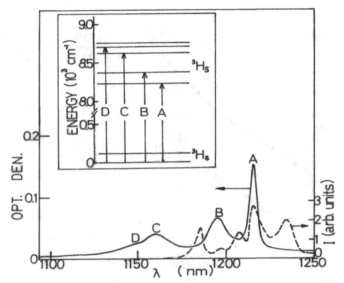

Figure 5. Emission spectrum, in the 1100-1250 nm region, of Tm³⁺doped LiNbO₃ crystal excited with 478 nm laser at 296 K, together with the $^3H_6 \rightarrow ^3H_5$ absorption band (the scale is shown at the left side). Inset shows the energy level assignment of the fine structure of the absorption band. OPT. DEN.: optical density, I: emission intensity in arbitrary units slope appears at almost the same power (about 5 W) for both emission bands

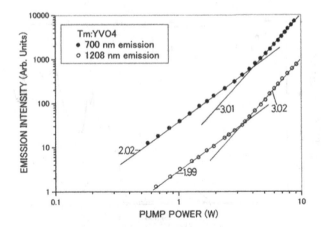

Figure 6. Intensities of 700 and 1208 nm emission of Tm³⁺doped YVO₄ plotted against the pump power of 798 nm high power laser diode. Straight lines are guided by eyes. Number (s) means the slope of the straight line.

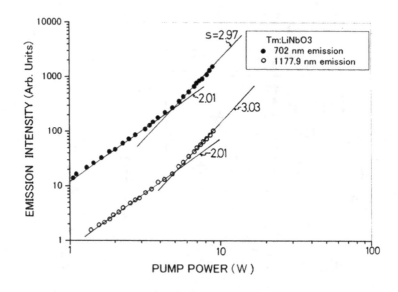

Figure 7. Intensities of 702 and 1177.9 nm emission of Tm^{3+}-doped LiNbO$_3$ plotted against the pump power of 798 nm high power laser diode. Straight lines are guided by eyes. Number means the slope (s) of the straight line.

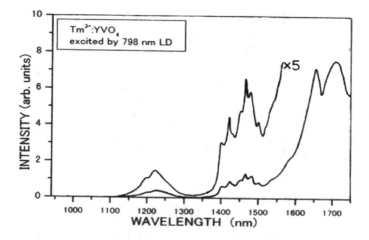

Figure 8. Emission spectra of Tm^{3+}-doped YVO$_4$ crystal excited with the 798 nm LD2 laser of 10 W at 296 K.

Fig.8 shows the emission spectrum of Tm^{3+} doped YVO$_4$ excited with the 298 nm laser of 10 W. From comparison of Fig.8 with Fig.2, we found that the emission band line shape of 1400-1520 nm region at 2 W is quite similar to the line shape at about 10 W, but line shape of the emission bands in 1600-1750 and 1150-1300 nm regions at low pump power is different from the line shape at high pump power. For example, two peaks are observed at 1655 and 1710 nm for the emission band in 1600-1750 nm region. Fig.9 shows the log-log plotted pump power dependence for 1466 and 1728 nm emissions. The two emissions are observed to have the same slope of 1.44. The 1466 and 1728 nm wavelengths were selected to investigate the power dependence of the emission bands appeared in the 1400-1520 nm and 1600-1750 nm regions.

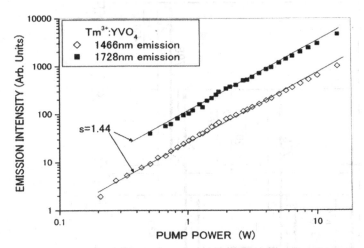

Figure 9. Intensities of 1466 and 1728 nm emission of Tm^{3+}doped YVO$_4$ plotted against the pump power of 798 nm laser diode at 296 K. Straight lines are guided by eyes. Number means the slope of the straight line.

3. Discussion

YVO$_4$ crystal is tetragonal belonging to space group I4$_1$/amd (D$_{4h}^{19}$). The dopant rare-earth ions are substituted for Y^{3+} ion sites. The local site symmetry is D$_{2d}$ and is surrounded by eight O^{2-} ions. LiNbO$_3$ crystal belongs to the space group R3c(C$_{3v}^6$). Trivalent rare-earth ions such as Nd^{3+} and Tm^{3+} are substituted for Li$^+$ ions in LiNbO$_3$ [26]. Li$^-$ ion at the octahedral site in LiNbO$_3$ is surrounded by six O^{2-} ions. When the trivalent rare-earth ion enters Li$^+$ octahedral site, the rare-earth ion does not have the O$_h$ site symmetry but it has a lower site symmetry because of difference of ionic radius between rare-earth and Li$^+$ ions and because of charge compensating vacancies located close to rare-earth ion. When we compare the Tm^{3+} absorption spectrum of LiNbO$_3$ with that of YVO$_4$, we found that each of the absorption bands consists of several sharp lines in LiNbO$_3$, while it consists of broad lines in YVO$_4$ and gives rise to broad absorption band. This indicates that although the local site symmetry of Tm^{3+} in $LiNbO_3$ is deformed from the octahedral one, its symmetry does not deviate from the octahedral largely and it is much higher than the D_{2d} symmetry of the case of YVO_4.

Energy level diagram of Tm^{3+}, which was estimated from the absorption spectrum of YVO_4, is given in Fig.10. The same diagram is also given for $LiNbO_3$.

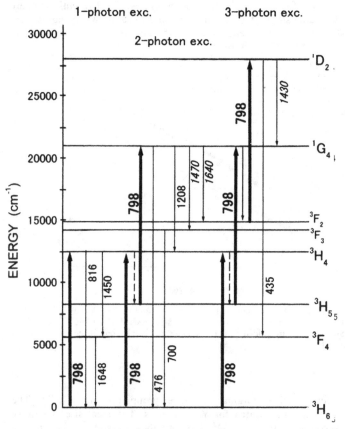

Figure 10. Energy level diagram of Tm^{3+} ion in YVO_4 crystal and down- and up-conversion luminescence processes under the one-, two- and three-photon excitations with 798 nm laser diode. In the two-photon process, the emission bands appeared in the one-photon process are also observed, but they are not shown in this figure. The same is true for the three-photon process. Broken line indicates the nonradiative multiphonon relaxation. Number 798 means the wavelengths of the excitation light, while numbers 816, 1450, etc. mean the wavelength at the center of each luminescence band. Italic number such as *1470* means a peak wavelength of emission band expected by the $^1G_4 \rightarrow {}^3F_3$ transition although the emission band is not clearly observed experimentally because of overlapping with the other emission bands

As the cases of various Tm^{3+}-doped crystals and glasses, the emission bands observed in 1400-1510 and 1600-1750 nm regions in both crystals are attributed to the $^3H_4 \rightarrow ^3F_4$ and $^3F_4 \rightarrow ^3H_6$ transitions, respectively, and the 816 nm emission band is attributed to the $^3H_4 \rightarrow ^3H_6$ transition as shown in Fig.10. Emission band due to the $^3F_4 \rightarrow ^3H_6$ transition has a peak in a region of 1800-1900 nm in most Tm^{3+}-doped crystals and glasses. Our spectrophotometer has a measurement limit at 1750 nm. The bands shown in Fig.2 are the high energy components of the $^3F_4 \rightarrow ^3H_6$ emission band. The 435 and 476 nm up-converted emission bands are attributed to the $^1D_2 \rightarrow ^3F_4$ and $^1G_4 \rightarrow ^3H_6$ transitions, respectively. The up-converted $^1D_2 \rightarrow ^3H_6$ emission band is expected to appear at about 360 nm by the three-photon excitation, but it is not observed by the present study because of a sensitivity limit of our spectral analyzer at the wavelength region below 400 nm.

As seen in Fig.5, the 1216 nm emission band of $LiNbO_3$ and the 1208 nm emission band of YVO_4 are located close to the absorption band due to the $^3H_6 \rightarrow ^3H_5$ transition. Thus one might attribute these emission bands to the $^3H_5 \rightarrow ^3H_6$ transition, which occurs after the electrons excited to the $^3H_4$ state by the 798 nm pumping laser are non-radiatively relaxed to the $^3H_5$ state. In fact the $^3H_5 \rightarrow ^3H_6$ emission band is observed at 1230 nm in YCl_3 crystal [27], Ga_2S_3-GeS_2-La_2S_3 glasses [28], and $LaCl_3$ [29] by excitation into the $^3H_4$ state with 800 or 794 nm laser. However, such a down-converted $^3H_5 \rightarrow ^3H_6$ emission has not been observed by excitation into the $^3H_4$ or $^3F_3$ state in various materials including $LiNbO_3$ and YVO_4. Nunez et al tried to observe the $^3H_5 \rightarrow ^3H_6$ emission by excitation into the $^3H_4$ state with 800 nm laser, but they could not observe it [19]. We irradiated 1150 nm He-Ne laser (of 5 mW) to the crystal to excite the $^3H_5$ state directly, but we observed only the 1648 nm $^3F_4 \rightarrow ^3H_6$ emission band and we could not observe the 1208 nm emission band. Therefore we have to abandon the assignment of $^3H_5 \rightarrow ^3H_6$ transition for the 1216 nm emission band of $LiNbO_3$ and the 1208 nm emission band of YVO_4.

The maximum phonon energy of YVO_4 host crystal is 890 cm^{-1} [30] or 980 cm^{-1} [17]. The phonon energy of $LiNbO_3$ is 880 cm^{-1} [19]. Taking into account that the $^3H_5$-$^3F_4$ energy gap is about 2600 cm^{-1}, this gap can be bridged by three or four phonons, indicating that the probability of nonradiative multiphonon $^3H_5 \rightarrow ^3F_4$ relaxation is much higher than the probability of radiative $^3H_5 \rightarrow ^3H_6$ transition. On the other hand, the radiative $^3H_5 \rightarrow ^3H_6$ transition is possible in YCl_3, Ga_2S_3-GeS_2-La_2S_3 and $LaCl_3$ because they have much low phonon energy which makes non-radiative multi-phonon $^3H_5 \rightarrow ^3F_4$ relaxation difficult.

Neither the 1216 nm emission band of $LiNbO_3$ nor the 1208 nm emission band of YVO_4 shows linear pump-power dependence but shows the non-linear quadratic and cubic dependences. This result indicates that these emission bands are not caused by one-photon excitation but by two- or three-photon excitation. From the energy level diagram, the 1208 nm emission band is attributable to the $^1G_4 \rightarrow {}^3H_4$ transition and the 700 nm emission band to the $^3F_3 \rightarrow {}^3H_6$ transition (see Fig.10). We observed the quadratic and cubic pump power dependences for the 700 and 1208 nm emission bands as shown in Fig.6. This indicates that these emission bands are caused by two- and three-photon excitations at relatively lower and higher pump powers, respectively, as mentioned in the following paragraph.

The absorption of 798 nm light gives rise to populate the $^3H_4$ excited state followed by fast nonradiative relaxation to the $^3H_5$ state. Then a second 798 nm photon is absorbed, bringing the Tm^{3+} ion from the populated $^3H_5$ state to the $^1G_4$ excited state   (i.e. excited state absorption) from which 1208 nm emission takes place and at the same time from which radiative transition to the $^3F_2$ and $^3F_3$ states followed by 700 nm emission takes place. Thus, the two-step excitation (i.e. two photon process) gives rise to 700 and 1208 nm emission. At the same time, it is conceivable that the two-step excitation also gives rise to the 1450 and 1648 nm emission bands. At higher pump powers above 4W, it is suggested that a third 798 nm photon is absorbed   to give rise to transition from the $^3F_2$ state to the $^1D_2$ state followed by radiative relaxation to the $^1G_4$ state, resulting in the 700 and 1208 nm emission bands. In this case these emission bands are produced by three-step excitation process (i.e. three-photon process). The 702 and 1216 nm emission bands of $LiNbO_3$ correspond to the 700 and 1208 nm emission bands in YVO_4, respectively, thus the same excitation and relaxation processes are given for Tm^{3+} in $LiNbO_3$.

The 700 nm emission of YVO_4 and the 702 nm emission of $LiNbO_3$ are called up-conversion because they are produced by the excitation with lower photon energy. According to the definition of up-conversion, the 700 and 702 nm emissions are up-conversion. However, the 1208 nm emission of YVO_4 and the 1216 nm emission of $LiNbO_3$ are not up-conversion although they are also produced multi-photon process, because their photon energies are lower than the excitation energy. They are called up-conversion-pumped emission [10,31]. As shown in Fig.10, the up-conversion-pumped emission is emission from energy level which lies above the energy level excited with pumping radiation and its wavelength is longer than the wavelength of pump radiation.

As shown in Fig.9, the intensity of the 1466 emission does not exhibit linear power dependence when   it is produced by high power laser but   it exhibits nonlinear power dependence (i.e. 1.44-power dependence). The same is observed for the 1728 nm emission. This is explained as follows. Emission near 1466 nm is due to the $^3H_4 \rightarrow {}^3H_6$ transition. This $^3H_4 \rightarrow {}^3H_6$ emission appears by not only the one-photon excitation with 798 nm radiation but also two- and three-photon excitation with increasing pump power. Emission by the former process is much stronger than the emission by the latter processes since the number of electrons relaxed to the $^3H_4$ state is much larger in the former process than   in the latter ones. The contribution of the latter processes is not so large, therefore quadratic or cubic power dependence is not obtained, but an intermediate power dependence such as 1.44-power dependence is obtained. Regarding the 1728 nm emission due to the $^3F_4 \rightarrow {}^3H_6$ transition, the same is also true.

We observed that the line shape of the down-converted emission bands around 1700 and 1200 nm at low pump power below 2 W is different from that at high power above 8 W, but no difference is observed for the 1450 nm emission band (Fig.8). Two reasons are suggested for this observation. One reason (called Reason A) is an overlap with the other emission bands. The transition from the $^1G_4$ state to the $^3F_2$ state gives rise to an emission band around 1640 nm as shown in Fig.6, and it overlaps with the 1640 nm $^3F_4 \rightarrow ^3H_6$ emission band. The $^1G_4 \rightarrow ^3F_3$ and $^1D_2 \rightarrow ^1G_4$ transitions give rise to emission bands around 1470 and 1530 nm, respectively, and they overlap with the 1450 nm $^3H_4 \rightarrow ^3H_6$ emission band. Emission band which overlaps with the 1208 nm $^1G_4 \rightarrow ^3H_4$ emission band is not expected from the energy diagram. Therefore the change of the emission line shape around 1700 and 1450 nm is expected but no change is expected for the line shape around 1200 nm. This does not agree with the observation, indicating the Reason A is not acceptable. The other reason (called Reason B) is a self-absorption by overlap of emission band with absorption band. The 1640 nm $^3F_4 \rightarrow ^3H_6$ emission is re-absorbed by the $^3H_6 \rightarrow ^3F_4$ transition which gives rise to absorption band in 1600-1900 nm region, and the 1208 nm $^1G_4 \rightarrow ^3H_4$ emission is absorbed by the $^3H_6 \rightarrow ^3H_4$ transition which gives rise to absorption band in 1150-1250 nm region (Fig.5). Such a self-absorption would be stronger at low pump power than at high power. As a result, the emission bands around 1700 and 1200 nm are expected to show a different line shape depending on pump power. On the other hand, the 1450 nm $^3H_4 \rightarrow ^3H_6$ emission is not re-absorbed, resulting in no change in the line shape around 1450 nm. Therefore the Reason B is acceptable to explain the observed result.

4. Acknowledgments

We thank Mr. Wang Xiaoming for kindly supplying Tm-doped YVO$_4$ crystals which were used in the present paper. The present work is partially supported by the Grant-in-Aid for Scientific Research (C) from the Japan Society for Promotion of Science.

5. References

1. Hauglie-Hanssen, G. and Djeu N. (1994) Further investigations of a 2 μm Tm^{3+}:YVO$_4$ laser, *IEEE J. Quant. Electr.* **30**, 275-279.
2. Bourdet, G.L. and Lescroart, G. (1998) Theoretical modeling and design of a Tm^{3+}:YVO$_4$ microchip laser, *Opt. Commun.* **149**, 404-414.
3. Bourdet, G.L. (2000) Gain and absorption saturation coupling in end pumped Tm^{3+}:YVO$_4$ and Tm^{3+},Ho:YLF CW amplification, *Opt. Commun.* **173**, 333-340.
4. Stoneman, R.C. and Esterowitz, L. (1995) Efficient 1.94 μm Tm^{3+}:YALO laser, *IEEE J. Selected Topics Quant. Electr.* **1**, 78-81.
5. Tanabe, S., Feng, X. and Hanada, T. (2000) Improved emission of Tm-doped glass for a 1.4-μm amplifier by radiative energy transfer between Tm^{3+} and Nd^{3+}, *Opt. Lett.* **25**, 817-819.
6. Hayashi, H., Tanabe, S. and Hanada, T. (2001) 1.4 μm band emission properties of Tm^{3+} ions in transparent glass ceramics containing PbF$_2$ nanocrystals for S-band amplifier, *J. Appl. Phys.* **89**, 1041-1045.
7. Bourdet, G.L., Lescroart, G. and Muller, R. (1998) Spectral characteristics of 2 μm

microchip Tm^{3+}:YVO_4 and Tm,Ho:YLF lasers, *Opt. Commun.* **150**, 141-146.

8. Saito, H., Chaddha, S., Chang, R.S.F. and Djeu, N. (1992) Efficient 1.94 μm Tm^{3+} laser in YVO_4 host, *Opt. Lett.* **17**, 189-191.

9. Zayhowski, J.J., Harrison, J., Dill C. and Ochoa, J. (1995) Tm:YVO_4 microchip laser, *Appl. Opt.* **34**, 435-437.

10. Allain, J.Y., Monerie, M. and Poignant, H. (1990) Blue upconversion fluorozirconate fiber laser, *Electr. Lett.* **26**, 166-168.

11. Antipenko, B.M., Voronin, S.P. and Privalova, T.A. (1990) New laser channels of the Tm^{3+} ion, *Opt. Spectrosc.* **68**, 164-166.

12. Caird, J.A., DeShazer, L.G. and Nella, J. (1975) Characteristics of room-temperature 2.3 μm laser emission from Tm^{3+} in YAG and $YalO_3$, *IEEE J. Quant. Electr.* **11**, 874-880.

13. Wu, M.Q., Yang, W.Q., Chen, J.K. and Zhuang, J. (1998) Absorption spectrum and parameters calculation of Tm^{3+} in YVO_4 crystal, *SPIE* **3549**, 68-73.

14. Huang, L.L., Yie, L.H. and Shen, W.Z. (1997) Spectral characteristic parameters of Tm^{3+} in YVO_4 crystal (in Chinese), *J. Zhejiang Univ.* **31**, 147-153.

15. Ermeneux, F.S., Goutaudier, C., Moncorge, R., Cohen-Adad, M.T., Bettinelli, M. and Cavalli, E. (1997) Growth and fluorescence properties of Tm^{3+} in YVO_4 and Y_2O_3 single crystals, *Opt. Commun.* **8**, 83-90.

16. Ohta, K., Saito, H. and Obara, M. (1993) Spectroscopic characterization of Tm^{3+}:YVO_4 crystal as an efficient diode pumped laser source near 2000 nm, *J. Appl. Phys.* **73**, 3149-3152.

17. Ermeneux, F.S., Goutaudier, C., Moncorge, R., Sun, Y., Cone, R.L., Zannoni, E. and Cavalli, E. (2000) Multiphonon relaxation in YVO_4 single crystal , *Phys.Rev.B* **61**, 3915-3921.

18. Wortman, D.F., Leavitt, R.P. and Morrison, C.A. (1974) Analysis of the ground configuration of Tm^{3+} in YVO_4, *J. Phys. Chem. Solids* **35**, 591-593.

19. Nunez, L., Tocho, J.O., Sanz-Carcia, J.A., Rodriguez, E., Cusso, F., Hanna, D.C., Tropper, A.C. and Large, A.C. (1993) Optical absorption and luminescence of Tm^{3+}-doped $LiNbO_3$ and $LiNbO_3$ (MgO) crystals, *J. Lumin.* **55**, 253-263.

20. Johnson, L.F. and Ballman, A.A. (1969) Coherent emission from rare earth ions in electrooptic crystals, *J. Appl. Phys.* **40**, 297-302.

21. Tsuboi, T. (2000) Luminescence of Tm^{3+} ion in $LiNbO_3$ crystal *J. Electrochem. Soc.* 147, 1997-2001.

22. Grubb, S.G., Benner, K.W., Cannon, R.S. and Hummer, W.F. (1992) CW room emperature blue upconversion fibre laser, *Electron. Lett.* **28**, 1243-1245.

23. Sanders, S., Waarts, R.G, Mehuys, D.G. and Welch, D.F. (1995) Laser diode pumped 106 mW blue upconversion fiber laser, *Appl. Phys. Lett.* **67**, 1815-1817.

24. Paschotta, P., Moore, N., Clarkson, W.A., Topper, A.C., Hanna, D.V. and Maaze, G. (1997) 230 mW of blue light from a thulium-doped upconversion fiber laser, *IEEE J. Selected Topics Quant. Electr.* **3**, 1100-1111.

25. McAleavey, F.J., O'Gorman, J., Donegan, J.F., MacCraith, B.D., Hegarty, J. and Maze, G. (1997) Narrow linewidth, tunable Tm^{3+}-doped fluoride fiber laser for optical based hydrocarbon gas sening, *IEEE J. Selected Topics Quant. Electr.* **3**, 1103-1111.

26. Lorenzo, A., Loro, H., Munoz Santiuste, J.E., Terrile, M.C., Boulon, G., Bausa, L.E. and Garcia, Sole J. (1997) RBS/channeling to locate active ions in laser materials: application to rare-earth activated $LiNbO_3$, *Opt. Materials* **85**,55-63.

27. Ganem, J., Schmidt, P. and Bowman, S.R.(2000) Cross relaxation in a Tm^{3+}doped low-phonon energy laser material, in H. Injeyan, U. Keller and C. Marshal (eds.), *Advanced Solid State Lasers*, OSA TOPS 34, pp. 536-541.

28. Kadono, K., Yazawa, T., Shojiya, M. and Kawamoto, Y. (2000) Judd-Ofelt analysis and luminescence property of Tm^{3+} in Ga_2S_3-GeS_2 glasses, *J. Non-Crystal. Solids* **274**, 75-80.

29. Soga, K., Wang, W.Z., Feng, X., Riman, R.E., Brown, J.B. and Mikeska, K.R. (2001) Optical properties of rare-earth doped lanthanum chloride prepared by reactive atmosphere processing, Report at the 2001 annual meeting, Basic Research Division of the American Ceramic Society

(Indianapolis,   USA),BSD-8P;
http://www002.upp.so-net.ne.jp/ksoga/slideshow/acers2001.pdf.

30. Capobianco, J.A., Kabro, P., Ermeneux, F.S., Moncorge, R., Bettinelli, M. and Cavalli, E. (1997) Optical spectroscopy, fluorescence dynamics and crystal-field analysis of Er^{3+} in YVO_4, *Chem. Phys.* **214,** 329-340.

31. Pollack, S.A., Chang, D.B. and Moise, N.L. (1986) Up-conversion-pumped infrared erbium laser, *J. Appl. Phys.* **60,** 4077-4086.

VALENCE STATE STABILITY IN SrTiO₃ DOPED WITH ME/RE-IONS

N.A. KULAGIN[1], J. DOJCILOVIC[2], D. POPOVIC[2]

[1]*Kharkiv National University for Radioelectronics, av. Shakespeare 6-48, 61045 Kharkiv, Ukraine. E-mail: nkulagin@bestnet.kharkov ua.*
[2]*Physics Department, University of Belgrade, Academski Str. 12-16, 11000 Beograd, SRY*

Abstract. A study of the electronic structure, dielectric and spectral parameters of pure and doped strontium titanate single crystals at temperature ranging from 10 to 300 K is presented. Change in the oxidation state of Ti^{+4} ions toward Ti^{+3} as well as change in the compound stoichiometry and formation of clusters with Ti^{+3} and Me^{+3} ions lead to different properties of SrTiO₃ crystals. The charge transfer, the tunnelling current, the change of the oxidation states and anomalies in spectral properties for the crystals doped with Me and/or RE ions are discussed in detail.

Keywords: electronic state, dielectric constant, temperature dependence, strontium titanate.

1. Introduction

Strontium titanate single crystal, SrTiO₃, has been supporting a steady interest for a long time. This crystal presents the perovskites structure ABO_3 [1-4]. According to the experimental data already published, several phase transitions (PT), at temperatures: 10, 30, 65 and 105 K were evidenced in SrTiO₃ [1-6]. It is well known that the structural PT at T = (105 ± 2) K is a displacive-type one and below T = 105 K, a change of the crystalline symmetry, from cubic to tetrahedral, is observed. The PT is triggered by the rotation of the TiO_6 oxide octahedra with the Ti^{+4} ion in the centre of the grouping.

It should be noted, that the ferroelectric or anti-ferroelectric properties of this crystal [1, 2] are still not well-understood, but the temperature dependence of the dielectric constant, ε_0, is described by the Curie-Weiss law in the first approximation [1-3]:

J.-C. Krupa and N.A. Kulagin (eds.), Physics of Laser Crystals, 187–199.
© 2003 *Kluwer Academic Publishers. Printed in the Netherlands.*

$$\varepsilon_0(T) = \frac{C}{T - T_C}, \tag{1}$$

where T_c - is the PT temperature and C, is the Curie constant.

Up to now, different parameter values for $SrTiO_3$ were reported in literature. For instance, the crystalline lattice parameter was measured at $a = 3.9051$ Å and $a = 3.9002$ Å [7-8]. In the same way, the dielectric constant value varies from 200 to 360 units [5, 7, 9] and anomalous properties for different $SrTiO_3$ samples such as photochromism and dichroism were also evidenced [10-13].

A detailed investigation of the structure of $SrTiO_3$ reveals distortion of the perovskite structure above and below T_c [4, 14]. The structure variation of the investigated samples can result in different parameter values characterizing the crystals. The crystal growing conditions are certainly responsible of the anomalous dielectric and spectral behaviour of $SrTiO_3$ [8, 15-16] because of changes in stoichiometry and oxidation state of Ti^{+n} ions. The oxidation state of Ti^{+n} ions in pure and doped strontium titanate grown by different methods have been reported in [17-18]. Interesting results were obtained for $SrTiO_3$ single crystal presenting macro-defects, the so-called "blue colour" crystal, where 20 % of titanium ions have changed oxidation state [8].

It should be noticed that variations in the Sr/Ti ratio were also observed in the range of 5 to 20 % for the pure and doped crystals respectively. We propose that the presence of Ti^{+3} ions can give rise to a local electronic level in the forbidden band and consequently leads to the appearance of conductivity in the crystal. Results presented in [9, 18] show once more that distortion in the crystalline lattice can induce unusual properties in $SrTiO_3$.

ESR investigations of Me ions: Fe, Cr, Co, Ni and structural defects in $SrTiO_3$ single crystals were conducted thoroughly by K.Muller and co-workers [18-19]. In [21-22] abnormal temperature dependence of ESR spectra for definite $SrTiO_3$ single crystals was studied. For samples with abnormal ESR behaviour, the reference signal at $T = (110 \pm 3)$ K is firstly vanishing and it appears again after 600-1200 s of cooling at 77 K. It should be noted that this phenomenon could not be observed in the same sample when the experiment is repeated.

Other optical properties, such as dichroism and birefringence also vary after cooling the doped crystals at 77 K during 600 s [22-23]. Details of the temperature dependence on the optical and dielectric properties of $SrTiO_3$ were reported in [24].

Since results for ESR, dielectric and optical parameters in strontium titanate are contradictory, we have carried out a new study of the properties of pure and doped $SrTiO_3$ single crystals with the full control of i) the stoichiometry, ii) the crystallographic parameters, iii) the element concentration and iv) the oxidation state of the regular and doping ions. The temperature dependences of ε_0 and $\tan\delta$ for $SrTiO_3$ crystals doped with ions of the iron group elements, Me ions, and/or lanthanide group, were investigated in the temperature range:10-300 K.

The results reported in [25] suggest that the substitution Me^{+3} / Ti^{+4} in the doped crystals (and the related change of stoichiometry) results in the appearance of a large amount of oxygen vacancies. For investigation of the oxidation state of the ions and the crystal stoichiometry, the valence shift of the X-ray lines were used [5, 25].

2. Samples and Techniques.

The experimental techniques were already published in [22-23]. Pure, un-doped (but containing uncontrolled impurities) and doped $SrTiO_3$ single crystals were grown by the Verneuil technique. ESR experiments were performed on a Varian spectrometer with $\lambda = 3$ cm and 8 mm. The concentration of the paramagnetic ions into the samples was less than 10^{-5} wt % and the total concentration of other impurities was less than 10^{-4} wt %.

Pure crystals are optically transparent from 395 nm to higher wavelengths (see Fig.1). and the value of ε_0 is near 360 units. Un-doped crystals are containing Fe^{+3}, Cr^{+3} and other ions with a concentration around 10^{-4} to 10^{-3} wt %. The following uncontrolled dopants were found in the samples by spectral investigation: Ca, Mg, Si, Al etc. with a concentration around 10^{-3} wt %. The ratio between strontium and titanium is: 1:1.2 and the value of ε_0 for the un-doped samples varies from 250 to 300 units. Optical spectra of most of the un-doped samples contain bands at 430 and 520 nm.

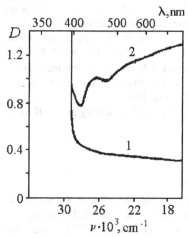

Figure 1. Oprical absorption spectra of pure $SrTiO_3$ (1) and "blue" one (2)

The optical and luminescence properties of the un-doped "blue color" crystals with intense blue color, were studied in [15-16]. Value of ε_0 for "blue crystals" is (120 ± 5) units and the optical spectra consist in very intense optical bands at 430 nm and 520 nm.

Strontium titanate samples doped with V, Cr, Mn, Fe, Co and Ni ions were grown by the standard method. The concentration of the Me ions into the crystals varies from 10^{-2} to 10^{-1} wt %. For example, $SrTiO_3$:Mn single crystals showed 0.05 - 0.1 wt % of Mn^{+n} ions. In these samples Fe^{+3} and Cr^{+3} ions were also present at concentrations: C_{Fe} - 10^{-3} wt% and C_{Cr} - 10^{-4} wt % as uncontrolled impurities. The total amount of the impurities was less than 10^{-2} wt %. Crystals doped with Cr, Co, Ni ions contain similar concentration of controlled and uncontrolled impurities. Value of ε_0 for the doped samples is ranging from 200 to 280 units.

$SrTiO_3$:V:Fe samples contained the following doping ions: 0.05 to 0.1 wt % of Fe^{+3} ions (ESR data) and 0.08 to 0.1 wt % of V ions (optical data). The optical bands due to impurities at 430 and 520 nm were studied for all the doped samples [15, 25]. The latter samples showed an abnormal temperature dependence in ESR experiment and non-linear optical properties. Value of ε_0 for the doped samples is 200 – 280 units.

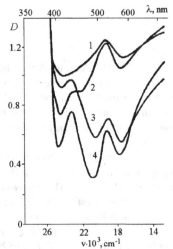

Figure 2. Optical absorption spectra of SrTiO$_3$ doped with Sm (1), Pr (2), Nd (3) and Tm (4)

Crystals doped with rare earth ions: Pr, Nd, Sm, Tm contain RE impurities with concentration 10^{-2} to 10^{-3} wt % and uncontrolled ones: Ca, Mg, Cr and Fe with concentration of 10^{-4} to 10^{-3} wt %. Optical absorption spectra are presented in figure 2. It should be noticed that the optical spectra for the crystals doped with Pr, Nd and Tm are similar to the ones recorded for the "blue color" crystals: intense optical absorption bands at 430 and 520 nm but the optical absorption spectra of crystals doped with Sm contains the second band, only.

Two sets of SrTiO$_3$ single crystals doped with Cr, Mn, Fe and V impurities with concentration C = $4*10^{-3}$- $5*10^{-2}$ wt % were studied. For the first set no peculiarities were observed in ESR or optical spectra but for the second one (SrTiO$_3$:V:Fe), the spectroscopic parameters, the temperature dependence of ESR and the optical data were different [23-24]. The dielectric constant, ε_0, of the samples at 300 K varied from 120 to 360 units and a change in the crystallographic parameter, a, was also observed: For example, a = 3.9064 Å for the pure crystal and a = 3.9002 Å for the "blue crystal" [12-13, 24].

In [21-22] was studied the strong influence of a "soft" annealing at 600 K in air during 1800 - 3600 s.

3. Results of the Spectral Investigations.

The optical absorption spectra of pure or doped with Me and RE ions SrTiO$_3$ single crystals were studied in [15-18]. The Me ions with unfilled 3d-shell occupy Ti^{+4} sites. The main oxidation state of V, Cr and Fe ions, according to [26] is Me^{+3} but for the ions at the end of the iron group, Co and Ni for example, Me^{+2} oxidation state is observed. Figure 3 shows the well known optical absorption spectrum of Ni^{+2} ions in octahedral position. The optical absorption spectra of SrTiO$_3$ samples doped with Me ions correspond to the standard spectra of Me^{+3} or Me^{+2} ions in oxides [26]. The shift of the band-edge in the doped crystals is due to a charge-transfer process [8]. It should be noted that the formation of Me^{+3} and Me^{+2} oxidation state induces a large amount of oxygen vacancies.

Me^{+4} ions were observed by optical and ESR techniques after a high-temperature treatment in oxygen atmosphere.

Doped RE ions with unfilled 4f-shell in $SrTiO_3$ crystals occupy Sr^{+2} sites. Different ions, such as Sm or Tm, may be introduced as RE^{+2}. It should be noticed that the ionic radii of RE^{+3} ions (0.82 – 0.97 Å) differ from Sr^{+2} one (1.12 Å) [27] and that may be the reason of the small concentration of these ions embedded in the crystals. In contrary, the ionic radius of Sm^{+2} is similar to Sr^{+2} one (1.13 Å). The optical spectra of $SrTiO_3$ crystals doped with RE (Fig. 2) show bands at 430 and 520 nm but for Sm, the first band is absent as was already mentioned. The intense bands at 430 and 520 nm prevent the study of the weak f-f transitions in rare earth ions. General data for the samples studied are listed in Table 1.

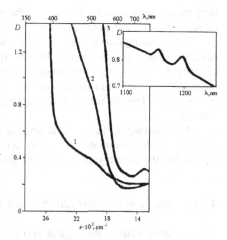

Figure 3. Optical absorption spectra of $SrTiO_3$:Ni with different concentration of Ni.

TABLE 1. Properties of $SrTiO_3$ single crystals

Crystal\ Properties	$C(Me^{+3})$	ε	Optical absorption, nm,	Luminescence	C^* 10^4 K	$C(Ti^{+3})$ %
$SrTiO_3$- standard	10^{-5}	360	-	-	9.9	-
$SrTiO_3$ – un-doped	10^{-3}	280	430,520	450	15.0	5±2
$SrTiO_3$ -blue	10^{-3}	120	430,520	450	-	30±6
$SrTiO_3$:Mn	10^{-2}	220	430,520	-	33.0	20±7
$SrTiO_3$:Fe:V	10^{-2}	220	430,520, 620	-	45.6	20±7

The optical absorption bands at 430 and 520 nm were observed for all the un-doped and doped crystals [8] but with different intensity. The intensity of the first band is very high in the "blue crystal". An extra optical absorption band with a maximum at 620 nm, was observed in the second set of samples. The intensity of this optical line first increases during 1800 s of cooling at 77 K but starts to decrease [1] after 3600 s [22-23] to reach a minimum value for the optical absorption coefficient equal to $k = 1.35*10^{-5}$ mm at 620 nm

ESR spectra of Me ions, Cr^{+3}, Fe^{+3} and $Mn^{+2, +4}$ were studied at 77 - 300 K. Cr^{+3} ions with $3d^3$ configuration in $SrTiO_3$ and Mn^{+2} with $3d^5$ configuration occupy the ortho-rhombohedric distorted Ti- sites in the crystalline lattice. The Mn^{+4} ions with $3d^3$ configuration substitute for Ti ions in the cubic positions as well as Fe^{+3} ions with $3d^5$ configuration. The defect: $Fe^{+3} - V_{O^{-2}}$ (Fe^{+3} ion with neighbouring oxygen vacancy) has an axial symmetry according to K. Muller et al. [19-20].

The temperature dependence of ESR spectra at T= 300 - 77 K for the first set of doped crystals is similar to the results reported by K. Muller [19]. For these crystals, ESR signal is vanishing during 30 - 60 s when approaching T_c. Below PT temperature, the classic ESR spectra of the doping ions in the rhombohedric phase of strontium titanate [19] was observed.

For doped crystals of the second set, we have observed the following anomaly (which is similar to data reported in [22-23]): ESR spectra vanish at T = (110 ± 10) K during 600-900 s. This phenomenon does not depend on the impurities in a wide range of concentration of the doping ions. We did not observe this effect in the pure crystals and in crystals with a small concentration of uncontrolled impurities (C < 10^{-5} wt %).

After cooling of the doped crystals at 77 K during 600 - 1200 s, the spectra of impurity in the rhombohedric environment appears. When the experiment was repeated on the same crystal, only the classic picture was observed

4. Investigation of the Oxidation State of Ti^{+n} - Ions.

The oxidation state of the regular and doping ions plays an important role in the optical and dielectric properties of strontium titanate. Oxidation states of Ti and Sr ions were studied by using the valence shift of X-ray lines (VSXRL method) described in detail in books [5, 25] and papers [17-18]. Moreover, VSXRL method was also used for estimation of the sample stoichiometry. In short, VSXRL method is based on the dependence of K, L, M,...etc absorption-edge energy upon the ion oxidation state[25]. The theoretical dependence of energy E_x on the valence of the ion n is the following [25]:

$$E_x = a_0 + a_1*n + a_2*n^2 + a_3*n^3,$$

$$(2)$$

where $a_0 > a_1 > a_2 > a_3$.

The valence shift (VS) of X ray line: $\Delta E_x = E_x^n - E_x^{n-1}$ is determined by the difference in energy of X ray absorption-edge between ions with oxidation state "n" and "$n-1$". At the first step, the dependence of ΔE_x may be represented by a square function of n. For electronic transitions from Me^n to Me^{n-1} in Me and RE ions, the energy of the X-ray edge increases by 1eV, approximately [5]. For example, the theoretical energy value for $TiK_{\alpha1}$-line, for Ti^{+3} ($3d^1$), for the $1s_{1/2}3d^1 > 2p_{3/2}3d^1$ electronic transition is $E(TiK_{\alpha1}) = 4508.80$ eV, and VS ($Ti^{+3} > Ti^{+4}$) = -0.86 eV [16]. The experimental value of $E(TiK_{\alpha1})$ is E = 4510.84 eV for metallic Ti, in good agreement with the experimental value reported in [28].

We have also determined the energy of $TiK_{\alpha1}$-line for Ti^{+3} ion via a detailed study of the spin-orbit interaction and calculation of the full electronic level scheme of $n'l'^{-1}nl^N$ configurations. By this way, the energy of $TiK_{\alpha1}$-line is: $E(Ti\ K_{\alpha1}) = 4509.81$ eV and VS ($Ti^{+3} > Ti^{+4}$) = 1.87 eV [15,16].

It should be noticed, that the energy of the X-ray lines of Me or RE-ions decreases when the oxidation state of the 3d- or 4f-ion increases. For ions with unfilled ns- or np- shell, the opposite dependence was observed [5]. The VS values for s- and p-ions are smaller than the ones for d- or f-ions.

Eq.(2) can be used for an unambiguous determination of the oxidation state of an ion in solids and the relative concentration of ions inducing an oxidation state change can be estimated by using the next relation [8]:

$$C(Me^{+n}) = \frac{\Delta E_{texper}}{\Delta E_{theor}},$$ (3)

where ΔE_{exp} - is the experimental VS value for the ion in the crystal and $\Delta E_{theor} = E_x^n - E_x^{n-1}$ is the theoretical one. For the calculation of $C(Me^{+n})$, a decomposition of the profile of $MeK_{\alpha 1}$-line into two or three Lorentz functions was used.

Use of Eq.(3) supposes that the half-width, Γ, of the $TiK_{\alpha 1}$-line for all the investigated samples is the same. The experimental error was estimated by comparing the half-width of the $CrK_{\alpha 1}$ line for ruby recorded with our set-up and with a double-crystals spectrometer leading respectively to: $\Gamma_1 = 2.05$ eV and $\Gamma_2 = 1.96$ eV [28]. Accordingly, the relative error $\rho = \Delta\Gamma/\Delta E_x$ and minimum error δ are: $\Delta\Gamma = 0.09$ eV, $\rho = 1.07 \cdot 10^{-5}$, and $\delta = 0.09$ eV. The half-width variation of $TiK_{\alpha 1}$-line in $SrTiO_3$ samples during the experiments was less than 5 %. It corresponds to the accuracy of our concentration measurements, in the first approximation.

According to Eq.(2), the energy of $TiK_{\alpha 1}$ line increases with the number of 3d-electrons. For Sr-ions we must observe a similar change of energy of $SrK_{\alpha 1}$-line along the valence of Sr-ions. The valence shift of $TiK_{\alpha 1}$-line and $SrK_{\alpha 1}$-line was determined with a

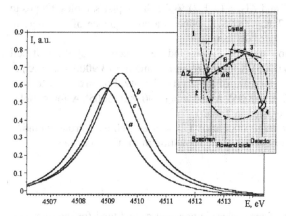

Figure 4. Profile of $TiK_{\alpha 1}$ line for pure (a) "blue color" (b) and $SrTiO_3$:V:Fe crystals (c) Insertion: Scheme of microanalysor: 1-gun, 2 - sample, 3 - crystal-analysor and 4 - counter.

micro-analyser. A simplified scheme of the device is shown in Fig.4 (insertion). An electron gun (1) provides 10 nA electron beam send on a 10 μm in diameter probe spot. (2). The emitted radiations from the excited ions are dispersed by the crystal-analyser (3) and the monochromatic radiation intensity is recorded by the counter (4). [17-18]. The pure $SrTiO_3$ crystal response was used as a reference in all experiments carried out The experimental recordings are shown on figure.4. Curves A and B correspond to the pure and "blue color" sample respectively. Curve C corresponds to the different doped crystals, for example $SrTiO_3$:V:Fe. A shift of $TiK_{\alpha 1}$-line for the "blue crystal" and doped samples toward higher energy region is easily observed. It corresponds to the transition of Ti^{+4} toward the Ti^{+3} oxidation state. Experimental profiles for $TiK_{\alpha1}$-lines were decomposed by two or three Lorentz functions. The maximum value of VS for $TiK_{\alpha1}$-line in the pure sample and "blue crystal" is: $\Delta E(TiK_{\alpha1}) = (0.56 \pm 0.11)$ eV. For $SrTiO_3$:Mn and $SrTiO_3$:V:Fe , VS was measured at: $\Delta E = (0.37 \pm 0.13)$ eV.

The study of the $SrTiO_3$:RE samples has been carried out in the same way. The shift of the $TiK_{\alpha1}$-line for $SrTiO_3$:Nd and $SrTiO_3$:Tm in comparison with the "blue crystal" is equal to $\Delta E = (0.35 \pm 0.12)$ eV. $SrTiO_3$:Sm sample provides similar data.

The effective number of 3d-electrons and the relative concentration of Ti^{+3} ions were determined via the experimental values of ΔE_{exp}. Eq.(3)leads to $C(Ti^{+3})$= (30 ± 6) % for the "blue" sample and C = (20 ± 7) % for the doped ones. Increase of Ti^{+3} concentration is accompanied by an increase of the $TiK_{\alpha1}$-line intensity up to ΔI_{Ti} = (12 ± 2) % for the "blue crystal' and ΔI_{Ti} = (8 ± 2) % for the doped samples.

A detailed investigation of VS of $TiK_{\alpha1}$-line was carried out by using a high resolution double-crystals X-ray spectrometer. This technique was described in [18]. The value of VS for "blue crystal" is $\Delta E = (0.45 \pm 0.08)$ eV.

It is worthy noticing that since any shift of the $SrK_{\alpha1}$ line was observed, the oxidation state of Sr ions remains the same in all samples. But the intensity of $SrK_{\alpha1}$ line for the "blue crystal" and the doped samples (mainly with RE) decreases: ΔI_{Sr} = (6 ± 2) % for "blue crystals" and ΔI_{Sr} = (4 ± 2) % for the doped samples. Opposite variation of the intensity of $TiK_{\alpha1}$ and $SrK_{\alpha1}$-line is a consequence of a change in Sr/Ti ratio and the stoichiometry of the samples.

Changes in the oxidation state of Ti-ions induce a decrease of the unit cell parameter a: $a = (3.9064 \pm 0.0005)$ Å for the reference and $a = (3.9002 \pm 0.0005)$ Å for the "blue" samples and an increase of the crystal density. For strontium titanate doped with Mn: $a = (3.9045 \pm 0.0005)$ Å and $a = (3.9038 \pm 0.0005)$ Å for samples doped with V and Fe. Data are listed in Table 1.

We may conclude that a large amount of Ti^{+3} ions as well as a strong variation of the stoichiometry are a consequence of the formation of $Sr_xTi_yO_z$ crystals.

5. Dielectric Investigations.

The temperature dependence of the dielectric constant ε_0 and tanδ for the pure, un-doped and doped crystals with Mn, Fe, and V ions on the one hand and Nd, Sm and Tm on the other hand has been investigated. For dielectric measurements a C-bridge 75D (1 MHz) apparatus was used and the temperature was controlled down to 10 K with a 22c-Kriodin (Lake Shore Cryotronics Inc.) device. The relative error attached to the dielectric constant value is 1 to 2 %. The accuracy of the temperature determination is 0.2 to 0.5 K. Measurements of ε_0 and tanδ for all samples were

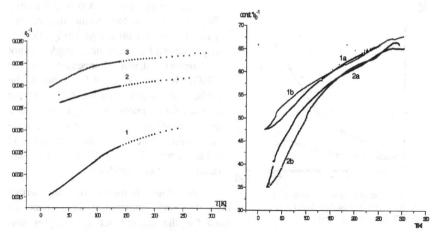

Figure 5. Temperature dependence of ε_0^{-1} for pure SrTiO$_3$ (1), SrTiO$_3$:Mn (2) and SrTiO$_3$:Fe:V (3) and. SrTiO$_3$:V:Fe samples:(curve 2a, cooling 2b, heating) after annealing.

Figure 6. Temperature dependence of const*ε_0 for SrTiO$_3$:Mn (curve 1a, cooling, 1b, heating)

performed between 300 and 10 K with a temperature scanning step of 1.0 K/min at high temperature and 0.1 K/min from 120 K down to 10 K. The experimental technique is described in [18, 24].

The temperature dependence of $\varepsilon_0(T)$ at T> T$_c$ is described by Eq.(1) with a relatively high accuracy. In figure 5, the dependence of the $\varepsilon_0^{-1}(T)$ for pure and doped SrTiO$_3$ samples in original oxidation state are shown. At room temperature, the values of ε_0 for different samples in original state differ very strongly: $\varepsilon_0 = 240$ - 360. At low temperatures (T< 77 K), a strong increase of $\varepsilon_0(T)$ was observed for the pure samples accordingly to the classic theory, but for the doped ones, only a "soft" growth of $\varepsilon_0(T)$ was recorded. Figure 5 also shows a break in $1/\varepsilon_0(T)$ variation for 1 MHz at T =107-112 K. for the pure samples (curve 1) and at T$_c$= (108± 2)K for SrTiO$_3$:V:Fe (curve 3). For SrTiO$_3$:Mn crystal (curve 2), T$_c$ is equal to (112 ± 4) K in agreement with Raman's spectroscopy data [11].. At T$_c$ = (109 ± 3) K, we have observed a change in ε_0 and tanδ temperature dependences which is related to the structural PT. The experimental data recorded for SrTiO$_3$:V:Fe shows peculiar points in the curve $1/\varepsilon_0(T)$ at the following temperatures: 190, 145, 90, 75, 60 and 35 K.

In figure 6, the temperature dependence on ε_0 in SrTiO$_3$:Mn (curve 1, a: cooling stage, b: heating) and for SrTiO$_3$:V:Fe samples:(curve 2, a: cooling, b: heating) after annealing is presented. The variation of $\varepsilon_0^{-1}(T)$ shows that for the first set of samples in the original oxidation state, the thermal hysterisis is relatively weak and does not change drastically (decrease of 8 -10% only) after a thermal treatment at 600 K. For SrTiO$_3$:V:Fe crystals (the second set), a relatively strong hysterisis phenomenon is observed after the same thermal treatment. It may result from a strong change in the oxidation state and structural constants of strontium titanate.

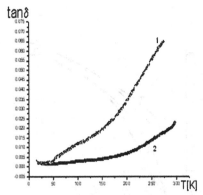

tanδ

T[K]

Figure 1. Temperature dependence of tanδ for SrTiO₃:Fe:V crystals before (1) and after (2) thermal treatment

The thermal treatment at 600 K during 1800 s changes the temperature dependence of ε_0 and tanδ for all samples (Fig. 7). The well observed PT in original samples do not show up in the thermally annealed ones. For SrTiO₃:Mn and SrTiO₃:V:Fe samples, the temperature dependence on the dielectric constant shows anomaly of $\varepsilon_0(T)$ at 35 K: a strong growth of $\varepsilon_0(T)$ at T < 40 K during the cooling cycle of the original and thermally treated SrTiO₃:V:Fe but a smooth variation for the annealed one.

A strong influence of the thermal annealing and temperature dependence on *tanδ* for the second set of samples was observed. (Fig.7). It is important to notice that tanδ value at 300 K is three time higher than the one obtained after a "soft" annealing stage of these samples. The *tanδ*(T) dependence for SrTiO₃:V:Fe crystals (see Table) is similar to the classic behavior of semiconductors.

Analysis of the temperature dependence of the dielectric constants shows different anomalies under 35 K during the cooling of SrTiO₃:Mn and SrTiO₃:V:Fe samples: a strong growth of $\varepsilon_0(T)$ at T < 40 K during the cooling of the virgin SrTiO₃:V:Fe and smooth temperature dependence during the heating stage. In addition, different values of the constant C in Eq.(1) were obtained: $C = 9,9*10^4$ K for the pure crystal, $C = 3,3*10^5$ K for SrTiO₃:Mn and $C = 4,56*10^5$ K for SrTiO₃:V:Fe.

Thermal treatments change the temperature dependence of ε_0 and *tanδ* for the samples of the both series. For crystals doped with Mn ions, the temperature dependence of ε_0 is more linear after treatment than before and PT which is well observed for original samples is not recorded in crystals thermally annealed.

The introduction of rare earth ions via $RE^{+3} \rightarrow Sr^{+2}$ substitution changes the SrTiO₃ crystalline lattice parameter according to the difference in ion-radii. Consequently, SrTiO₃ single crystals doped with a large amount of RE ions cannot be grown. The ionic radii of RE^{+2} and Sr^{+2} are more close but we have no confirmation of the $RE^{+2} \rightarrow Sr^{+2}$ substitution. Data on the temperature dependence of dielectric parameters of SrTiO₃:RE samples are ambiguous. For crystals, such as SrTiO₃:Pr or

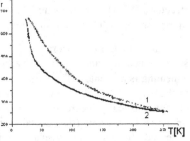

T[K]

Figure 8. Temperature dependence of ε_0 for SrTiO₃:Tm under heating (1) and cooling (2) of the crystal.

$SrTiO_3$:Tm (with different concentration of impurity), the results are similar to data obtained for Me doped samples. In figure 8, data for $SrTiO_3$:Tm with $C_{Tm} = 5 \cdot 10^{-3}$ wt % are plotted. It is easy to see that $\varepsilon_0(T)$ variation is smoother than for Me doped crystals. The profile analysis has provided anomalies at 107 K and 50 K during the heating cycle.

6. Discussion.

The study of the stability of the valence state in the pure and doped $SrTiO_3$ crystals, has emphasized the presence of Ti^{+3} in un-doped and doped with Me and/or RE ions crystals . Change in Ti^{+n} ion oxidation state results in the formation of a large concentration of oxygen vacancies leading to a violation of the sample stoichiometry. Deficiency of strontium oxide and emanation of oxygen during the crystal growth are the main causes of the change of the different properties, in particular, density, optical behavior and temperature dependence of the dielectric and spectral parameters of $SrTiO_3$: for the pure crystal, $a = (3.9064 \pm 0.0005)$ Å and the oxidation state of the main part of Ti ions is Ti^{+4} but for the "blue crystal" the concentration of Ti^{+3} ions is $C = (30 \pm 6)$ % and $a = (3.9002 \pm 0.0005)$ Å. As a consequence of decreasing a and Sr/Ti ratio, the crystal density increases.

For the doped samples, the unit cell parameter a is less than for the "blue crystal and even lesser than for the pure one. We may conclude that oxygen vacancies into strontium titanium perovskite have been ordered accordingly to the results reported in [13]. In any case, a single crystal which includes around 20 % of vacancy must be ordered. It should be noticed that titanium oxide that generates the structure of the $SrTiO_3$ crystal may present different crystallographic structures depending on Ti/O ratio (from TiO to Ti_2O_5).

The oxidation state of Me ions substituted for Ti^{+4} sites, in general, Me^{+3}, is determined by the relationship between the ionization energy of Me^{+n} and the Madelung's constant. In other words, α_M depends on the number of oxygen ions (or vacancies) in the neighborhood. The oxidation state of RE ions substituted for Sr^{+2} ions, is determined by similar rules.

The concentration of Ti^{+3} ions correlates well the dielectric properties as well as the optical parameters, in particular, the intensity of the optical bands at 430 and 520 nm. The increase of the absorption band intensity at 620 nm in $SrTiO_3$:V:Fe together with anomaly in ESR parameters denote a change in the electronic structure of strontium titanate. A correlated anomaly in the temperature dependence of ε_0 and $tan\delta$ confirms this conclusion.

Therefore, the main reason of changes in spectral and dielectric properties of doped strontium titanate is related to the presence of Ti^{+3} ion and an increase of the oxygen vacancy concentration. Indeed, Ti^{+3} defect creates a local energy level in the forbidden band of strontium titanate which changes the electronic structure of the crystal. the concentration. The formation of $Ti^{+3}Me^{+3}O_x$ clusters may also play a role in the property changes.

The introduction of impurities of the iron and lanthanum group elements with "+3" oxidation state into $SrTiO_3$ single crystals lead to the formation of Ti^{+3} ions and a "smoothing" of $\varepsilon_0(T)$ dependence over temperature. This phenomenon has been observed for $SrTiO_3$:Mn and $SrTiO_3$:Tm samples. Other samples, such as $SrTiO_3$:V:Fe show anomaly in the temperature dependence of the dielectric properties with critical points which were determined.

High value of the refractive index n = 2.4 (similar to diamond), non-linear optical properties and photochromism in some un-doped and doped SrTiO3 samples are opening a new area of technological developments for theses compounds offering a large possibility of ionic substitution.

7. Acknowledgement.

Authors are very grateful to A. Sandulenko for samples and interesting discussion and to S. .Spacovic for assistance during the experiments.

8. References

1. Smolensky, S.A., Bokov, V.A., Isupov, V.A. (1985) *Ferrielectrics and Anti-Ferrielectrics,* Nauka Press, Leningrad.
2. Brus, A.R., Cowley, A. (1981) *Structural Phase Transitions,* Taylor and Francis, London.
3. Blinc, R., Zeks, B (1974) *Soft Modes in Ferroelectrics and Antiferroelectrics,* N.Holland Publ.Co., Amsterdam.
4. van Benthem, K., Elsusser, C., French, K.P. (2001) Bulk electronic structure of $SrTiO_3$: experiment and theory, *J. Appl. Phys.***90**, 6156-6164.
5. Kulagin, N., Sviridov, D., (1990) *Introduction to Doped Crystals Physics,* High School Publ. Kharkov.
6. Luxin, Cao, Sozontov, E., Zegenhagen, J. (2000) Cubic to tetragonal phase transition of $SrTiO_3$ under epitaxial stress: an X ray backscattering study, *phys.stat.sol.* **181**, 387-404
7. Grechushnikov, B. (1968) *Materials for IR Optics,* Nauka, Moscow.
8. Kulagin, N. (1983) Defects and colour centers of $SrTiO_3$ single crystals, *Phys.Sol.State* **25**, 3392- 3397.
9. Dojcilvic, J., Kulagin, N. (1996) Anomalies of phase transition and dielectric properties of $SrTiO_3$ crystals . *Phys,Sol.State* **38**, 2012-2017.
10. Konstantinova, A., Stepanov, A.,.Korostel, L., et al. (1993.) Investigations of abnormal dichroism and double-refraction of $SrTiO_3$ crystals, *Crystallog.Reports* **38**. 194-200.
11. Korostel, L., .Zhabotinsky, E., Kostantinova, A. et al. (1994) Investigation of change of defects structure of strontium titanate under 77 K by optical methods, *Crystallography. Reports* **39**. 1092-1097.
12. Konstantinova, A., Korostel, L., Kulagin, N. (1995) Estimation of nonstoichiometry of strontium titanate, *Crystallogr. Reports* **40**, 312-317.
13. Konstantinova, A., Korostel, L., Sulanov, C. (1998) Anomaly of optical properties $SrTiO_3$ crystal related to real structure, *Crystal. Reports* **43**, 903-905
14. Kvjatkovskyy, O.A. About nature of ferroelectric properties of solid solutions $Sr_{1-x}AxTiO_3$ and $KTa_{1-x}Nb_xO_3$ (2002) *Phys.Sol.State* **44**, 1087-1092
15. Kulagin, N., Podus, L., Apanasenko, A., Zajtseva, Ju (1981) Study of the defects of gowth in sngle cystals of sronsium ttanate *Ukr. J. .Phys.*, **26**, 309-312.
16. Kulagin, N., Landar, S., Litvinov, L., Tolok, V. (1981) Spectra and radiation stableness of $SrTiO_3$ single crystals doped with 3d and 4f ions, *Sov.Optics Spectr.* **50**, 888-892.
17. Kulagin, N., Ozerov, M. (1993) Electronic states of Ti^{+3} sons in $SrTO_x$ and TiO_y single crystals, *Phys. Sol. State*, **35**, 2472-2478.

18. Kulagin, N., Dojcilovic, J. (1999) Change of the valency of Ti^{+n} ions and dielectrical properties of strontium titanate single srystals, *Physica* **B269**, 49-59.
19. Muller, K.A. (1981) Paramagnetic point and pair defects in oxide perovskites. *J.Physique* **42**, 551-557.
20. Muller, KA, Blazey, KW, Kool, TW (1993) Tetrahedrally coordinated Cr^{+5} in $SrTiO_3$, *Sol.Stat.Comm.* **85**, 381-384.
21. Kulagin, N., Sandulenko, V., Korostel, L. (1993) Phase transition in $SrTiO_3$ crystals: a new approach, *Abstract book of VIII Intern. Ferroelectricity. Meet.*, Gaithersburg, 428
22. Kulagin, N., Korostel, L., Ozerov, M., Konstantinova, A., Volkov, V.(1993) Electronic structure of doped clusters and dielectric properties of strontium titanate, *Bull.of Russian Acad.Sci. Phys.* **57**, 203-205.
23. Kulagin, N., Korostel, L., Sandulenko, V., Konstantinova, A. (1994) Local structure of doped clusters in oxide crystals, *Bull Russian Acad.Sci. Phys.* **58**, 21-24.
24. Kulagin, N., Dojcilivic, J., Popovic, D. (2001) Low temperature dielectrical properties of $SrTiO_3$ single crystals and valency of ions, *Cryogenic*, **41**, 745-750.
25. Kulagin, N.A., Sviridov, D.T. *Methods of Calculation of Electronic Structure forFree and Doped Ions*, Nauka, Moscow
26. Sviridova, R.K., Sviridov, D.T., Smirnov, Ju.F. (1976) *Optical Spectra of Ions of Transition Metals in Crystals*, Nauka, Moscow.
27. Dieke, G.H. *Spectra and Energy Levels of Rare Earth Ions in Crystals*, J.Wiley, N.Y.
28. Barinsky, L.B., Nefedov, V.I. (1966) *X Ray Sspectroscopy for Derermination of Chanrge of Atoms and Molecules*, Nauka, Moscow.

OPTICAL TRANSPARENCY OF QUARTZ AND FULLERITE C_{60} AT HIGH POWER INFRARED LASER IRRADIATION

I.I. GERU, D.M. SPOIALA, I.T. DIHOR
*State University of Moldova, Superconductivity
and Magnetism Laboratory, 60 Mateevici str.,
MD2009 Chisinau, Moldova*

Abstract. The increase of the optical transparency of quartz and fullerite thin films under the effect of high power non-resonance IR laser irradiation was found. It is shown that laser induced optical transparency at power density of radiation close to destruction cut-off of the materials takes place. For fullerite C_{60} thin films this effect in UV-VIS range arises at thicknesses >600 Å. The laser induced red shift of 642.8 cm^{-1} line and splitting of 674.3 cm^{-1} line in C_{60} fullerite thin films was noticed.

The destruction of C_{60} fullerite thin films under effect of high power IR laser radiation taking into account the "cumulative" effect is discussed. Two types of self-organizing structures caused by interaction with high power IR laser radiation with C_{60} fullerite thin films were discovered.

1. Introduction

The investigation of interaction of high power infrared laser irradiation with optical materials presents interest taking into account the necessity of creation (design) of a new high power lasers. At present there is much experimental data about the destruction of optical media by hypershort pulses (or consequences of pulses) of laser radiation. The theory of physical processes which take place in such interaction is developed also [1, 2].

The processes leading to laser destruction of the optical elements in case of continuous regime of irradiation were investigated to a less extend. In particular there are a few theoretical works. It is caused by insufficient experimental data and by great number of laser destruction physical mechanisms. In the real crystals and films these mechanisms are interconnected and are displayed simultaneously.

In cases of semiconductors and dielectrics the mechanisms of influence of laser radiation on materials depend on correlation between the photon energy quantum $\hbar\omega$ and the energy gap E_g. When photon energy quantum $\hbar\omega$ is higher then energy gap E_g ($\hbar\omega \geq E_g$) the transformation of surface properties of the semiconductors or dielectrics under the effect of high power laser irradiation arises. At inverse limit case when the photon energy quantum is smaller as compared with the energy gap E_g ($\hbar\omega \leq E_g$), the volume properties of semiconductors or dielectrics change. In both cases the interaction of the laser radiation with materials leads to creation of new defects. Moreover, the laser radiation stimulates the processes of migration and

J.-C. Krupa and N.A. Kulagin (eds.), Physics of Laser Crystals, 201–217.
© 2003 *Kluwer Academic Publishers. Printed in the Netherlands.*

redistribution of the point defects, impurities and their complexes which were present in the crystals before the irradiation.

In the present work only the case when the photon energy quantum is smaller than the energy gap E_g in conditions close to the destruction cut-off is studied. The investigated materials were fullerite C_{60} and quartz under the effect of neodymium lasers ($\lambda=1.06$ μm) in continuously and pulse regimes. Alongside with the experimental data the new theoretical results concerning the non-linear non-resonance interaction of high power infrared laser irradiation with substance are presented, too.

2. Formation of Volume Defects in Semiconductors under the Effect of High Power Laser Irradiation

In [3-6] the creation of the volume defects under the effect of high power laser irradiation in binary and ternary semiconducting single crystals of the IV-VI compounds, in silicon doped by phosphor (Si:P), as well as in III-V single crystals were investigated. The concentration of carriers was changed within limits $n = 10^{15}$ -10^{20} cm^{-3}. The irradiation was carried out under the room temperature. During the irradiation under the room temperature the temperature of the crystals was 473K. The CO and CO_2 lasers with $\lambda = 5 - 6$ μm and $\lambda = 10.6$ μm were used. The density of power radiation was below cut-off of the destruction of samples.

For clarification of the regularity of laser-stimulated processes taking place in the crystal lattice, the optical spectra of reflection and transparency, the spectra of electron paramagnetic resonance as well as the electro-physical, photoelectrical and Roentgen-structural properties of the crystals were investigated. It is shown that the velocity of the defects generation depends on density of laser radiation power, on energy of the photons and on impurities concentration.

It is known that for semiconducting materials the deviation from stoichiometry as well as the formation in the semiconducting matrix of different inclusions such as microdefects, complexes and clusters is characteristic. The stability of these inclusions is determined by the energy of atomisation, which characterises the strength of the bound of the complex and matrix of the solvent. Specifically for impurities of II, III, and V groups in IV-VI compounds the atomization energy is higher than the atomization energy of the semiconducting matrix [7, 8].

In [6] the generation of volume defects in some semiconductors under the effect of the infrared (IR) laser radiation ($\hbar\omega < E_g$) has been investigated. It was shown that the rate of defects generation depends on the density of laser power, laser length wave and impurities concentration. The spectral dependence of the reflection coefficient before and after the irradiation is characterized by a typical minimum. This minimum corresponds to plasmon frequency for given concentration n of carriers with m_c effective mass: $\omega_p=(4\pi e^2 n/m_c\varepsilon_\infty)^{1/2}$, where ε_∞ is high frequency dielectric constant. The carriers concentrations determined from ω_p in a satisfactory correlation with Hall effect data obtained under the same conditions of irradiation.

For irradiated surface of samples the plasmon minimum (ω_{p1}) at low power of radiation W moves to a long wave range with increase of W. The plasmon minimum ω_{p2} corresponding to non irradiated opposite surface of the sample stays unchanged with increasing of W. With increasing of the density of radiation power the shift of both plasmon minima ω_{p1} and ω_{p2} in a long wave range occur. But the shift of ω_{p2} plasmon minimum corresponding to no irradiated surface of the sample takes place with on insignificant time delay. This delay reduces after achievement of the range where the spectral position of the plasmon minimum ω_{p1} changes weakly. In the course of time the plasmon minimum ω_{p2} reaches the spectral position ω_{p1}. Hereinafter on the spectral position of both plasmon minima ($\omega_{p1}=\omega_{p2}$) remains the same. The concentration of free carriers reaches a weakly dependence on duration of irradiation too.

With increasing of the density of radiation power the position of plasmon minimum ω_{p1} begins to shift to a long wave range again. In addition, the concentration of free carriers decreases.

These data show that generation of defects under the effect of laser radiation takes place on irradiated surface during irradiation, the front of generation moves towards of propagation of the light. All main peculiarities of laser annealing in case when $\hbar\omega < E_g$ can be explained taking into account the following effects [6]:
- the generation of single proper or impurity atoms as a result of dissociation of inclusions;
- the subsequent motion of these atoms in the crystal through formation of activated states with its localization in sites of crystalline lattice;

The most probable mechanism of creation of single atoms at stationary laser annealing consists in existence of temperature gradient on boundary between inclusion and matrix. This temperature gradient is due to direct absorption of the radiation by inclusion and formation of the termodiffusion beam. Under some condition this beam can be predominant in comparison with diffusion one [7, 8]. The time of termodiffusion carrying-out of the substance from inclusion is a 1.5-2.0 order of magnitude smaller in comparison with the diffusion one. The time of termodiffusion carrying-out of the substance from inclusion under the effect of IR laser radiation decreases with increasing of the power of radiation.

3. Influence of High Power IR Laser Radiation on Quartz: laser Induced Transparency in IR Range

Quartz was one of the first materials to be extensively investigated in the infrared range [9]. Because of its scientific and technological importance it continues to be the frequent subject of investigation.

Saksena [10] gave the interpretation of the infrared and Raman spectra of the quartz on the basis of a group theoretical analysis of the lattice vibrations. He finds that the 24 normal modes of vibration of zero wave vector can be classified according to their symmetry properties as follows: 4 non-degenerate totally symmetric vibrations (class A) active only in Raman spectra, 4 non-degenerate vibrations (class B) active only in infrared E ray, and 8 doubly degenerate vibrations (class E) active in the infrared O ray and Raman spectra.

We have investigated the infrared spectrum of α-quartz close to strong absorption line at 3648 cm^{-1} pertaining to class E (Fig.1).

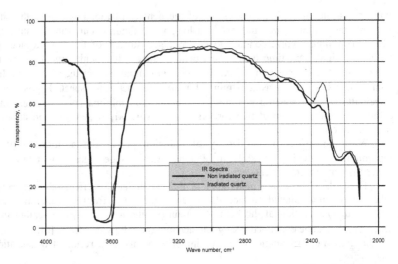

Figure 1. The infrared spectrum of α-quartz before and after high power IR laser irradiation.

On Fig.1 the thin line represents the IR spectrum of the α-quartz in absence of IR laser radiation and the thick line corresponds to IR spectrum of α-quartz after three 40 kJ pulses of GOS-1001 laser (duration of each pulse being τ=1 mm) with intervals between pulses equal to 5 min. We see the increase of optical transparency of quartz under the effect of high power laser radiation, which can be caused by laser annealing of the irradiated material. The increase of optical transparency is due to the movement of impurity atoms under the effect of laser radiation from irradiated surface to the opposite one during laser annealing. This effect is an irreversible effect since the IR spectra of irradiated quartz samples stay unchanged for a long time (many months).

4. Optical Transparency of Fullerite C_{60} in UV-VIS Range Induced by High Power IR Laser Radiation

The fullerite C_{60} thin films on glass, quartz, KCl and KBr substrates were obtained by vacuum evaporation ($\sim 10^{-5}$ torr) of soot containing molecules C_{60} [11]. The soot was prepared using the classical method of evaporation of graphite electrodes (with diameters of 6 mm and length of 200 mm) in electric arc in helium atmosphere [12].

The optical properties of fullerite C_{60} thin films in UV-VIS range were investigated using M40 spectrometer. The UV-VIS spectra of these films contain 5 absorption lines (Fig.2) corresponding to optical transitions $h_u \leftrightarrow t_{1u}$, $h_u \leftrightarrow t_{1g}$, $(h_g+g_g) \leftrightarrow t_{1u}$, $h_u \leftrightarrow h_g$ and $(h_g+g_g) \leftrightarrow t_{2u}$ in order to increase the photon energies. The transition $h_u \leftrightarrow t_{1u}$ (HOMO↔LUMO) at 1.8 eV close to fundamental absorption edge is an electric dipole forbidden transition whereas the remaining transitions are electric dipole allowed transitions. Figure 3 shows the existence of electric dipole forbidden transitions in fullerite C_{60} thin films with the thickness of d = 600, 800, 1000 and 1800 Å under the room temperature. In Tabl.1 the corresponding values of energy gap $E_g^{dir}(h_u, t_{1u})$ for these thin films are presented. We see that, with exception of the case d=600 Å, the

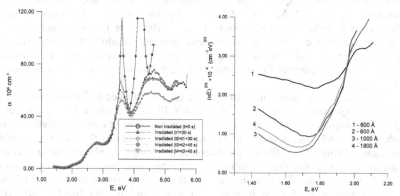

Figure 2. The UV-VIS spectrum of C_{60} fullerite thin film before and after irradiation (α is absorption coefficient).

Figure 3. The electric dipole forbidden transitions in fullerite C_{60} thin films.

energy gap for fullerite thin films of different thickness is practically the same. The value of $E_g^{dir}(h_u, t_{1u})$ for the film with thickness d = 600 Å is caused by more important contribution of mechanical strains which exist on the boundary substrate-film in this case.

TABLE 1. Energy gap E_g^{dir} for C_{60} fullerite thin films.

Type of transiton	Energy gap E_g^{dir}, eV			
	d=600 Å	d=800 Å	d=1000 Å	d=1800 Å
$h_u \rightarrow t_{1u}$	1.57	1.72	1.70	1.67
$h_u \rightarrow t_{1g}$	2.37	2.32	2.32	2.35
$h_g + g_g \rightarrow t_{1u}$	-	-	3.33	-
$h_u \rightarrow h_g$	-	-	3.97	-

For fullerite C_{60} film with thickness d = 2.93 μm the indirect dipole forbidden optical transitions in 1.38 - 2.34 eV range were also found (0.92, 1.11 and 1.72 eV). The corresponding values of optical phonons are 2623.7, 2571.1 and 3024.5 cm^{-1}. These values are higher than the maximum value for energy of phonons in fullerite C_{60}. In the second order IR spectra of fullerite C_{60} the combination modes 2676, 2559 and 2998 cm^{-1} were observed [13]. Therefore three types of indirect dipole forbidden optical transitions can be caused by participation of two phonons in each transition with observance of corresponding selection.

Under the effect of continuous IR laser radiation (laser LTN-101) with density of photons flow W = 8.2·10^{22} photons/cm^2·s on fullerite film with thickness d = 600 Å during the time t = 30 s (three irradiations with durations $t_1 = t_2 = t_3 = 10$ s and the interval $\Delta t \gg t_i$ between irradiations) the electron spectrum remains unchanged. At increasing of film thickness (d = 800 Å) with preservation of the regime of irradiation the strong radiational effects arise (Fig.2). In this case the transitions $h_u \rightarrow t_{1g}$ (HOMO→LUMO+1) in first approximation are not affected by radiation whereas the optical transparency increases essentially for $(h_g+g_g) \rightarrow t_{1u}$ (HOMO-1→LUMO+2) and $h_u \rightarrow h_g$ (HOMO→LUMO+3) transitions. Moreover, the transitions $h_u \rightarrow t_{1u}$ (HOMO→LUMO) are also not affected by laser radiation whereas for fullerite C_{60} nanostructures in water (fullerene C_{60} aggregates in water with concentration c = 0.3 mg/ml) the laser induced red shift of HOMO→LUMO line was observed [14]. The increasing of the irradiation duration leads to increasing of the optical transparency for indicated allowed transitions in an all studied spectral range (Fig.2).

Thus, the non-reversible increasing of the C_{60} thin films optical transparency in the UV-VIS range under the effect of high power IR laser irradiation is a cumulative effect. The accumulation of the laser interaction consequences on the fullerite occurs both in space and time. Namely, the IR laser irradiation causes the increase in optical transparency only of C_{60} films with thickness d = 600 Å. This increasing becomes more evident at in the time of irradiation. On the other hand, the accumulation in time of the radiation effects occurs only for the case of experimental reveal of their spatial accumulation. So the increasing of irradiation time of the C_{60} film with thickness d = 600 Å does not lead to the change of optical transparency.

One of causes of the optical transparency increasing in the UV-VIS range of C_{60} thin films under IR laser irradiation can be the homogenization of point defects caused by this interaction. Due to small values of linear absorption coefficient of the IR radiation with wavelength λ = 1.06 μm in the C_{60} fullerite, the effect of optical increasing under high power IR laser irradiation is a nonlinear effect. This irreversible increasing of the optical transparency is caused by laser annealing of the point defects, containing in the fullerite before irradiation as well as the point defects arisen during the irradiation process. The irreversible structural modifications of the fullerite thin films are more evident in case of thick films ($d \geq 1$ μm). However, because of great values of absorption coefficient for dipolar-allowed optical transitions $(h_g+g_g) \leftrightarrow t_{2g}$ (HOMO-1↔LUMO+2), $h_u \leftrightarrow g_u$ (HOMO↔LUMO+3), these transitions, in spite of their sensibility to high power IR laser irradiation (see Fig.2), can not be detected for fullerite films with thicknesses $d > 1$ μm.

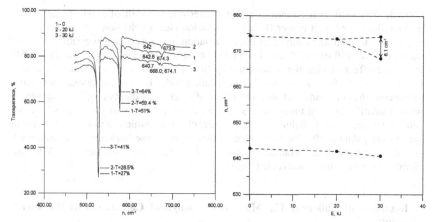

Figure 4. The laser induced transparency of C_{60} fullerite thin film (d=2.93 µm) in 450 - 750 cm^{-1} range.

Figure 5. The red shift of 642.8 cm^{-1} line and splitting of 674.3 cm^{-1} line under the effect of IR laser radiation on C_{60} fullerite film.

The infrared spectra of C_{60} fullerite deposited on KCl and KBr substrates were studied using SPECORD-75 IR spectrometer in the 400 - 4000 cm^{-1}. These spectra have four vibrational intramolecular modes at 526, 576, 1178 and 1428 cm^{-1}. These modes are characterized by irreducible representation F_{1u} type of the symmetry point group I_h of the C_{60} molecule. The fact that first order vibrational modes of fullerite coincide with vibrational modes of free state C_{60} molecule proves the toughness of the chemical bonds between carbon atoms of C_{60} molecule in comparison with chemical bonds between carbon atoms of neighbourhood C_{60} molecules of fullerite. The influence of IR radiation of GOS-1001 laser on C_{60} fullerite films was studied in 400 - 4000 cm^{-1} range. The increasing of optical transparency was established in the IR range when the optical density of photons increases. The IR spectra of C_{60} fullerite film with thickness d = 2.93 µm in 450 - 750 cm^{-1} range are presented in Fig.4. Curve 1 corresponds to non- irradiated film and curves 2 and 3 correspond to irradiated film for different values of laser excitation energy (E_1 = 0, E_2 = 20 kJ, E_3 = 30 kJ). The increasing of optical transparency was noticed at increasing of E from 0 to 20 kJ. The decrease in optical transparency at energy E increasing from 20 to 30 kJ is caused by absorption of IR light in the destroyed fullerite "islands" under influence of high power radiation (transition "fullerite-carbon" and explosive evaporation of fullerite from the substrate). The mentioned transformations take places only in vicinity of some spottinesses which leads to local destruction of fullerite films under the influence of laser irradiation. For the rest the film keeps the high quality. This is confirmed by the behaviour of 526 and 576 cm^{-1} vibrational modes. As can be seen from Fig.4, these lines shift to low frequencies range at increasing of energy from 0 to 30 kJ. In spite of the fact that the values of these spectral shifts are small, the weak shift effect of $\omega_1(F_{1u})$ = 526 cm^{-1} and $\omega_2(F_{1u})$ = 576 cm^{-1} vibrational intramolecular modes to low frequencies ("red shift") under the effect of high power IR laser radiation can be considered as experimentally established effect. It was multiply noticed, lines shift in the same direction. Moreover, the shift of small intensity lines under influence of high power IR laser irradiation (see a behaviour of 642.8 cm^{-1} and 674.3 cm^{-1} lines under action of IR laser irradiation in Fig.4) was also established. Fig.5 shows the red shift of 642.8 cm^{-1} line and splitting of 674.3 cm^{-1} under the effect of IR laser. The affect of vibrational structure by high power IR laser irradiation is a weak effect which manifests at high levels of excitation under limit of radiational destruction of material (but in vicinity of its). The effect of electric field action of electromagnetic wave is small because of the symmetry centre existence in C_{60} molecule. The 642.8 cm^{-1} and 674.3 cm^{-1} lines belong to C_{70} molecule part which according to Raman spectroscopy data is 10-15% in the studied thin films. These lines are more sensitive to the influence of laser irradiation than $\omega_1(F_{1u})=526$ cm^{-1} and $\omega_2(F_{1u})=576$ cm^{-1} vibrational modes of C_{60} molecule (Fig.5).

5. Rotations of Fullerene C_{60} Molecules: Analytical Calculation of the Inertia Moments

It is known [15, 16] that molecules C_{60} form the fullerite crystal with two temperatures phases. One phase is orientation-disordered with face centred cubic lattice ($T > 249K$, space group of symmetry is T_h^3 ($P2_1/Fm3$)). Another phase is orientation ordered with simple cubic lattice ($T < 249K$, space group of symmetry is T_h^6 ($P2_1/a\overline{3}$)). In orientation-disordered phase the almost free rotation of C_{60} molecules with the period $T = 9.1 \cdot 10^{-12}$ s takes place. This value only three times exceeds the period of rotation of fullerene molecules in free state [15]. Almost free rotation of fullerene molecules in solid state is caused by strong intramolecular carbon-carbon interactions in comparison with the intermolecular ones. In order to confirm this conception we have calculated analytically the moments of inertia of molecule C_{60} for the cases of its rotations around C_2, $C_3(S_6)$ and $C_5(S_{10})$ axes of symmetry. For the moment of inertia $I(C_2)$ corresponding to rotation of fullerene C_{60} molecule around C_2 axis of symmetry we get:

$$I(C_2) = \left\{ 6 + 8\frac{R^2}{a^2}\left[\cos^2\alpha_2 \sin^2\gamma_2 + \sin^2\delta_2 + \sin^2\theta_2 + 2\right] - \right.$$
$$- 2\sin^2\delta_2 + 8\sin^2\theta_2 + +16\left[1 - \left(\frac{1}{2} + \frac{\sqrt{5}}{10}\right)\left(1 - \cos\frac{2\pi}{5}\right)^2\right] + \quad , \quad (1)$$
$$\left. + 8\left[\frac{R^2}{a^2} + \left(\frac{1}{2} + \frac{\sqrt{5}}{10}\right)\left(1 - \cos\frac{2\pi}{5}\right)^2 - 1\right](\sin^2\beta_2 + \sin^2\varphi_2)\right\} m_c a^2$$

where m_c is the mass of carbon atom, R and a are the radius of molecule C_{60} and the side of hexagon ($R = 0.357$ nm, $a = 0.142$ nm [17]), and angles α_2, β_2, γ_2, δ_2, φ_2 and θ_2 are:

$$\alpha_2 = \arcsin\frac{a}{2R},$$

$$\beta_2 = \alpha_2 + \arccos\frac{2R^2 - a^2}{2R\left\{R^2 - a^2\left[1 - \left(\frac{1}{2} + \frac{\sqrt{5}}{10}\right)\left(1 - \cos\frac{2\pi}{5}\right)^2\right]\right\}^{1/2}},$$

$$\gamma_2 = \arccos\frac{2R^2\cos^2\alpha_2 - 3a^2}{2R^2\cos^2\alpha_2},$$

$$\delta_2 = \alpha_2 + \arccos\frac{2R^2 - a^2\left\{\left[\left(\frac{1}{2} + \frac{\sqrt{5}}{10}\right)^{1/2} + \left(\frac{1}{4} + \frac{\sqrt{5}}{10}\right)^{1/2}\right]^2 + \frac{1}{4}\right\}}{2R\left(R^2 - \frac{a^2}{4}\right)^{1/2}},$$

$$\varphi_2 = \gamma_2 + \arccos\left\{\frac{R(1 + \cos^2\alpha_2)}{2\cos\alpha_2\left\{R^2 - a^2\left[1 - \left(\frac{1}{2} + \frac{\sqrt{5}}{10}\right)\left(1 - \cos\frac{2\pi}{5}\right)^2\right]\right\}^{1/2}} - \right.$$

$$\left. \frac{a^2\left\{1 - \left(\frac{1}{2} + \frac{\sqrt{5}}{10}\right)\left(1 - \cos\frac{2\pi}{5}\right)^2 + \left[\cos\frac{2\pi}{5}\left(\frac{1}{2} + \frac{\sqrt{5}}{10}\right)^{1/2} + \left(\frac{1}{4} + \frac{\sqrt{5}}{10}\right)^{1/2}\right]^2\right\}}{2R\cos\alpha_2\left\{R^2 - a^2\left[1 - \left(\frac{1}{2} + \frac{\sqrt{5}}{10}\right)\left(1 - \cos\frac{2\pi}{5}\right)^2\right]\right\}^{1/2}}\right\}$$

(2)

$$\theta_2 = \delta_2 + \arccos\frac{R^2 - a^2}{\left[\left(R^2 - a^2\right)\left(R^2 - \frac{a^2}{4}\right)\right]^{1/2}}.$$

The inertia moment $I(C_3)$ at rotation of C_{60} molecule around $C_3(S_6)$ axis

$$I(C_3) = 12\left[\frac{5}{4} + \frac{R^2}{a^2}\left(\sin^2\beta_3 + \frac{1}{2}\sin^2\gamma_3 + \frac{3}{2}\sin^2\gamma_3\right) + \right.$$

is:
$$\left. + \left(\frac{R^2}{a^2}\cos^2\alpha_3 + \frac{3}{4}\right)\sin^2\delta_3\right]m_c a^2 \tag{3}$$

where
$$\alpha_3 = \alpha_2,$$

$$\beta_3 = \alpha_3 + \arccos\frac{2R^2 - a^2}{2R^2},$$

$$\gamma_3 = \frac{R\cos\alpha_3}{\left(R^2\cos^2\alpha_3 + \frac{3}{4}a^2\right)^{\frac{1}{2}}} +$$

$$+ \arccos\frac{R^2(1+\cos^2\alpha_3) - a^2\left\{\left[\left(\frac{1}{2}+\frac{\sqrt{5}}{10}\right)^{\frac{1}{2}} + \left(\frac{1}{4}+\frac{\sqrt{5}}{10}\right)^{\frac{1}{2}}\right]^2 - \frac{3}{4}\right\}}{2R\left(R^2\cos^2\alpha_3 + \frac{3}{4}a^2\right)^{\frac{1}{2}}}, \tag{4}$$

$$\delta_3 = \arccos\frac{R\cos\alpha_3}{R^2\cos^2\alpha_3 + \frac{3}{4}a^2} + \arccos\frac{2\left(R^2\cos^2\alpha_3 + \frac{3}{4}a^2\right) - 3a^2}{2\left(R^2\cos^2\alpha_3 + \frac{3}{4}a^2\right)}.$$

$$\varphi_3 = \gamma_3 + \beta_3 - \alpha_3.$$

Analogically for moment of inertia $I(C_5)$ at rotation of C_{60} molecule around C_5 axis we have

$$I(C_5) = 10\left[\frac{3}{2} + \frac{\sqrt{5}}{10}\sin^2\beta_5 + 2(\sin^2\beta_5 + \sin^2\delta_5)\times\right.$$

$$\left. \times\left(\frac{R^2}{a^2}\cos^2\alpha_5 + \frac{\sqrt{5}}{10} + \frac{1}{4}\right)\right]m_c a^2 \tag{5}$$

where
$$\alpha_5 = \arcsin\frac{a}{R}\left(\frac{1}{2} + \frac{\sqrt{5}}{10}\right)^{\frac{1}{2}},$$

$$\beta_5 = \alpha_5 + \arccos\left[1 - \frac{a^2}{2R^2}\right],$$

$$\gamma_s = \text{arctg}\frac{a\left(\frac{1}{4}+\frac{\sqrt{5}}{10}\right)^{\frac{1}{2}}}{R\cos\alpha_s} + \arccos\left[1-\frac{3a^2}{2\left[R^2\cos^2\alpha_s+a^2\left(\frac{1}{4}+\frac{\sqrt{5}}{10}\right)\right]}\right], \quad (6)$$

$$\delta_s = \beta_s + \arccos\frac{R^2\left(1+\cos^2\alpha_s\right)-a^2\left\{2\left[\left(\frac{1}{2}+\frac{\sqrt{5}}{10}\right)\left(\frac{1}{4}+\frac{\sqrt{5}}{10}\right)\right]^{\frac{1}{2}}+\frac{1}{2}+\frac{\sqrt{5}}{10}\right\}}{2R\left[R^2\cos^2\alpha_s+a^2\left(\frac{1}{4}+\frac{\sqrt{5}}{10}\right)\right]}.$$

According to [18] the numeric values for $I(C_2)$, $I(C_3)$ and $I(C_5)$ of inertia moment are $I(C_2) = 0.992 \cdot 10^{-43}$ kg·cm^2, $I(C_3) = 0.995 \cdot 10^{-43}$ kg·cm^2 and $I(C_5) = 1.001 \cdot 10^{-43}$ kg·cm^2. Since C2, C3 and C5 are not orthogonal axes (the angle between C_3 and C_5 is 77.55°) $I(C_2)$, $I(C_3)$ and $I(C_5)$ are not considered as the principal values for tensor of inertia moments. These values were obtained in the basis of orthogonal axes C_2^x, C_2^y and C_2^z: $I_{xx} = 1.016 \cdot 10^{-43}$ kg·cm^2, $I_{yy} = 1.029 \cdot 10^{-43}$ kg·cm^2, $I_{zz} = 1.042 \cdot 10^{-43}$ kg·cm^2. We note that in simple model of homogeneous distribution of the mass of fullerene molecule on its spherical surface the moment of inertia is $I_0 = 1.016 \cdot 10^{-43}$ kg·cm^2. This confirms the model of spherical bidimensional quantum well for molecule C$_{60}$ proposed by Rotkin and Suris [19].

The obtained values for the inertia moments of fullerene molecule lead to small values of the energy of rotation quantum ($E_{rot}/\hbar = 166$ MHz). Therefore the vibration-rotation structure of the energy spectrum of fullerene molecule was only observed in radio-microwave range [13].

6. The Destruction of Fullerite C$_{60}$ by High Power Laser Radiation

As it was shown in Part 4, the influence of high power laser radiation brings to the increase in optical transparency of the fullerite C$_{60}$ in the UV-VIS and IR spectral ranges. This occurs at high levels of optical excitation of fullerite but not sufficiently high for its laser destruction. In other words, the observed laser induced increasing of the fullerite optical transparency in the large spectral range takes places in the radiation regime close to the cut-off of destruction of material, but below it. If the dose of irradiation exceeds the cut-off of material radiational stability, the irradiated material begins to destroy, and this occurs first of all in the places of spottinesses accumulation. The beginning of such destruction of C$_{60}$ thin films is noticed in the IR spectra at $E_3 = 30$ kJ laser excitation energy, which in condition of our experiments gives the value of falling electromagnetic irradiation power on area unit of irradiated surface of 37.7 W/cm^2 (Fig.4, curve 3). Namely, the laser destruction products of fullerite stipulate the optical transparency decreasing in the range of unceasing background and IR lines of weak intensity at increasing of laser excitation energy from $E_2 = 20$ kJ to $E_3 = 30$ kJ (respectively at increasing of falling irradiation power on unit area from 25.2 W/cm^2 to 37.7 W/cm^2). Herewith the small areas of radiational destruction (in comparison with the area of irradiated film surface) are well visible in optical microscope.

The measurement of electrical conductivity of studied C_{60} fullerite thin films shows, that it has poor conductivity ($\sigma < 10^{-10}\ \Omega^{-1}\cdot cm^{-1}$). So we consider that fullerite is a transparent dielectric material under the effect of high power IR laser radiation. The material destruction occurs under the thermal influence of laser irradiation, which strongly differs from ordinary heating by external heat source. In particular, in multiphase systems (in our case there are 85% of C_{60} phase and 15% of C_{70} phase) the thermal influence can warm selectively one of the phases, and this leads to the change of chemical transformation velocity in such systems [20]. Thereby, heat influence of non-resonance laser irradiation leads to changing of the chemical reaction velocity in the transparent medium. In turn, the products of chemical transformation, which can be different in case of slow and quick chemical reactions, also absorb non-resonance laser radiation. The laser destruction of the transparent dielectrics is accompanied by process of heat propagation, which can be described quantitatively.

Let the heat conducting thin film sample with thermal conductivity coefficient κ and specific thermal capacity C_v have thickness l ($0 < z < l$) and is infinite in the film plane: $\rho = (x, y)$. The function $T(\rho, t)$ describing the temperature distribution in the film limits and its temporary evolution satisfies the thermal conductivity equation:

$$C_V \frac{\partial T(\rho,z,t)}{\partial t} - \kappa \Delta T(\rho,z,t) = Q, \tag{7}$$

where Q is the heat source power in unit volume. We will expect that sample is thermo isolated and warms only at the expense of absorption of external radiation in volume, which is switched on in the time moment $t = 0$ (before it was at T_0 temperature). If the sample was infinite, the temperature distribution in the time moment t would be determined by the formula:

$$T(\mathbf{r},t) = T_0 + \frac{1}{C_V (4\pi\chi)^{\frac{3}{2}}} \int_0^t \frac{dS}{S^{\frac{3}{2}}} \int d^3\mathbf{r}'\, e^{-\frac{(r')^2}{4\chi S}} Q(\mathbf{r} - \mathbf{r}', t - S), \tag{8}$$

where $\chi = \kappa/C_v$ is a temperature conductivity coefficient. The solution correctness of (8) can be verified by substitution of it in the initial Eq.(7). The solution (8) cannot be used for describing of the local laser destruction of fullerite experiment (see Part 4). In this case we need to find the solution of heat conduction Eq.(7) in the limited volume. This problem in turn is reduced to the problem for infinite volume with additional fictitious heat source. Herewith heat conduction of sample leads to frontier conditions:

$$\left.\frac{\partial T}{\partial z}\right|_{z=0} = 0, \qquad \left.\frac{\partial T}{\partial z}\right|_{z=l} = 0. \tag{9}$$

Considering the frontier conditions (9), the solution of heat conduction equation in the limited volume for temperature Fourier-component $\Theta(\mathbf{k}, z, t)$ is

$$Q(\mathbf{k},z,t)=\frac{1}{C_V\sqrt{4\pi\chi}}\int_0^t\frac{dS}{\sqrt{S}}e^{-2k^2S}\int_0^t dz'\sum_{n=-\infty}^{+\infty}e^{\frac{(z'+2nl)^2}{4\chi S}}\times$$
$$\times\left[P(\mathbf{k},z-z',t-S)+P(\mathbf{k},z'-z,t-S)\right] \tag{10}$$

where $P(\mathbf{k},z-z',t-t')$ and $P(\mathbf{k},z'-z,t-S)$ are Fourier transformants of function $Q(\mathbf{\rho-\rho'},z-z',t-S)$ and $Q(\mathbf{\rho-\rho'},z'-z,t-S)$. The Eq.(10) is the exact solution of heat conduction equation in the limited volume in form of integral containing the time overdue heat source. We neglect the influence of heat change with ambient on the basis that burning of sample takes place during a short time. Then the exact solution of Eq.(1) with initial conditions $T(\mathbf{\rho},z,0)=T_s$, where T_s is stationary temperature in the end of the stage preceding to destruction, and with frontier conditions (9) have the following view:

$$T(\mathbf{\rho},z,t)=T_0+\frac{W}{C_V(4\pi\chi)^{3/2}}\int_{\max(0,t-t_0)}^t\frac{dS}{S^{3/2}}e^{-\frac{\rho^2}{4\chi S}}\times$$
$$\times\sum_{n=-\infty}^{+\infty}\left\{e^{-\frac{[z-v(t-s)+2nl]^2}{4\chi S}}+e^{-\frac{[z+v(t-s)+2nl]^2}{4\chi S}}\right\} \tag{11}$$

We proceed to the following dimensionless variables and parameters for tractability:

$$\mathbf{R}=\frac{\mathbf{\rho}}{2\sqrt{\chi t}},\qquad \tau=\frac{t}{t_0},\qquad \alpha=\frac{C_v(4\pi\chi)^{\frac{1}{2}}\sqrt{t_0}\,E}{k_BW},$$

$$\widetilde{T_1}(\mathbf{R},\tau)=\frac{\alpha k_B}{E}T(\mathbf{\rho},0,t),\qquad \widetilde{T_2}(\mathbf{R},\tau)=\frac{\alpha k_B}{E}T(\mathbf{\rho},l,t). \tag{12}$$

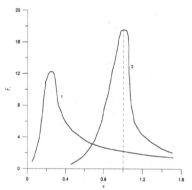

Figure 6. The temperature dependences $\widetilde{T_1}$ and $\widetilde{T_2}$ vs. time in $R \ll 1$ positions

where E is the activation energy of oxidation, k_B is Boltzman constant.

The dependences $\widetilde{T_1}$ and $\widetilde{T_2}$ vs. time in $R \ll 1$ positions calculated on the basis of Eqs.(11) and (12) are presented in Fig.6. The types of temperature modification on film input side (irradiated side) and output ones are different. The temperature on input side of film near to the laser beam centre achieves the maximum almost in the beginning of $(0, t_0)$ interval.

7. Self-organizing Processes under the Effect of High Power IR Laser Irradiation of Fullerite C_{60} Thin Films

Alongside with the increase on the optical transparency of the fullerite C_{60} films (see Part 4) the formation of microstructures is possible under the effect of high power IR laser radiation. These structures are formed from products of laser stimulated thermo-chemical reactions in conditions close to material destruction cut-off. Under other identical conditions the form and concentration of these finite structures depend on technology of fullerite C_{60} thin films deposition and on their thicknesses.

In Fig.7 the TEM image ($\times 11000$) is shown for non-irradiated fullerite C_{60} thin film (400 Å) sublimated on graphite thin film (300 Å) which, in its turn, is deposited on NaCl crystal substrate. We see that surface of fullerite film have a relatively homogenous structure.

After its irradiation by one 1 ms pulse of GOS-1001 IR laser with power density P = 62.9 W/cm^2 the finite discrete structures of quasi-spherical form with radius about 0.5 μm and concentration $0.55 \cdot 10^{10}$ cm^{-2} arise. Such structures appear due to the existence of strong non-equilibrium states of fullerite which is due to non-linear interaction with high power laser radiation. This interaction leads to self-organization and formation of the dissipative structures.

The dissipative structures of quasi-spherical form arises (Fig.8) as a consequence of laser induced thermal effect and laser stimulated thermo-chemical reactions. The products of chemical transformation, in their turn, absorb the electromagnetic radiation.

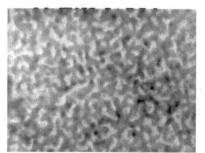

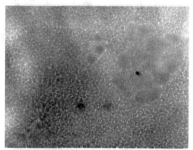

Figure 7. TEM image (×11000) of the non-irradiated C_{60} fullerite thin film.

Figure 8. TEM image (×11000) of the irradiated C_{60} fullerite film (d=400 Å) with quasi-spherical dissipative structure.

In the regions of the film with compositional and structural spottinesses the laser stimulated thermal effect is more evident. This is caused by the increase of the optical absorption coefficient. On Fig.9 the TEM image (×10000) of such region for the same fullerite C_{60} film is presented. We see that the dimensions of the dissipative structure remain the same but their homogeneity strongly changed. The bright region inside of

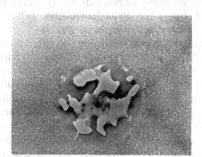

Figure 9. TEM image (×11000) of the irradiated C_{60} fullerite film (d = 400 Å) with quasi-spherical dissipative structure in the spottiness region.

the dissipative structure has shown the absence of substance in this region. This effect can be caused by laser stimulated chemical reaction of oxidation of the fullerite with formation of CO and CO_2 gases at higher temperature in comparison with previous case.

The increase in the thickness of fullerite C_{60} film from 400 to 700 Å leads to formation of the quasi-square (in the profile) structures at the same condition of irradiation. These structures are localized, probably, in graphite close to interface between graphite and NaCl substrate. In Fig.10 and 11 two TEM images (×11000 and ×22000, respectively) for two regions of the fullerite C_{60} film of 700 Å thickness are presented. The linear dimensions of quasi-square structures are about 0.6 - 2 μm and their surface density is $7.2 \cdot 10^5$ cm^2. Formation of such structures can be caused by cumulative effect discussed in Part 6. As a consequence of this effect the increase of the temperature of interface graphite-NaCl substrate takes place with the some delay in comparison with increasing of temperature close to irradiated fullerite surface. This delay is the more the more is the sum of fullerene and graphite films thicknesses. Immediately after end of the delay the sharp increase in temperature arises in comparison with increasing of temperature of the irradiated surface. In these conditions the growth of new phases is possible.

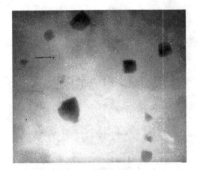

Figure 10. TEM image (×11000) of irradiated C$_{60}$ fullerite film (d=700 Å) with multiple quasi-square structures (a~0.4-1.0 μm).

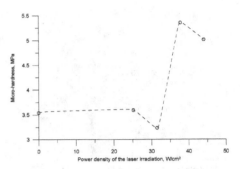

Figure 12. The dependence of C$_{60}$ fullerite thin film micro-hardness from laser power density.

Figure 11. TEM image (×22000) of irradiated C$_{60}$ fullerite film (d=700 Å) with single quasi-square structure (a~1.5 μm).

The existence close to the interface graphite-NaCl substrate of the condensation centres with quasi-square symmetry and the growth on its basis of microstructures does not contradict the cubic symmetry of NaCl crystal. However, the final conclusions about "cumulative" nature of laser stimulated microstructures in form of distorted octahedrons or square prisms localized in graphite film near interface graphite-NaCl substrate must be made only after supplementary investigation. Specifically, the supplementary information about the effect of laser effect on fullerite C$_{60}$ can be obtained from micro-hardness data. In Figure 12 the dependence of micro-hardness from laser power density is presented. We see that the micro-hardness have minimal value at power density W_1 = 31.4 W/cm^2, maximal value at W_2 = 37.7 W/cm^2 and decreases at future power density increasing. These results suggest the origin of hardening to two photon ($2\hbar\omega$=2.34 eV) induced polymerization of C$_{60}$ molecules in fullerite thin films in 31.4-37.7 W/cm^2 density power range.

8. Acknowledgements

This work is supported partially by Supreme Council for Science and Technology Development of Republic of Moldova, Grant 01016C.

9. References

1. Gromovoj, Yu.S., Plyatsko, S.V., Sizov, F.F., Korovina, L.A. (1990) The site and pseudodonor character of manganese in single-crystal lead telluride, *J. Phys. Condens. Matter.* **2**, 10391-10401.
2. Plyatsko, S.V., Gromovoj, Yu.S., Kadyshev, S.K., Klimov, A.A. (1994) Conversion of intrinsic and extrinsic defects by laser light in lead selenide and its solid solutions, *Semiconductors* **28**, 83-86.
3. Gromovoj, Yu.S., Plyatsko, S.V. (1994) The 1st Int. Conf. on Matter. Sci. of Chalcogenide and Diamond-Structure Semiconductors, Abstr. Booklet, Chernivtsi, **2**, 143.
4. Plyatsko, S.V. (2000) Generation of bulk defects in some semiconductors by laser radiation in the transparency region of the crystal, *Semiconductors* **34**, 983-1102.
5. Gurevich, L.V., Karachevtsev, G.V., Kondrat'ev, V.N. et al. (1974) *Energies of Breaking of the Chemical Couplings. Potentials of Ionisations and Electron Affinity*, Nauka, Moscow.
6. Smorodina, T.A., Sheftal', N.N., Tsuranov, A.P. (1986) *Inclusion of Impurity Centres in Crystalline Layer of Semiconductor*, Nauka, Leningrad.
7. Grigor'ev, N.N., Kudykina, T.A., Plyatsko, S.V., Sizov F.F. (1988) Laser-induced thermodiffusion of semiconductor inhomogeneities in the spectral transparency region, *Semicond. Sci. Technol.*, **3**, 951-957.
8. Grigor'ev, N.N., Kudykina, T.A., Plyatsko, S.V., Sizov, F.F. (1988) Heating of inclusions by intense laser radiation in transparent narrow-Gap IV–VI semiconductors, *Infr. Phys.* **28**, 307-310.
9. Spitzer, W.G., Kleinman, D.A. (1961) Infrared lattice bands of quartz, *Phys. Rev.* **121**, 1324-1335.
10. Saksena, B.D., (1940) *Proc. Indian Acad. Sci.*, **12A**, 93.
11. Manaila, R., Geru, I., Spoiala, D., Dihor, I., Devenyi, A. (1999) Structure and defects in thin C_{60} films, *Fullerene Science and Technology* **7**, 59-75.
12. Krätschmer, W., Fostiropoulos, K., Huffman, D.R. (1990) The infrared and ultraviolet absorption spectra of laboratory-produced carbon dust: Evidence for the presence of the C_{60} molecule, *Chem. Phys. Lett.* **170**, 167-170.
13. Dresselhaus, M.S., Dresselhaus, G., Eklund P.C. (1996) *Science of Fullerenes and Carbon Nanotubes*, Academic Press, N.Y.
14. Geru, I.I., Dihor, I.T., Andrievsky, G.V. (1999) Laser induced red shift of HOMO-LUMO E_g of the fullerite C_{60} nanostructures in water, *The 4th Biennial International Workshop in Russia "Fullernes and Atomic Clusters" IWFAC'99*, St. Petersburg, Russia, Book of Abstracts, 336.
15. Eletskii, A.V., Smirnov, B.M. (1995) Fullerenes and carbon structures, *Physics-Uspekhi*, **38**, 935-964.
16. Grushko, Yu.S., Ganja, Yu.V., Kovalev, M.F. et al. (1993) Raman diffusion of light in the high temperature phase of C_{60} fullerite, *Phys.Sol. State* **35**, 980-986.
17. Eletskii, A.V., Smirnov, B.M. (1993) Fullerenes, *Uspekhi Fizicheskikh Nauk*, **163**, 33-60 (in Russian).
18. Geru, I.I., Dihor, I.T. (1997) Radiofrequency resonance of rotating fullerene molecule, *Specialized Colloque AMPERE/RAMIS'97, Abstracts of XVII Conference on Radio- and Microwave Spectroscopy*, Poznan, 57.
19. Rotkin, V.V., Suris, R.A. (1994) Spherical quantum well model of the C_{60} molecule, *Mol. Materials*, **4**, 87-94.
20. Lifshitz, I.M., Slezov, V.V. (1958) About kinetics of supersaturated solid solution diffusion disintegration, *JETP*, **35**, 479-492.

AN OPTIMAL WAY FOR COMPUTATION OF NON-LINEAR OPTICAL PROPERTIES OF CRYSTALS AND LARGE CONJUGATED MOLECULES WITHIN CLUSTER MODELS

Yu. F. Pedash, A. Yu. Semenov

Research Institute for Chemistry of Kharkiv V.N. Karazin National University, Svobody Square 4, 61077 Kharkiv, Ukraine.

E-mail: pedash@univer.kharkov.ua

Abstract: Cluster models are of importance in quantum-mechanical computations of Nonlinear Optical (NLO) properties of macroscopic objects such as crystals and large conjugated molecules, since they preserve important electronic structure features, topology and symmetry of the nuclear frame. In a cluster approach the NLO properties such as polarizability (α), first (β) and second (γ) hyperpolarizability tensors are computed for a series of clusters large enough for further extrapolation of the calculated values to infinite cluster dimensions. For realistic models the most appropriate and acceptable is the finite field method, when α, β, γ are derived by differentiation of either induced dipole moment $d(\varepsilon)$ or total energy $E(\varepsilon)$ with respect to the intensity of a trial uniform electric field (ε). In this study, we apply non-finite-difference formulae of numerical differentiation with interpolating Lagrangian polynomials on regular even- and odd-point grids. Estimates of absolute errors (δ) are obtained for derivatives of any order, including the number of points in the grid, the ε value, the values of the higher derivatives of $d(\varepsilon)$ and $E(\varepsilon)$ with respect to ε. Errors in α, β, γ are minimized with an optimal ε that can be found *a priori* from model systems. Our approach is more than just a technical improvement: it yields the maximum size of a cluster that can be studied with a finite-field method. In particular, the dependence of the δ interval on the spatial length and extent of delocalization of valence electrons is established. Of special notice are the cases where the finite-field method fails due to divergence. It is found for the variational methods of quantum chemistry that other things being equal, the errors are decreased by two orders of magnitude by differentiation of $E(\varepsilon)$ rather than $d(\varepsilon)$, even though the order of derivative is lower by one in the latter case. New analytical expressions are presented for the higher order NLO properties of conjugated molecular systems.

1. Introduction

Investigation of Nonlinear Optical (NLO) properties of crystalline materials and extended conjugated molecules remains the focal point of theoretical and experimental interest as numerous reviews suggest [1-4]. Theoretical estimates of NLO susceptibilities often rely on quantum-mechanical computations of nonlinear response of representative microscopic clusters to the electric field of an incident light wave [5-7].

J.-C. Krupa and N.A. Kulagin (eds.), Physics of Laser Crystals, 219–243.
© 2003 *Kluwer Academic Publishers. Printed in the Netherlands.*

These clusters are selected to reproduce the essential features of electronic structure of a crystal, a polymer molecule or its important part, a conjugated chromophore fragment within a larger molecular species, e.g. dendrimer [6-11], etc. Cluster models are more appropriate than models with periodic boundary conditions when topology and symmetry of real species have to be reproduced. Theoretical estimates of microscopic susceptibilities become more reliable with larger cluster sizes. In this connection, of special interest is the accuracy of cluster calculations depending on spatial dimensions of the delocalized electronic cloud. As applied to this problem, cluster models of crystals and conjugated molecules are closely related. For instance, we can build the test set of conjugated chain hydrocarbon molecules treating them by the Huckel method. This π-electron model has its solid-state counterpart in the atomic crystal lattice donating one electron per atom into the valence zone, under the tight-binding condition [12,13]. Such simple systems avoid irrelevant details and seem most appropriate for analysis of the numerical approach.

Two basic procedures are commonly used to compute hyperpolarizabilities. These are the Finite-Field (FF) method and the Sum-over-States (SOS) method. [6-7] . The latter of two involves summation of a series derived from the perturbation theory and is quite demanding when applied to multielectron systems. The computational expense builds up due to a great number of excited states to be included in the summation, a need for a basis set control, a rather specific character of non-uniform equations that appear in most quantum chemical approaches, the problem of integration over a continuous spectrum, etc. In this study, we concentrate our attention on the FF method as the computationally simplest and equally applicable to numerous models of approximate wave functions and a variety of effective Hamiltonians.

The following are common definitions and conventions used in the paper.

An external electric field applied to a molecular system generates in it an induced dipole moment that increases with the field strength:

$$d_i(\varepsilon) = d_i(0) + \alpha_{ij}\varepsilon_j + \frac{\beta_{ijk}}{2!}\varepsilon_j\varepsilon_k + \frac{\gamma_{ijkl}}{3!}\varepsilon_j\varepsilon_k\varepsilon_l + ..., \qquad (1)$$

Here, d_i is a component of the dipole moment, α is the polarizability tensor, β and γ are the first and second hyperpolarizability tensors, ε_i is a component of the external field. The subscripts take values from the Cartesian coordinate set $\{x, y, z\}$. Summations are performed over subscripts that occur in pairs. Taylor expansion of the free energy $E(\varepsilon)$ is equivalent to the $d(\varepsilon)$ expansion:

$$E(\varepsilon) = E(0) - d_i\varepsilon_i - \frac{\alpha_{ij}}{2!}\varepsilon_i\varepsilon_j - \frac{\beta_{ijk}}{3!}\varepsilon_i\varepsilon_j\varepsilon_k - \frac{\gamma_{ijkl}}{4!}\varepsilon_i\varepsilon_j\varepsilon_k\varepsilon_l - ... \qquad (2)$$

The higher order polarizabilities will be also denoted by a general $E^{(k)}$, where k refers to the polarizability order.

Normally, hyperpolarizability tensors depend on frequencies of the applied field. This dependence is known as dispersion, and frequency dependent polarizabilities are referred to as dynamic polarizabilities. For instance, the polarizability tensor and the related refractive index depend on a single frequency, $\alpha(-\omega;\omega)$. The first hyperpolarizability tensor depends on two frequencies, $\beta(-\omega;\omega_1,\omega_2)$. The negative sign at ω corresponds to light emission, while the positive sign corresponds to light absorption. Nonlinear effects are described by these frequency dependent polarizability tensors and are most prominent when the applied frequencies ω_1, ω_2, and their sums $\omega_1 + \omega_2$ are close to the characteristic frequencies of molecular electronic and vibrational transitions. In particular, the linear electro-optic effect, also known as the Pockels effect, is described by the $\beta(-\omega;\omega,0)$ tensor. So-called optical rectification corresponds to the $\beta(0;\omega,-\omega)$ tensor. The second harmonic generation, $\beta(-2\omega;\omega,\omega)$, is a special case of the combined harmonic generation, $\beta(-(\omega_1+\omega_2);\omega_1,\omega_2)$. The 4th order tensor, i.e. the second hyperpolarizability, depends on three frequencies, $\gamma(-\omega;\omega_1,\omega_2,\omega_3)$. The $\gamma(-3\omega;\omega,\omega,\omega)$ tensor is responsible for the third harmonic generation. The optical Kerr effect is determined by the $\gamma(-\omega;\omega,0,0)$. The frequency-independent second harmonic generation corresponds to the $\gamma(-2\omega;\omega,\omega,0)$ tensor. The nonlinear polarizability at the basic frequency is given by the $\gamma(-\omega;\omega,\omega,-\omega)$ tensor. The electro-optic Kerr effect is described by the real part of the $\gamma(-\omega;\omega,\omega_1,-\omega_1)$ tensor [5-7, 14, 15].

Polarizability dispersion vanishes when the basic ω_1, ω, and combined $\omega_1 + \omega$ frequencies of the applied external field are far from the fundamental molecular frequencies. This far-from-resonance case gives static polarizabilities. It is important to note that estimates of static polarizabilities can vary substantially when evaluated from different frequency-dependent properties by extrapolation to zero frequency. In the light of a variety of notation seen in the literature (see e.g. [16-18]), the hierarchy of scaling factors in agreement with the definitions (1) and (2) is given below:

$$\beta(-\omega;\omega,0) \rightarrow 2\beta(0), \quad \beta(0;\omega,-\omega) \rightarrow (1/2)\beta(0), \quad \beta(-2\omega;\omega,\omega), \rightarrow (1/2)\beta(0)$$
$$\gamma(-3\omega;\omega,\omega,\omega) \rightarrow (1/4)\gamma(0), \quad \gamma(-\omega;\omega,0,0) \rightarrow 3\gamma(0), \tag{3}$$
$$\gamma(-2\omega;\omega,\omega,0) \rightarrow (3/2)\gamma(0), \quad \gamma(-\omega;\omega,\omega,-\omega) \rightarrow (3/4)\gamma(0), \quad \gamma(-\omega;\omega,\omega,\omega_1,-\omega_1) \rightarrow (3/2)\gamma(0).$$

In the three-dimensional space, $3n$ cartesian components are needed to define every n^{th} order tensor; for example, 9 components for the α tensor, 27 components for β, 81 for γ and so on. The static polarizability tensors obey the Kleiman symmetry relations [6] establishing that the tensor components are invariant with respect to every permutation of their indices. For example, in case of β the symmetry relations give:

$$\beta_{xyz} = \beta_{yzx} = \beta_{zxy} = \beta_{xzy} = \beta_{yxz} = \beta_{zyx}. \tag{4}$$

The symmetry reduces the number of independent components to 6 for α, 10 for β and 15 for γ.

Let consider static (hyper)polarizabilities obeying the Kleiman symmetry relations.

In practice, the tensor invariants (i.e. average values that are independent of the coordinate system) are used together with individual tensor components. The tensor invariants are: for polarizability

$$\overline{\alpha} = (\alpha_{xx} + \alpha_{yy} + \alpha_{zz})/3, \tag{5}$$

for the first hyperpolarizability

$$\beta_x = \beta_{xxx} + \beta_{xyy} + \beta_{xzz},$$
$$\beta_y = \beta_{yyy} + \beta_{yxx} + \beta_{yzz}, \tag{6}$$
$$\beta_z = \beta_{zzz} + \beta_{zxx} + \beta_{zyy}.$$

for the second hyperpolarizability

$$\bar{\gamma} = \left(\gamma_{xxxx} + \gamma_{yyyy} + \gamma_{zzzz} + 2(\gamma_{xxyy} + \gamma_{xxzz} + \gamma_{yyzz})\right)/5. \tag{7}$$

In the FF method, the hyperpolarizabilities are calculated either as the derivatives of d(ε) with respect to the strength of an infinitesimal trial electric field $\varepsilon=0$, or as the corresponding derivatives of energy, E(ε). The errors in the calculated values of the α, β, γ tensor components are dominated by the accuracy of the numerical differentiation according to (1) or (2). The following important practical questions immediately appear with the numerical differentiation, but are commonly overlooked by practitioners. *First*, what is the optimal field strength for the system under study? *Second*, what minimum number of computations with different fields ε is needed in order to achieve a specified precision? *Third*, what function should be differentiated to give better accuracy, dipole moment or energy? *Fourth*, what is the expected accuracy of the resulting hyperpolarizability values?

Clearly, the order of the numerical derivative that has to be computed is lower with d(ε) than with E(ε). Since the accuracy of numerical differentiation drops dramatically with every order, d(ε) appears preferable to E(ε). On the other hand, the energy is obtained far more accurately than the dipole moment from any variational quantum chemical procedure. Thus, an optimal algorithm for calculating hyperpolarizabilities should provide some compromise between the two numerical differentiation schemes.

The choice of the optimal ε value is motivated by two reasons. On the one hand, it cannot be too small, otherwise the machine-supported precision available for floating-point operations will be insufficient. On the other hand, ε cannot be too large, otherwise the higher order terms in (1) and (2) will be large and must not be omitted. The number of data points with a constant or an adjustable step determines the minimal order of ε in the residual term containing the higher order derivatives.

The numerical estimates of β, γ found in the literature usually give the number of digits far beyond the accuracy allowed by the employed computational schemes. For instance, in [6, 7, 19-21], where symmetrized finite-difference formulas are used, the β values are given to within 0.001 to 0.01 atomic units, and γ to within 0.01 to 0.1, representing 5 to 6 significant digits. The finite-difference formulas are terminated at the ε^2 terms, not allowing hyperpolarizability accuracy higher than several atomic units, as will be illustrated in the last part of the present work. A series of papers developing and using the MOPAC software for the numerical differentiation [22] employs finite-difference formulas with an odd number of central differences over 5 points, corresponding to the use of a Lagrange interpolation polynomial [23-25]. Ref.[6], Eq.(8d) derives βs from dipole moments with a substantial loss in accuracy and a residual term estimated to ε^2 rather than ε^4. One more deficiency of the simplest formulas considered here is in lose of intermediate points used in calculation of lower derivatives on going to higher derivatives (see e.g. [6, 19]).

It follows from the above discussion that the computational and machine precision related aspects of the FF method are worth a deeper consideration. Prior to deal with extended molecular systems, let us give a concise and instructive survey of the results on the higher order polarizabilities of the simplest model cases.

2. Taylor Series Expansion of Energy in External Field Strength and Higher Order Hyperpolarizabilities of the Simplest Quantum-chemical Systems

The material reviewed in this part contains some results that have not been published previously. It is presented in order to illustrate general trends in (hyper)polarizabilities as their order increases. The results will be used in the next part for a critical characterization of the FF method and estimation of its precision. It is expected that this compilation of data will be of interest on its own accord.

2.1 HARMONIC OSCILLATOR

The Schrödinger equation for a one-dimensional harmonic oscillator under a uniform electric field of strength ε_0 is expressed as

$$-\frac{\hbar^2}{2\mu}\frac{d^2\Psi_m}{dx^2} + \left(\frac{kx^2}{2} - e_0\varepsilon_0 x\right)\Psi_m = E_m\Psi_m,$$
(8)

where μ is the (reduced) mass of the system, e_0 is its charge, $-kx$ is the force present in the system without the external field. The solution of (8) is given by (9):

$$E_m = \left(m + \frac{1}{2}\right)\hbar\omega - \frac{e_0^2\varepsilon_0^2}{2\mu\omega^2}, \quad \omega = \sqrt{k/\mu}.$$
(9)

The only non-vanishing (hyper) polarizability is the polarizability α that is the same for all states

$$E_{xx}^{(2)} \equiv \alpha_{xx} = \frac{e_0^2}{k}.$$
(10)

The linear harmonic oscillator is a unique example of a system with vanishing hyperpolarizabilities (see also [26]). It seems reasonable that in realistic systems with restoring forces approximately linear with displacement from equilibrium, the quadratic and cubic nonlinearities are forced to be small to negligible. As soon as we give up the harmonic approximation, we immediately arrive at finite hyperpolarizabilities of different values for different excited states.

2.2 THE MORSE OSCLLATOR

The potential energy of the Morse oscillator is given as a function of displacement x from equilibrium by

$$V = V_0(e^{-2ax} - 2e^{-ax}),\tag{11}$$

where V_0 is the potential well depth and a is an adjustable parameter. Defining

$$g = \frac{2\sqrt{2mV_0}}{\hbar a}\tag{12}$$

the energy spectrum for the bound states is written as

$$E_n = -V_0[1 - \frac{2}{g}\left(n + \frac{1}{2}\right)]^2, \; n = 0,1,2\ldots,\ldots\tag{13}$$

The expression is valid for $n < \frac{1}{2}(g - 1)$. The range of $g < 1$ corresponds to the continuous spectrum.

Let consider three lowest levels of the Morse oscillator. The transition dipole moments taken from [27] are as follows:

$$X_{01} = \frac{1}{a} \cdot \frac{\sqrt{g-3}}{g-2}; \; X_{02} = -\frac{1}{a} \cdot \frac{1}{g-3}\sqrt{\frac{g-5}{(g-2)g}};\tag{14}$$

$$X_{12} = \frac{1}{a} \cdot \frac{1}{g-4}\sqrt{\frac{2(g-3)(g-5)}{g-2}}.$$

The transition energies according to (12) и (13) are:

$$E_{01} = \frac{a^2(g-2)}{2}; \; E_{02} = a^2(g-3); \; E_{12} = \frac{a^2(g-4)}{2}.\tag{15}$$

Extensive but routine transformations, whose details are omitted here, yield the following analytic expressions for α and γ:

$$\alpha = \frac{1}{a^4(g-2)}\left[\frac{4(g-3)}{(g-2)^2} + \frac{g-5}{(g-3)^3}\right],\tag{16}$$

$$\gamma = \frac{24\alpha}{a^6(g-2)}\left[\frac{8(g-5)}{(g-2)(g-4)^2} - \frac{8(g-3)}{(g-2)^3} - \frac{g-5}{(g-3)^4}\right].\tag{17}$$

2.3. Particle on a Straight Segment, Infinitely Deep Square Well

It is not straightforward in this case to obtain an analytic expression for an arbitrary-order correction $E^{(N)}$ to the energy of the ith stationary state in the presence of an external field. Instead, a reasonable estimate can be obtained by perturbation theory.

The exact solution for the non-perturbed system is:

$$E_i^0 = \frac{\hbar^2 \pi^2}{2\mu L^2} i^2 , \; |\Psi_i^0\rangle = \sqrt{2/L} \, \sin\left(\frac{\pi x}{L} i\right), \; i = 1,2,3,\ldots. \qquad (18)$$

Here, L is the length of the segment or the potential well oriented in x direction. The perturbation operator:

$$V = -e_0 \varepsilon_0 x, \qquad (19)$$

is given in the basis of $|\Psi_i^0\rangle$ by

$$\overline{V}_{ij} = \begin{cases} \frac{1}{2}, & \text{at } i = j \\ 0, & \text{at } i + j = 2n \\ \dfrac{ij}{\left(i^2 - j^2\right)^2}, & \text{at } i + j = 2n+1 \end{cases}, \quad \text{where n is an integer.}$$

Corrections to the energy $\overline{E}_i^0 = i^2$ are given by

$$E_i^{(k)} = -\frac{k \Omega^{4k-1} L^{3k-2}}{\pi^{4k-2} a_0^{k-1} e_0^{k-2}} S_i^{(k)}, \; k = 2,4,6,\ldots. \qquad (20)$$

with $a_0 = \hbar^2 / \mu e_0^2$ and the $S_i^{(k)}$ values listed in Tabl.1. Closed expressions for $E^{(2)}$ and $E^{(4)}$ were obtained by Ducuing [28]:

$$E_i^{(2)} \equiv \alpha_{xx}(i) = \frac{L^4}{12} \frac{L^4}{\pi^4 a_0 i^4} (15 - \pi^2 i^2), \qquad (21)$$

$$E_i^{(4)} = \gamma_{xxxx}(i) = \frac{L^{10}}{12 \pi^{10} a_0^3 e_0^2 i^{10}} (-1980 + 210\pi^2 i^2 - \pi^4 i^4). \qquad (22)$$

These formulas have been extensively used for analysis of scaling laws of both α and γ of linear polyenes within the metallic model.

TABLE 1. The coefficients $S_i^{(k)}$ in Taylor expansion of energy: The potential well.

I	$S^{(0)}$	$S^{(2)}$	$S^{(4)}$	$S^{(6)}$	$S^{(8)}$	$S^{(10)}$
1	1.0	$-1.65\ 10^{-2}$	$4.95\ 10^{-5}$	$-3.61\ 10^{-7}$	$3.39\ 10^{-9}$	$-3.60\ 10^{-11}$
2	4.0	$4.92\ 10^{-3}$	$-4.79\ 10^{-5}$	$3.66\ 10^{-7}$	$-3.41\ 10^{-9}$	$3.60\ 10^{-11}$
3	9.0	$2.93\ 10^{-3}$	$-1.54\ 10^{-6}$	$-4.70\ 10^{-9}$	$1.47\ 10^{-11}$	$-2.20\ 10^{-14}$
4	16.0	$1.79\ 10^{-3}$	$-6.15\ 10^{-8}$	$-2.23\ 10^{-10}$	$-4.80\ 10^{-14}$	$1.69\ 10^{-16}$
5	25.0	$1.19\ 10^{-3}$	$1.17\ 10^{-8}$	$-1.57\ 10^{-11}$	$-5.13\ 10^{-15}$	$7.09\ 10^{-19}$
6	36.0	$8.44\ 10^{-4}$	$9.15\ 10^{-9}$	$-1.52\ 10^{-12}$	$-3.86\ 10^{-16}$	$-2.45\ 10^{-20}$
7	49.0	$6.27\ 10^{-4}$	$4.91\ 10^{-9}$	$-1.72\ 10^{-13}$	$-3.66\ 10^{-17}$	$-2.68\ 10^{-21}$
8	64.0	$4.84\ 10^{-4}$	$2.58\ 10^{-9}$	$-1.69\ 10^{-14}$	$-4.35\ 10^{-18}$	$-2.60\ 10^{-22}$
9	81.0	$3.84\ 10^{-4}$	$1.40\ 10^{-9}$	$1.11\ 10^{-15}$	$-6.19\ 10^{-19}$	$-2.91\ 10^{-23}$
10	100.0	$3.12\ 10^{-4}$	$7.94\ 10^{-10}$	$2.03\ 10^{-15}$	$-1.00\ 10^{-19}$	$-3.822\ 10^{-24}$

TABLE 2. The coefficients $S_i^{(k)}$ in Taylor expansion of energy: Rigid rotor

N	$S^{(0)}$	$S^{(2)}$	$S^{(4)}$	$S^{(6)}$	$S^{(8)}$	$S^{(10)}$
0	0.0	-2.0	3.5	$-1.29\ 10^{+1}$	$5.96\ 10^{+1}$	$-3.09\ 10^{+2}$
1	1.0	1.67	-3.53	$1.29\ 10^{+1}$	$-5.96\ 10^{+1}$	$3.09\ 10^{+2}$
		$-3.33\ 10^{-1}$	$2.31\ 10^{-2}$	$-3.72\ 10^{-3}$	$7.64\ 10^{-04}$	$-1.77\ 10^{-4}$
2	4.0	$1.33\ 10^{-1}$	$3.21\ 10^{-2}$	$-2.14\ 10^{-03}$	$-9.16\ 10^{-4}$	$6.98\ 10^{-5}$
			$-2.35\ 10^{-2}$	$3.78\ 10^{-3}$	$-7.66\ 10^{-04}$	$1.77\ 10^{-4}$
3	9.0	$5.71\ 10^{-2}$	$2.73\ 10^{-4}$	$7.43\ 10^{-05}$	$-1.64\ 10^{-6}$	$1.77\ 10^{-8}$
				$-6.46\ 10^{-5}$	$1.99\ 10^{-06}$	$7.45\ 10^{-9}$
4	16	$3.17\ 10^{-2}$	$4.36\ 10^{-5}$	$1.98\ 10^{-07}$	$4.09\ 10^{-8}$	$-3.80\ 10^{-10}$
					$-3.79\ 10^{-08}$	$4.13\ 10^{-10}$
5	25	$2.02\ 10^{-2}$	$1.09\ 10^{-5}$	$1.85\ 10^{-08}$	$4.91\ 10^{-11}$	$7.77\ 10^{-12}$
						$-7.42\ 10^{-12}$
6	36	$1.40\ 10^{-2}$	$3.55\ 10^{-6}$	$2.76\ 10^{-09}$	$3.26\ 10^{-12}$	$4.96\ 10^{-15}$
7	49	$1.03\ 10^{-2}$	$1.39\ 10^{-6}$	$5.64\ 10^{-10}$	$3.42\ 10^{-13}$	$2.62\ 10^{-16}$
8	64	$7.84\ 10^{-3}$	$6.16\ 10^{-7}$	$1.44\ 10^{-10}$	$4.97\ 10^{-14}$	$2.13\ 10^{-17}$
9	81	$6.19\ 10^{-3}$	$3.02\ 10^{-7}$	$4.35\ 10^{-11}$	$9.16\ 10^{-15}$	$2.38\ 10^{-18}$
10	100	$5.01\ 10^{-3}$	$1.60\ 10^{-7}$	$1.50\ 10^{-11}$	$2.03\ 10^{-15}$	$3.39\ 10^{-19}$

2.4. Particle on a Circle, Rigid Rotor

The eigenstates and eigenvectors of a free particle rotating on a circle of length L are well-known:

$$E_m^0 = \frac{2\pi^2 h^2}{\mu L^2} m^2 \ , \ |\Psi_m^0\rangle = \frac{1}{\sqrt{2\pi}} \exp(im\varphi), \ m = 0, \pm 1, \pm 2, \dots . \quad (23)$$

The external electric field perturbation operator is

$$V = \frac{e_0 \varepsilon_0 L \cos(\varphi)}{2\pi} . \quad (24)$$

Due to the double degeneracy of the excited states (23), a more convenient basis is constructed with the odd and even wavefunctions

$$|\Psi_{m,1}^0\rangle = \frac{1}{\sqrt{\pi}}\cos(m\varphi), \quad |\Psi_{m,2}^0\rangle = \frac{1}{\sqrt{\pi}}\sin(m\varphi), \quad m = 0,1,2,... \quad (25)$$

The energy corrections are obtained by standard PT formulas. Omitting physical constants, the non-vanishing elements of the V matrix are:

$$\langle \Psi_{m,1}^0 | \overline{V} | \Psi_{l,1}^0 \rangle = \langle \Psi_{m,2}^0 | \overline{V} | \Psi_{l,1}^0 \rangle = 1,$$
$$\langle \Psi_{0,1}^0 | \overline{V} | \Psi_{1,1}^0 \rangle = \langle \Psi_{1,1}^0 | \overline{V} | \Psi_{0,1}^0 \rangle = \sqrt{2} \quad \cdot \quad l = m+1, \quad m \neq 0, \qquad (26)$$

Putting $\overline{E}_m^0 = m^2$ leads to

$$E_m^{(k)} = \frac{k! L^{3k-2}}{2^{3k-1}\pi^{3k-2}a_0^{k-1}e_0^{k-2}} S_m^{(k)}, \quad k = 2,4,6,.... \qquad (27)$$

Table 2 lists the normalized $S_m^{(k)}$ values. The mth level of the unperturbed system is split by the electric field already at the $2m^{th}$ order of PT.

The formulas for α and γ then become:

$$E_m^{(2)} \equiv \alpha_{xx}(m) = -\frac{2!L^4}{2^5\pi^4 a_0}\begin{cases}\dfrac{2}{4m^2-1}, m = 0,2,3,... \\ 5/3, -1/3, m = 1\end{cases}, \qquad (28)$$

$$E_m^{(4)} \equiv \gamma_{xxxx}(m) = -\frac{4!L^{10}}{2^{11}\pi^{10}a_0^3 e_0^2}\begin{cases}\dfrac{20m^2+7}{2(4m^2-1)^3(m^2-1)}, m = 0,3,4,... \\ 5/216, -763/216, m = 1 \\ \dfrac{11691}{364500}, -\dfrac{951}{40500}, m = 2\end{cases} \qquad (29)$$

The results suggest that Taylor (ε) expansion of energy for finite movement of a particle within a line segment or a circle is rather specific in each stationary state. In the ground state, the even-order derivatives alternate signs, while the odd-order derivatives vanish by symmetry. In the excited states, both values and signs of $E_i^{(k)}$ depend on k and i in a rather complicated manner. The dipole polarizability α of a line segment and a circle is positive only in the ground state and is negative in all excited states. Contrary to α, the second hyperpolarizability γ is negative in the ground state. Despite the fact that $E_i^{(k)}$ in both (20) and (27) increase as the same power of length L, the numerical values of $E_i^{(k)}$ may substantially differ for the line segment and the circle. Thus, for the ground state of the line segment (I) and the circle (II)

$$\frac{\overline{\alpha}(I)}{\overline{\alpha}(II)} = 5 - \pi^2/3 = 1.710132, \tag{30}$$

$$\frac{\overline{\gamma}(I)}{\overline{\gamma}(II)} = \frac{2^4}{21}\left(1980 - 210\pi^2 + \pi^4\right) = 3.651175, \tag{31}$$

$$\frac{\overline{\gamma}(I)}{\overline{\gamma}(II)} = 3.651175,$$

whereas for the highly excited states with $m \to \infty$:

$$\frac{\overline{\alpha}(I)}{\overline{\alpha}(II)} = \frac{4}{3}\pi^2 = 13.159473, \tag{32}$$

$$\frac{\overline{\gamma}(I)}{\overline{\gamma}(II)} = \frac{2^8}{15}\pi^4 = 1662.448487. \tag{33}$$

The ratios (30)–(33) describe to average tensor values that are related to the tensor components by

$$\overline{\alpha}(I) = \frac{1}{3}\alpha_{xx}, \quad \overline{\alpha}(II) = \frac{2}{3}\alpha_{xx}, \tag{34}$$

$$\overline{\gamma}(I) = \frac{1}{5}\gamma_{xxxx}, \quad \overline{\gamma}(II) = \frac{8}{15}\gamma_{xxxx}, \tag{35}$$

see (5) and (7).

It is anticipated that the treatment just presented will be useful for analysis of nonlinear optical properties, with specific emphasis on linear and cyclic systems. In particular, crystals and polyenes can be described by either a cyclic or a linear model, whose properties can be compared using the above results.

2.5 The Hydrogen Atom

Consider this illustrative example of the simplest quantum-mechanical systems. On the one hand, it gives an order of magnitude picture that should be expected for atomic hyperpolarizabilities. On the other hand, a sophisticated approach applied to a real species is demonstrated (see [29-35]and references therein).

The following is a brief history of the subject. The linear response of the hydrogen atom to an electric field was first determined by P. Epstein and S. Schwarzschild (1916) within the Bohr-Sommerfeld model, and confirmed by V. Pauli (1926) with the Schrödinger equation. The quadratic correction, i.e., polarizability was calculated in 1926 independently by E. Schrödinger, V.A. Fock, I. Veller, P. Epstein, and G. Ventzel.

The cubic field effect was treated first by Yu. Ishida and S. Hiyama in 1928. The 4th correction had been calculated in the parabolic coordinate system beginning from the pioneering work of K. Basu and further extended by J. Bekenstein and J Krieger in 1969 using a quasi-classical approximation. Unfortunately, the fundamental expressions for $E^{(4)}$ in these papers are erroneous. The correct expression for $E^{(4)}$ that is valid for an arbitrary state, was found by S.P. Alliluev and I.A. Malkin as late as in 1973. Higher order corrections to the ground state energy ($k = 6, 8$) date back to G. Sevell, who still overlooked minor errors. The correct energy expansion up to $E^{(10)}$ was given by L. Mendelsohn in 1968. Further corrections up to the 17th order were derived in parabolic coordinates by Ch. Silverstone (1978). A detailed convergence study for the energy series as a function of the field strength constitutes another valuable result of that work. The algorithmic scheme was rather entangled, and some "black-box" reference tables were needed. Finally, B. Adams (1992) was the first to present explicit formulas for the energy corrections for an arbitrary state up to $E^{(6)}$ in the ordinary H atom, and up to the 30th order, $E^{(30)}$ in the ground state of the two-dimensional counterpart of H atom. The results were obtained employing Lie algebraic techniques and the $S_0(2,1)$ group symmetry.

Therefore, only six of the first energy corrections are available at present in analytic form for all states of the H atom. More useful in practice, and also demonstrative, are the numerical values of $E^{(N)}$ for the H atom in the ground state taken from [31] and given below in Tabl.3.

$$E = E^0 - E^{(2)}\varepsilon^2 - E^{(4)}\varepsilon^4 - E^{(6)}\varepsilon^6 - .... \qquad (36)$$

TABLE 3. The coefficients $E^{(k)}$ in the energy expansion in series with respect to ε: the Hydrogen atom

K	$E^{(k)}$
0	-0.5
2	-2.25
4	-55.546875
6	-4907.771484375
8	-794236.926452636718
10	-194531960.466499329
12	-66263036523.6891709
14	-29924943988411.9395
16	-17346970495631198.5

2.6. Conjugated Molecules within the Huckel Model, or Clusters Donating One Electron per Center into the Valence Zone

As stated in the Introduction, investigation of the dependence of the NLO characteristics on electron delocalization constitutes one of the goals of the present work. The Hückel model is the simplest model of conjugated molecules with delocalized electrons, at the same time taking into account details of the chemical structure. Multiple atom clusters of quasi-one-dimensional and quasi-two-dimensional crystals can be constructed in a similar way. The Hückel approach emphasizes the topological aspect in the structure of the molecule/cluster under study. Due to simplicity of the method, a number of NLO properties can be determined by the energy expansion.

Consider first a diatomic AB with the atomic ionization potentials +I for A and −I for B. Let β represent the bond resonance integral, and l the bond length. Then the electric response characteristics can be represented by polynomials in the atomic charge

$$q = I \Big/ \sqrt{I^2 + \beta^2} :$$

$$E^{(1)} = -lq, \; E^{(2)} = \frac{l^2}{2I}\big(q - q^3\big), \; E^{(3)} = \frac{3l^3}{4I^2}\big(q^3 - q^5\big),$$

$$E^{(4)} = \frac{3l^4}{8I^3}\big(-q^3 + 6q^5 - 5q^7\big),$$

$$E^{(5)} = \frac{15l^5}{16I^4}\big(-3q^5 + 10q^7 - 7q^9\big), \tag{37}$$

$$E^{(6)} = \frac{15l^6}{32I^5}\big(3q^5 - 45q^7 + 105q^9 - 63q^{11}\big),$$

$$E^{(7)} = \frac{315l^7}{64I^6}\big(5q^7 - 35q^9 + 63q^{11} - 33q^{13}\big),$$

$$E^{(8)} = \frac{315l^8}{128I^7}\big(-5q^7 + 140q^9 - 630q^{11} + 924q^{13} - 429q^{15}\big),$$

$$E^{(9)} = \frac{2835l^9}{256I^8}\big(-35q^9 + 420q^{11} - 1386q^{13} + 1716q^{15} - 715q^{17}\big),$$

$$E^{(10)} = \frac{14175l^{10}}{512I^9}\big(7q^9 - 315q^{11} + 2310q^{13} - 6006q^{15} + 6435q^{17} - 2431q^{19}\big)$$

The value of q varies from 0 to 1 depending on I and β. However, the first three coefficients preserve their signs with any q. Their magnitudes are determined by the I/β ratio. Higher derivatives vanish at critical parameter values, and as k growths, $E^{(k)}$ become increasingly oscillating functions of q. For example, the 4th, 5th, 6th, and 7th derivatives of energy vanish at q equal to

$$1/\sqrt{5}, \; \sqrt{3/7}, \; \frac{1}{21}\sqrt{147 \pm 42\sqrt{7}}, \; \frac{1}{33}\sqrt{495 \pm 66\sqrt{15}}, \tag{38}$$

respectively. The $E^{(k)}$s of a diatomic can be evaluated numerically and their signs can be specified based on the bond length, dipole moment, and sign of the atomic charge.

Putting q = 0, a set of π-electron hyperpolarizabilities of the ethylene molecule is obtained:

$$E^{(2)} = \frac{1^2}{2|\beta|}, \quad E^{(4)} = -\frac{3l^4}{8|\beta|^3}, \quad E^{(6)} = \frac{45l^6}{32|\beta|^5},$$

$$E^{(8)} = -\frac{1575l^8}{128|\beta|^7}, \quad E^{(10)} = \frac{99225l^{10}}{512|\beta|^9}, \tag{39}$$

$$E^{(k)} = \frac{(-1)^{\frac{k}{2}+1} k!(k-2)!!^k}{2^{2k-2}(k/2)!(k/2-1)!|\beta|^{k-1}}.$$

The dependence of the electric response properties of the values of the resonance integral is an essential feature of (37) and (39).

The *butadiene* molecule considered in an equidistant configuration is characterized by the following response coefficients:

$$E^{(2)} = \frac{13\sqrt{51}^2}{10|\beta|}, \quad E^{(4)} = -\frac{51\sqrt{51}^4}{100|\beta|^3}, \quad E^{(6)} = -\frac{14247\sqrt{51}^6}{800|\beta|^5}, \tag{40}$$

$$E^{(8)} = \frac{1405341\sqrt{51}^8}{3200|\beta|^7}, \quad E^{(10)} = \frac{2042691211\sqrt{51}^{10}}{32000|\beta|^9}.$$

The following are the response coefficients of *benzene*:

$$E^{(2)} = \frac{1^2}{|\beta|}, \quad E^{(4)} = \frac{l^4}{4|\beta|^3}, \quad E^{(6)} = \frac{55l^6}{16|\beta|^5}, \quad E^{(8)} = -\frac{2765l^8}{192|\beta|^7},$$

$$E^{(10)} = -\frac{30625l^{10}}{256|\beta|^9}, \quad E^{(12)} = -\frac{1224877l^{12}}{1024|\beta|^{11}}, \tag{41}$$

$$E^{(14)} = \frac{76954377l^{14}}{4096|\beta|^{13}}, \quad E^{(16)} = \frac{809825266251^{16}}{16384|\beta|^{15}},$$

$$E^{(18)} = \frac{877483710853751^{18}}{65536|\beta|^{17}}, \quad E^{(20)} = -\frac{36092303795673125 \cdot l^{20}}{786432|\beta|^{19}}.$$

A linear chain of 30 atoms describes a 30-center polyene in the Hückel model.

Taking the x projection of the bond length l equal to $1.212436\,\overset{o}{A}$ and the bond resonance integral β of 2.274 eV the exact $E^{(k)}$ values in atomic units are:

$$
\begin{aligned}
E^{(2)} &= 4.946180 \cdot 10^4, & E^{(4)} &= 1.281121 \cdot 10^9, \\
E^{(6)} &= 3.244898 \cdot 10^{14}, & E^{(8)} &= 2.079066 \cdot 10^{20}, \\
E^{(10)} &= 1.022603 \cdot 10^{26}.
\end{aligned} \tag{42}
$$

The asymptotic $E^{(k)}$ values with increasing polyene length, $N \gg 1$, are given in the Hückel model by the formula

$$
E^{(k)} = \frac{N^{k+1}(k-1)!(el)^k}{|\beta|^{k-1}}. \tag{43}
$$

Comparison of the formulas (39) to (41) indicates that γ is fundamentally negative in ethylene and butadiene, and fundamentally positive in benzene. The properties of the energy expansion in ε are heavily governed by the π-electron shell topology. A sign-alternating pattern is observed for $E^{(k)}$ of ethylene, but not of butadiene or benzene. All three examples illustrate that the numerical coefficients at $l^k / |\beta|^{k-1}$ increase with the order of the derivative, k. A similar trend was found earlier with the Hydrogen atom, whereas with the square well and the ground state of the rigid rotor the opposite trend was observed: the coefficients at $L^{3k-2} / (a_0^{k-1} e_0^{k-2})$ decreased with the order k.

3. The Finite Field Method and Optimization of Parameters

The FF approach to the calculation of the molecular electric response properties is based on a numerical differentiation of dipole moment $d(\varepsilon)$ or energy $E(\varepsilon)$ with respect to an electric field, according to Eqs. (1) and (2). Both $E(\varepsilon)$ and $d(\varepsilon)$ are obtained as functions of ε_0 by addition to the fundamental one-electron Hamiltonian H^0 of a perturbation term describing the charge-field interaction:

$$
H(\varepsilon_0) = H^0 + e_0 \varepsilon_0 r + \sum_{i=1}^{M} z_i \varepsilon_0 R_i, \tag{44}
$$

In (44), e_0 and r, z_i and R_i are the charge and coordinate of electron and nuclei, respectively. We consider first the basic numerical differentiation procedure starting from the simplest case of one variable, in order to examine the sources of error and develop a general approach to optimization of the trial electric field strength. Then a computational scheme for evaluation of the electric response properties d_i, α_{ij}, β_{ijk}, γ_{ijkl}, etc. is discussed.

3.1. PITFALLS OF NUMERICAL DIFERENTIATION

As suggested by computational experience, numerical differentiation based on Lagrange polynomial interpolation is most convenient for the present purpose [36]. The simplest formulas with minimal errors use regular odd-point grids. Let the argument ε change in increments of ε_0, and the derivatives be calculated at $\varepsilon = 0$ based on a set of the $E(i\varepsilon)$ values, with $i = 0, \pm 1, \pm 2, \dots$. A general formula for the kth derivative then becomes:

$$E_L^{(k)} = \frac{1}{C_{k,L}\varepsilon^k} \sum_{i=-[L/2]}^{[L/2]} \left(a_{k,L}(i)E(i\varepsilon)\right) + R(k,L), \tag{45}$$

$$a_{k,L}(i) = (-1)^k a_{k,L}(-i). $$

where L is the overall number of points, $C_{k,L}$ and $a_{k,L}(i)$ are numerical coefficients at the Lagrange polynomial being differentiated, and $R(k,L)$ is the residual term:

$$R(k,L) = \begin{cases} B_{k,L}\ E^{(L)}(x)\ e^{L-k}, & \text{if } k = 2n+1 \\ B_{k,L}\ E^{(L+1)}(x)\ e^{L+1-k}, & \text{if } k = 2n \end{cases}, \quad -\varepsilon \le \xi \le \varepsilon .\tag{46}$$

The $a_{k,L}$, $B_{k,L}$, and $C_{k,L}$ values are given in Table 4. Let δ_i denote errors in the calculations of the energy values $E(i\varepsilon)$. The errors are bounded from above by Δ, such that for any i $|\delta_i| \le \Delta$. If

$$\left|E^{(k)}(\xi)\right| \le M_k, \tag{47}$$

then an upper bound on the error in the calculation of the kth derivative by the L point scheme may be written as:

$$d_{k,L}(e)\ J \begin{cases} A_{k,L}\ D\ e^{-k} + B_{k,L}\ M_L\ e^{L-k}, & \text{if } k = 2n+1 \\ A_{k,L}\ D\ e^{-k} + B_{k,L}\ M_{L+1}\ e^{L+1-k}, & \text{if } k = 2n \end{cases}, \tag{48}$$

with

$$A_{k,L} = \frac{1}{C_{k,L}} \sum_{i=-[L/2]}^{[L/2]} \left|a_{k,L}(i)\right|. \tag{49}$$

In Eq.(48), the first term decreases while the second term increases monotonically with increasing ε. An extremum point minimizing $\delta_{k,L}(\varepsilon)$ is determined by the condition $(\delta_{k,L}(\varepsilon))' = 0$. The optimum increment is thus

TABLE 4. Coefficients $a_{k,L}(i)$, $B_{k,L}$ and $C_{k,L}$ for the calculation of the kth order derivative based on L points using Eq. (45)

K	L	$C_{k,L2}$	$a_{k,L}(-5)$	$a_{k,L}(-4)$	$a_{k,L}(-3)$	$a_{k,L}(-2)$	$a_{k,L}(-1)$	$a_{k,L}(0)$	$B_{k,L}$
1	3	2					-1	0	1/6
	5	12				1	-8	0	-1/30
	7	60			-1	9	-45	0	1/140
	9	840		3	-32	168	-672	0	-1/630
	11	2520	-2	25	-150	600	-2100	0	1/2772
2	3	1					1	-2	1/12
	5	12				-1	16	-30	-1/90
	7	180			2	-27	270	-490	1/560
	9	5040		-9	128	-1008	8064	-14350	-1/3150
	11	25200	8	-125	1000	-6000	42000	-73766	1/16632
3	5	2				-1	2	0	1/4
	7	8			1	-8	13	0	-7/120
	9	240		-7	72	-338	488	0	41/3024
	11	30240	205	-2522	14607	-52428	70098	0	-479/151200
4	5	1				1	-4	6	1/6
	7	6			-1	12	-39	56	-7/240
	9	240		7	-96	676	-1952	2730	41/7560
	11	15120	-82	1261	-9738	52428	-140196	192654	-479/453600
5	7	2			-1	4	-5	0	1/3
	9	6		1	-9	26	-29	0	-13/144
	11	288	-13	152	-783	1872	-1938	0	139/6048
6	7	1			1	-6	15	-20	1/4
	9	4		-1	12	-52	116	-150	-13/240
	11	240	13	-190	1305	-4680	9690	-12276	139/12096
7	9	2		-1	6	-14	14	0	5/12
	11	24	5	-52	207	-408	378	0	-31/240
8	9	1		1	-8	28	-56	70	1/3
	11	3	-1	13	-69	204	-378	462	-31/360
9	11	2	-1	8	-27	48	-42	0	1/2
10	11	1	1	-10	45	-120	210	-252	5/12

$$e_{opt} = \begin{cases} \sqrt[L]{\dfrac{kA_{k,L}D}{(L-k)B_{k,L}M_L}} = C_1(k,L)\sqrt[L]{D/M_L}, & \text{if } k = 2n+1 \\[3mm] \sqrt[L+1]{\dfrac{kA_{k,L}D}{(L+1-k)B_{k,L}M_{L+1}}} = C_1(k,L)\sqrt[L+1]{D/M_{L+1}}, & \text{if } k = 2n \end{cases} \qquad (50)$$

Substitution of (50) into (48) gives (51):

$$d_{k,L}(e_{opt})J\begin{cases} L\sqrt[L]{\dfrac{A_{k,L}^{L-k}B_{k,L}^{k}D^{L-k}M_{L}^{k}}{k^{k}(L-k)^{L-k}}}=C_{2}(k,L)\sqrt[L]{D^{L-k}M_{L+1}^{k}}, & \text{if } k=2n+1 \\[4mm] (L+1)^{L+1}\sqrt{\dfrac{A_{k,L}^{L+1-k}B_{k,L}^{k}D^{L+1-k}M_{L+1}^{k}}{k^{k}(L+1-k)^{L+1-k}}}=C_{2}(k,L)^{L+1}\sqrt{D^{L+1-k}M_{L+1}^{k}}, & \text{if } k=2n \end{cases}$$

(51)

The values of C_1 and C_2 are given in Tab.5 - 6, correspondingly. A similar procedure was proposed for the first time by Pople (1968) and applied to the calculation of dipole polarizabilities by first order numerical differentiation of dipole moments with respect to an optimized field increment [37, 38].

3.2. Accuracy in the Calculations of the Electric Response and NLO Characteristics by Differentiation of Dipole Moment and Energy

The electric response characteristics can be derived either from energy (2) or from the induced dipole moment (1), provided that the Hellmann-Feynman theorem is fulfilled. Let us compare the accuracy of two approaches. The error Δ in energy Δ_E is well below the error in the wave function or, equivalently, in the density matrix, Δ_P.

Generally,

$$\Delta_E \approx \Delta_P^2,$$

(52)

as follows from any variational procedure used to obtain E. (Non-variational approaches are beyond the scope of this work). Provided that the same number of sampling points L if used for the evaluation of the kth order derivative of energy and the $(k-1)^{th}$ order derivative of dipole moment, substitution of (52) into (51) leads to the following error ratio $\delta_{k,L}^{E}/\delta_{k-1,L}^{P}$

$$\frac{\delta_{k,L}^{E}}{\delta_{k-1,L}^{P}}=\frac{C_{2}(k,L)\left(\Delta_{E}^{L+1-k}M_{L+1}^{k}\right)^{\frac{1}{L+1}}}{C_{2}(k-1,L)\left(\Delta_{P}^{L+1-k}M_{L+1}^{k-1}\right)^{\frac{1}{L}}}\approx$$

$$\approx \Delta_{E}^{\frac{(L+1-k)(L-1)}{2L(L+1)}}M_{L+1}^{\frac{L+1-k}{L(L+1)}}=\left(\Delta_{E}^{\frac{L-1}{2}}M_{L+1}\right)^{\frac{L+1-k}{L(L+1)}}.$$

(53)

The equality of the kth derivative of energy with the $(k-1)^{th}$ derivative of dipole moment $E^{(k)}(\varepsilon)=d^{(k-1)}(\varepsilon)$ is implied in (53). Several examples are considered below where γ is estimated by numerical differentiation over 7 and 9 sampling points. Let the characteristic computational error Δ be 10^{-12}.

TABLE 5. Coefficients $C_1(k,L)$ for the calculation of ε_{opt} by Eq. (50).

L	K									
	1	2	3	4	5	6	7	8	9	10
3	1.442	2.632								
5	1.623	2.493	1.783	2.402						
7	1.710	2.407	1.837	2.345	1.853	2.294				
9	1.762	2.349	1.869	2.303	1.885	2.264	1.880	2.232		
11	1.798	2.307	1.889	2.271	1.905	2.240	1.903	2.214	1.895	2.192

TABLE 6. Coefficients $C_2(k,L)$ for the calculation of $\delta_{k,L}(\varepsilon_{opt})$ by Eq. (51).

L	K									
	1	2	3	4	5	6	7	8	9	10
3	1.040	1.155								
5	1.156	1.287	1.324	1.442						
7	1.251	1.391	1.552	1.764	1.602	1.755				
9	1.330	1.473	1.733	2.023	2.050	2.374	1.894	2.076		
11	1.397	1.539	1.886	2.239	2.415	2.904	2.662	3.104	2.195	2.402

The ground state of the *straight segment* of length L

$$\delta_{4,7}^E \Big/ \delta_{3,7}^P \approx 5.7 \cdot 10^{-4} L^{\frac{11}{7}}, \quad \delta_{4,9}^E \Big/ \delta_{3,9}^P \approx 1.2 \cdot 10^{-4} L^{\frac{28}{15}}. \tag{54}$$

The *Hydrogen atom:*

$$\delta_{4,7}^E \Big/ \delta_{3,7}^P \approx 1.5 \cdot 10^{-2}, \quad \delta_{4,9}^E \Big/ \delta_{3,9}^P \approx 6.2 \cdot 10^{-3}. \tag{55}$$

The *ethylene molecule* ($\beta = -2.274\text{eV}$, $l = 1.4 \overset{\text{o}}{\text{A}}$). $\tag{56}$

$$\delta^E_{4,7}\Big/\delta^P_{3,7} \approx 3.2\cdot10^{-3}\frac{l^{4/7}}{|\beta|^{1/2}}=1.9\cdot10^{-2}, \tag{57}$$

$$\delta^E_{4,9}\Big/\delta^P_{3,9} \approx 9.0\cdot10^{-4}\frac{l^{2/3}}{|\beta|^{3/5}}=7.6\cdot10^{-3}. \tag{58}$$

The butadiene molecule:

$$\delta^E_{4,7}\Big/\delta^P_{3,7} \approx 4.4\cdot10^{-3}\frac{l^{4/7}}{|\beta|^{1/2}}=2.6\cdot10^{-2}, \tag{59}$$

$$\delta^E_{4,9}\Big/\delta^P_{3,9} \approx 1.2\cdot10^{-3}\frac{l^{2/3}}{|\beta|^{3/5}}=1.0\cdot10^{-2}. \tag{60}$$

30-center polyene with the 1.212436 Å projection of the bond length on the molecular axis (longitudinal component of γ)

$$\delta^E_{4,7}\Big/\delta^P_{3,7} \approx 7.6\cdot10^{-2},\quad \delta^E_{4,9}\Big/\delta^P_{3,9} \approx 3.4\cdot10^{-2}. \tag{61}$$

The error ratios just found suggest that in atoms and compact molecules more accurate results are obtained by differentiation of energy as a function of applied field. The conclusion is also valid for quasi-one-dimensional medium-length π-electron systems. On the other hand, $E^{(k)}$ of elongated systems is strongly governed by the extent of electron delocalization. Lowering of the order of the derivative is preferable in such cases, with differentiation of the induced dipole moment expected to produce more accurate results.

In Hückel *polyenes with a large number of atoms*, the asymptotic values for the derivatives can be shown to obey

$$\delta^E_{k,L} \approx \left(\Delta_E^{\frac{L+1-k}{L+1}}\,\Delta_P^{\frac{L+1-k}{L}}\,N^{\frac{(L+2)(L+1-k)}{L(L+1)}}\right)\delta^P_{k-1,L}, \tag{62}$$

N is the number of carbon centers. The coefficient at $\delta^P_{k-1,L}$ in (62) grows with increasing N.

It is readily seen from (62) that the δ^E/δ^P ratio rapidly decreases with increasing accuracy of the energy values.

3.3 Examples of Optimization of the Numerical Differentiation Step

As follows from Eq. (50) and the compilation of $C_1(k,L)$ in Table 5, the optimal value of the perturbation step depends on the number of sampling points used in the numerical differentiation, and only slightly changes with the order of the derivative k due to the k-dependence of $C_1(k,L)$.

Several examples are given below, with all quantities expressed in atomic units. A realistic machine accuracy Δ of 10^{-14} is used. It is easily reachable in practical calculations.

The ground state of the *straight segment* of length L. The optimal increment ε_{opt} and the corresponding error in the dipole polarizability can be written based on Eqs.(50) and (51) as:

$$\varepsilon_{opt}(5)=0.21L^{-8/3}, \qquad \delta(5)\leq1.8\cdot10^{-12}L^{16/3},$$
$$\varepsilon_{opt}(7)=0.65L^{-11/4}, \qquad \delta(7)\leq1.9\cdot10^{-13}L^{11/2}, \qquad (63)$$
$$\varepsilon_{opt}(9)=1.20L^{-14/5}, \qquad \delta(9)\leq5.9\cdot10^{-14}L^{28/5},$$

where the values in the brackets denote the number of points used in the numerical differentiation.

The second hyperpolarizability is characterized by the following errors:

$$\delta(5)\leq2.6\cdot10^{-10}L^{32/3}, \quad \delta(7)\leq3.4\cdot10^{-12}L^{11}, \quad \delta(9)\leq3.1\cdot10^{-13}L^{56/5}. \qquad (64)$$

The optimum values of the electric field increment remain nearly the same as for polarizability. The optimal increment asymptotically decreases $\varepsilon_{opt}\to0$ as $L\to\infty$. It is important to note that the error in polarizability increases with L more rapidly than the polarizability itself. An even more pronounced effect of this kind is observed for hyperpolarizability. Indeed,

$$\alpha_{xx}=4.3890\cdot10^{-3}L^4, \quad \gamma_{xxxx}=-4.2643\cdot10^{-6}L^{10}. \qquad (65)$$

that should be compared with the L scaling of the errors in (63) and (64). Fortunately, the relative errors contain an N power scaling factor, facilitating the calculation of both α and γ for fairly large N. For example, in the square well, the accuracy of better than 1% can be achieved for polarizability with $L\leq1.5\cdot10^6$ a.u. and for hyperpolarizability with $L\leq20000$ a. u.

The Hydrogen atom. No problems appear with the numerical calculation of polarizability. For hyperpolarizability the optimal values of the field increment $\mathcal{E}_{opt}$ and the corresponding error δ in atomic units are:

$$\mathcal{E}_{opt}(5)=9\cdot10^{-4},\ \delta(5)\leq0.7,$$
$$\mathcal{E}_{opt}(7)=2\cdot10^{-3},\ \delta(7)\leq0.03, \tag{66}$$
$$\mathcal{E}_{opt}(9)=3\cdot10^{-3},\ \delta(9)\leq0.007,$$
$$\gamma=1333.125.$$

These data illustrate the order of magnitude for both perturbation and error expected in atomic systems with the FF method.

The butadiene molecule described by (a) the Hückel model and (b) full configuration interaction (full-CI) with zero differential overlap (ZDO) within the Pariser-Parr-Pople (PPP) approach.

The analytical values for $E^{(k)}$ derived in the previous section can be used to estimate the ε_{opt} and δ_{opt} values for the second hyperpolarizability γ. These estimates are compared with the results obtained by direct numerical differentiation of energy from a variational minimization of the Hückel functional using the Jacobi diagonalization procedure:

Estimated: Directly computed:

$$\mathcal{E}_{opt}(5)=3\cdot10^{-4},\ \delta(5)\leq40;\qquad \mathcal{E}_{opt}(5)=1.3\cdot10^{-4},\ \delta(5)=1;$$
$$\mathcal{E}_{opt}(7)=9\cdot10^{-4},\ \delta(7)\leq0.5;\qquad \mathcal{E}_{opt}(7)=4.0\cdot10^{-4},\ \delta(7)=0.01; \tag{67}$$
$$\mathcal{E}_{opt}(9)=2\cdot10^{-3},\ \delta(9)\leq0.04;\qquad \mathcal{E}_{opt}(9)=0.5\cdot10^{-3},\ \delta(9)=0.003.$$

These data show that the developed numerical procedure for the optimization of ε and δ is rather efficient.

Higher order electric response coefficients are not readily available from calculations including high-level electron correlation effects, such as full CI. In this case, the optimal electric field strength can be determined from the fact that $E^{(k)}$ must be independent of ε. The optimal perturbation value of ε is then chosen to correspond to the flat region ($E^{(k)}(\varepsilon)=$const) on the numerically computed $E^{(k)}(\varepsilon)$ curve. For instance, the following γ_{xxxx} values are obtained by numerical differentiation over 7 sampling points. The results given in atomic units are

$\varepsilon=0.0004$	$\gamma_{xxxx}=43264.996$
$\varepsilon=0.0006$	$\gamma_{xxxx}=43268.038$
$\varepsilon=0.0008$	$\gamma_{xxxx}=43268.538$
$\varepsilon=0.0010$	$\gamma_{xxxx}=43268.626$

$$\varepsilon = 0.0020 \qquad \gamma_{xxxx} = 43268.909$$
$$\varepsilon = 0.0030 \qquad \gamma_{xxxx} = 43270.315$$

The variation of γ with ε is smallest with ε between 0.0008 and 0.001. Within this range, the hyperpolarizability γ changes only by 0.1, with an error of less than 0.5. Such accuracy seems quite satisfactory, so we can use the value $\varepsilon=0.00097$ from the above range. The step of 0.00097 is and is not far from $\varepsilon_{opt}=0.0004$ found earlier with the Hückel method. The Hückel method therefore can be used for estimation of optimal ε for many-atomic molecules, where full CI computations are hardly accessible.

30-center polyene in the Hückel method. An approach analogous to that used with butadiene above gives:

Estimated:	Directly computed:

$$\varepsilon_{opt}(5)=4\cdot10^{-5},\ \delta(5)\leq1\cdot10^{5}; \qquad \varepsilon_{opt}(5)=3.5\cdot10^{-5},\ \delta(5)=82000;$$
$$\varepsilon_{opt}(7)=1\cdot10^{-4},\ \delta(7)\leq3\cdot10^{3}; \qquad \varepsilon_{opt}(7)=8.5\cdot10^{-5},\ \delta(7)=3; \qquad (68)$$
$$\varepsilon_{opt}(9)=2\cdot10^{-4},\ \delta(9)\leq2\cdot10^{2}; \qquad \varepsilon_{opt}(9)=1.5\cdot10^{-4},\ \delta(9)=18.$$

The ε_{opt} value is 5 times smaller in this case, indicating that the step size should be lowered for systems with large electron delocalization.

Asymptotic estimates for polyenes with large, even number of atoms N>>1 give within the Hückel model:

$$\varepsilon_{k,L}^{E} \approx \left(\Delta_{E} N^{-(L+2)} \right)^{\frac{1}{L+1}}, \qquad (69)$$

$$\delta_{k,L}^{E} \approx \left(\Delta_{E}^{L+1-k} N^{K(L+2)} \right)^{\frac{1}{L+1}}. \qquad (70)$$

Both (69) and (70) confirm that the higher order NLO characteristics of long multicenter chains can be safely computed by numerical differentiation with an optimal step. Contrary to the case of square well, the γ value of polyenes increases more rapidly with chain length than the error.

The second order hyperpolarizabilities γ of π-conjugated molecules are, therefore, available with the accuracy of 1 a.u. or better, if the differentiation procedure is performed over 7 sampling points with an optimal ε ranging between 0.00001 and 0.0001 a.u. The α values are available with a much better accuracy under the same numerical conditions.

3.4. Algorithm for the Calculation of Hyperpolarizability Tensors

Suppose that the system under study is located in the XY plane. Then the one electron Hamiltonian of the system under the influence of a field ε_0 has the form [39]:

$$H = H^0 + \varepsilon(mX + nY), \quad n,m = 0, \pm 1, \pm 2, \pm 3, \tag{71}$$

with $\varepsilon = e_0 \varepsilon_0$. Computation of the energy for all possible values of n and m produces 49 energy values E(m,n) forming a 7×7 matrix. The rows and columns of the matrix are indexed by n and m, respectively. The n and m values are arranged in increasing order: -3, -2, -1, 0, 1, 2, 3. The components of the $|K\rangle$ vectors are defined by the numerical coefficients $a_{k,7}(i)/C_{k,7}$ given in Tabl.4. An additional auxiliary vector $|0\rangle$ is introduced:

$$|0\rangle = \begin{pmatrix} 0 \\ 0 \\ 0 \\ 1 \\ 0 \\ 0 \\ 0 \end{pmatrix}, \quad |1\rangle = \frac{1}{60\varepsilon} \begin{pmatrix} -1 \\ 9 \\ -45 \\ 0 \\ 45 \\ -9 \\ 1 \end{pmatrix}, \quad |2\rangle = \frac{1}{180\varepsilon^2} \begin{pmatrix} 2 \\ -27 \\ 27 \\ -490 \\ 27 \\ -27 \\ 2 \end{pmatrix}, \dots \tag{72}$$

With these definitions, the electric response and NLO characteristics are computed by the following formulas:

$$d_x = \langle 0|E|1\rangle, \quad d_y = \langle 1|E|0\rangle,$$

$$\alpha_{xx} = \langle 0|E|2\rangle, \quad \alpha_{yy} = \langle 2|E|0\rangle, \quad \alpha_{xy} = \langle 1|E|1\rangle,$$

$$\beta_{xxx} = \langle 0|E|3\rangle, \quad \beta_{yyy} = \langle 3|E|0\rangle, \quad \beta_{xyy} = \langle 2|E|1\rangle, \quad \beta_{yxx} = \langle 1|E|2\rangle \tag{73}$$

$$\gamma_{xxxx} = \langle 0|E|4\rangle, \quad \gamma_{yyyy} = \langle 4|E|0\rangle, \quad \gamma_{xxyy} = \langle 2|E|2\rangle, \quad \gamma_{xyyy} = \langle 3|E|1\rangle$$

$$\gamma_{yxxx} = \langle 1|E|3\rangle.$$

4. Conclusion

The analytic formulas for polarizability and hyperpolarizability of the Morse oscillator are presented. The numerical auxiliary tables are given for the calculation of higher order nonlinear susceptibilities of the square well and rigid rotor. The differences in the NLO characteristics of a particle moving on a line (square well) and a circle (rigid rotor) are established for an arbitrary length of allowed motion.

The new analytical expressions in the π-electron approximation are obtained for polarizability and higher order hyperpolarizabilities of a general AB diatomic, ethylene, butadiene, and benzene. The susceptibilities are shown to depend strongly on the parameters of the model, including vanishing and sign reversal. The obtained results are used for testing and selection of optimal numerical differentiation algorithms in the calculations of NLO properties.

A detailed error analysis is carried out for the Lagrange-polynomial-based numerical differentiation of the molecular energy functional with respect to the electric field strength. The optimal strength of the applied field for the numerical differentiation is obtained.

The accuracy of the calculated electric response and NLO properties is shown to increase by two orders of magnitude if numerical differentiation is applied to the energy rather than to the induced dipole moment.

It is established for the free electron motion in a square well of length L that the finite field method works only with relatively small L. As L grows, unavoidable errors accumulate much faster than the corresponding growth of the NLO property. For large L the error exceeds the calculated observable, with or without optimization of the trial field. This fact places serious restrictions on the numerical calculation of NLO properties of ensembles of loosely bound electrons, for instance, in valence shells of ionic crystals.

5. References

1. Rosenberg, W.R., Fejer, M.M. (Eds) (1998) *Advanced Solid State Lasers* , OSA: Washington.
2. Koechner, W. (1996) *.Solid-State Laser Engineering,* Springer Verlag, Berlin.
3. Nicogosyan, D.N. (1997) *Properties of Optical and Laser-Related Materials,* J. Wiley, Chichester.
4. Nalwa, H.S., Miyata, S. (Es) (1997) *Nonlinear Optics of Organic Molecules and Polymers,* CRC Press, Boca Raton.
5. Nalwa, H.S., Miyata, S. (Es) (1989) *Nonlinear optics of organics and semiconductors,* pringer: Berlin, 1989.
6. Chemla, D.S., Zyss, J. (1987) *Nonlinear optical properties of organic molecules and crystals, vol.1,* Academic Press, Orlando.
7. Chemla, D.S., Zyss, J. (1987) *Nonlinear optical properties of organic molecules and crystals, vol.1,* Academic Press, Orlando.
8. Pereverzev, Yu.V., Prezhdo, O.V. (2000) Mean-field theory of acentric order of dipolar chromophores in polymeric electro-optic materials, *Phys. Rev.* E 62, 8324-8334.
9. Pereverzev, Yu.V., Prezhdo, O.V., Dalton, L.R. (2001) Mean-field theory of acentric order of dipolar chromophores in polymeric electro-optic materials. Chromophores with displaced dipoles, *Chem. Phys. Lett* 240, 328-335.
10. Prezhdo, O.V. (2002 Assessment of theoretical approaches to the evaluation of dipole moments of chromophores for non-linear optics, *Adv. Materials* 14, 597-600.
11. Pereverzev, Yu.V., Prezhdo, O.V., Dalton, L.R. (2002) A model of phase transitions in the system of electro-optical dipolar chromophores subject to an electric field, *J. Chem. Phys.*117, 3354-3360.
12. Davison, S.G., Levine, J.D. (1970) *Surface States* In: Solid State Physics. Advances in Research and Applications. 25. Academic Press, N.Y.& London. .
13. Bassani, F., Paravichini, G.P. (1980) *Electronic states and optical transitions in solids,* Pergamon Press, N.Y.
14. Sutherland, R.L. (1996) *Handbook of Nonlinear Optics,* Dekker, NY.
15. Mukamel, S. (1997) *Principles of Nonlinear Optical Spectroscopy,* Oxford University Press, N.Y.

16.   Mestechkin, M.M., Whyman, G.E. (1993) Time-dependent Hartree-Fock theory: finite-field technique for calculation of the Pockels and Kerr effects and generation of harmonics, *Chem. Phys. Lett.* **214**, 144-148.

17.   Langhoff, P.W., Epstein, S.T., Karplus, M. (1972) Aspects of Time-dependent perturbation theory, *Rewiews of Modern Phys* **44**, 602-644.

18.   Orr, B.J., Ward, J.F. (1971) Perturbation theory of the non-linear optical polarization of an isolated system, *Mol. Phys.* **20**, 513-526.

19.   Lu, Y.-J., Lee, S.-L.(1994) Semiempirical calculations of molecular polarizabilities and hyperpolarizabilities of polycyclic aromatic compounds, *Chem. Phys* **179**, 431-444.

20.   Jacquemin, D. Champagne, B., André, J.-M., Kirtman, B. (1996) Exploratory Pariser-Parr-Pople investigation of the static first hyperpolarizability of polymethineimine chains, *Chem. Phys.* **213**, 217-228.

21.   Calaminici, P., Jug, C., Kuster,A.M. (1998) Density funmctional calculations of molecular polarizabilities and hyperpolarizabilities , *J. Chem. Phys.* **109**, 7756-7763.

22.   Stewart, J.J.P.(1995) *MOPAC: A General Molecular Orbital Packag. Version 7.2*, QCPE, Indiana University, Bloomington.

23.   Bartlett, R.J., Purvis, G.D. (1979) Molecular hyperpolarizabilities. I. Theoretical calculations including correlation, *Phys. Rev. A* **20**, 1313-1322.

24.   Williams, G.R.J. (1987) Finite field calculations of molecular polarizability and hyperpolarizability for organic π-electron systems, *J. Mol. Struct.* **151**, 215-222.

25.   Kurtz, H.A., Stewart, J.J.P., Dieter, K.M. (1990) Calculation of the nonlinear optical properties of molecules, *J. Comput. Chem* **11**, 82-87.

26.   Galitskii, B.M., Karnakov, B.M., Kogan, V.I. (1981) *Problem book in quantum mechanics*, Nauka, Moscow.

27.   Ogenko, B.M., Rosenbaum, B.M., Chuiko, A.A. (1991) *Theory of vibrations and reorientations in the surface atom groups*, Naukova Dumka, Kiev.

28.   Rustagi, K.C., Ducuing, J. (1974) Third-order optical polarizability of conjugated organic molecules, *Opt. Commun.* **10**, 258-261.

29.   Landau, L.D., Lifshitz, E.M.(1974) *Quantum mechanics. Nonrelativistic theory*, Nauka, Moscow.

30.   Sevell, G.L. (1949) Stark effect for a hydrogen atom in its ground state, *Proc. Cambridge Phil. Soc.* **45**, 678-679.

31.   Mendelsohn, L.B. (1968) High-order perturbation theory for a one-electron ion in a uniform electric field, *Phys. Rev* **176**, 90-95.

32.   Bekenstain, J.D., Kriegen, J.B.(1969) Stark effect in hydrogenic atoms: comparison of fourth-order perturbation theory with WKB approximation, *Phys. Rev.* **188**, 130-139.

33.   Alliluev, S.P., Malkin, I.A. (1974) On Stark effect calculations on the Hydrogen atom with allowance for dynamical symmetry O(2,2)xO(2) *JETF* **66**, 1283-1294.

34.   Silverstone, H.J. (1978) Perturbation theory of the Stark effect in hydrogen to arbitrarily high order, *Phys. Rev.* **A18**, 1853-1864.

35.   Adams, B.G. (1992) Unifield treatment of high-order perturbation theory for the Stark effect in a two- and three-dimensional hydrogen atom, *Phys. Rev.* **A46**, 4060-4064.

36.   Berezin, I.S., Zhidkov, N.P.(1996) *Computational methods. Vol. 1*, Nauka, Moscow.

37.   Pople, J.A., McIver, J.W., Ostlund, N.S.(1968) Self-consistent perturbation theory. I. Finite perturbation methods, *J. Chem. Phys.* **49**, 2960-2964.

38.   Pople, J.A., McIver, J.W., Ostlund, N.S. (1968) Self-consistent perturbation theory. II. Nuclear-spin coupling constants, *J. Chem. Phys.* **49**, 2965 - 2970.

39.   Pedash, Yu.F., Semenov, A.Yu. (1993) Electric response properties of π-electronic envelopes of polyenes, radicals and polymethines in full configuration interaction method, *Teor. Eksperim. Khimiya* **29**, 338-342.

MODIFICATION OF SOLID SURFACE BY A COMPRESSION PLASMA FLOW

M.M. KURAICA[1,2], V.M. ASTASHYNSKI[3], I.P. DOJCINOVIC[1], J. PURIC[1,2]

[1]*Faculty of Physics, University of Belgrade, P.O. Box 368, 11001 Belgrade, Yugoslavia*

[2]*Center for Science and Technology Development, Obilicev Venac 26, Belgrade, Yugoslavia*

[3]*Institute of Molecular and Atomic Physics, National Academy of Sciences of Belarus, F. Scaryna pr. 70, 220072 Minsk, Belarus*

Abstract. The main characteristics of several types of quasi-stationary plasma accelerators applicable in solid surface modifications have been described. The emphasis is on their applications in (i) obtaining the materials with improved qualities, (ii) creation of sub-micron highly oriented structures on silicon single crystal due to compression plasma flow action. The influence of external magnetic field on dimensions of created structures has been also observed and discussed

1. Introduction

Broad opportunities for modification of material surface properties are opened up by plasma methods based on the exposure of solid surfaces to plasma flows action. As indicated by our investigations, for effective material modifications it is necessary that power parameters of plasma flows ensure the rapid heating of treated surface and maintaining necessary temperature levels until completion of physical-chemical transformations in surface layer without both warming the underlying bulk and disturbing its structure and properties [1-4]. However, the obtaining of plasma flows with such parameters is a very difficult task. That is why the investigations of plasma flows action on solid surfaces were restricted so far (i) high energy pulse action (several tens microseconds long) or (ii) that of higher duration but lower intensity and relatively low velocity of plasma flows ($10^5 \div 10^6$ cm/s).

J.-C. Krupa and N.A. Kulagin (eds.), Physics of Laser Crystals, 245–255.

2. Quasi-stationary Plasma Accelerators Dedicated to Generation of High-energy Compression Plasma Flows

Nowadays only the quasi-stationary plasma accelerators of a new generation operating in the regime of ion current transfer are of great interest for creating high energy plasma flows [5-8]. In these systems, acceleration of plasma is accompanied by its compression due to interaction between longitudinal components of electric current and its azimuthal magnetic field. As a result, created at the outlet of the plasma accelerator discharge device is the compression plasma flow with plasma parameters much higher than those in the discharge device [9-10]. For solid surface modifications we use following acceleration units:

2.1. QUASI-STATIONARY PLASMA ACCELERATOR OF MAGNETOPLASMA COMPRESSOR TYPE (MPC) OF COMPACT GEPMETRY

Energy supply of MPC accelerator is a capacitor bank with total energy up to 15 kJ (C_0=1200µF, U_0=5 kV). The electrical discharge device of MPC with outer electrode 5 cm in diameter and 12 cm in length is mounted in the vacuum chamber with 30 x 30 x 150 cm dimensions (Fig. 1).

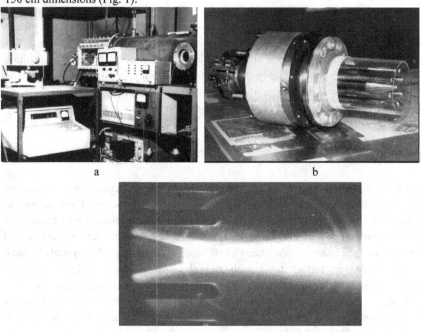

a   b

c

Figure 1. a – experimental setup of MPC-CG; b – photo of discharge device; c – compression plasma flow

MPC can operate in different working gases and their mixtures. Working gas can be introduced using electromagnetic pulse valve which provides mass rates in a range from 3 to 12 g/s. MPC can operate in the regime of "residual" gas at different pressures from 100 to 100 000 Pa. The energy efficiency coefficient of the MPC is 70%. At the outlet of MPC, the compression plasma flow 1-2 cm in diameter and 10-12 cm in length is created. Plasma parameters of compression plasma flow depending on the regime and the working gas type (hydrogen, helium, argon, nitrogen or their mixtures), can be varied in the following ranges [10-12]:

- discharge duration: 100-140 µs;
- peak current: 50 –120 kA;
- plasma velocity (3-7 x 10^6 cm/s);
- plasma electron density: $10^{16} - 10^{18}$ cm^{-3}, and
- plasma temperature: $1 - 4$ eV.

The input of powder into MPC discharge has been provided for diagnostic and technological purposes [13].

2.2. TWO STAGE QUASI-STATIONARY HIGH-CURRENT PLASMA ACCELERATOR (QHPA) OF P-50M TYPE.

Contrary to traditional plasma accelerator, the QHPA represents a two-stage plasmadynamic system with the magnetic shielding of accelerating channel elements. It operates in the ion current transfer mode and provides ion-drift acceleration of magnetized plasma [6]. One version of such accelerator with «transparent» (made of rods) passive anode and semiactive cathode transformers (electrodes) was named "QHPA P-50M" (P-semiactive; 50-characteristic scale equal to outer electrode diameter in centimetres) [14]. The magnetic shielding of elements of a QHPA main accelerating channel is performed in a self-consistent manner by currents flowing along the anode and cathode transformer rods.

The total energy stored in QHPA P-50M system attains 215 kJ. Discharge device of QHPA 120 cm long and 50 cm in outer diameter is located in vacuum chamber with dimensions of 0,8x0,8x4 m (Fig. 2).

At the outlet of inner (cathode) transformer, the compression plasma flow 3-5 cm in diameter and 30-40 cm long is formed (Fig. 2e). Plasma parameters of compression plasma flow depending on the accelerator operating regime and gas or gas mixture type can be changed within the following ranges [7, 15-17]:

- discharge duration: 500 µs;
- peak current: 200-450 kA;
- plasma velocity: 70-200 km/s;
- plasma electron density: $10^{16} - 10^{18}$ cm^{-3};
- plasma temperature: 10-15 eV.

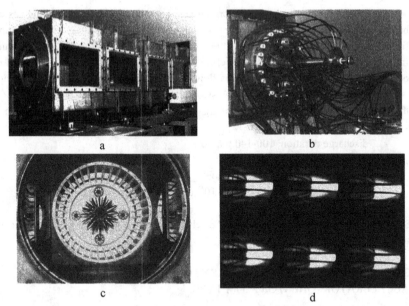

Figure 2. General views of the experimental setup (a,b), discharge device, taken along the device axis (c), and compression plasma flow (d)

2.3. MEDIUM-POWER MPC-Yu WITH PENERABLE ELECTRODES

MPC-Yu (Fig. 3) is a simplified one-stage version of QHPA with reduced dimensions [18]. Nevertheless, all of the QHPA basic properties and advantages are preserved. This type of accelerator generates relatively compact compression plasma flows with parameters comparable to those obtained in a two-stage QHPA P-50M.

Figure 3. General views of MPC-Yu (a) and compression plasma flow (b)

2.5. EROSIVE PLASMADYNAMIC SYSTEM FOR COMPRESSION PLASMA FLOWGENERATION IN AIR AT ATMOSPHERIC PRESSURE

Developed one- and two-stage erosive plasmadynamic systems (Fig. 4) for compression plasma flow generation make it possible to obtain in the air at atmospheric pressure the compression erosive plasma flows of desired composition, determined mainly by material of the inner electrode. Total energy supply of such erosive accelerator reaches up to 30 kJ. The discharge device is located in the chamber of 0,6 x 0,6 x 1,5 m dimensions which enables studies in any gas within wide range of pressures (from 10^{-2} Pa up to several atmospheres).

Generated in such a system is compression plasma flow 1-2 cm in diameter and 12-16 cm in length. It maintains very high stability throughout the lifetime of discharge. The plasma parameters of compression flow depend on the design of discharge device and on initial operating conditions of erosive system. The parameters can be changed within following ranges [19]:

- discharge duration: 100-150 μs;
- peak current: 70-200 kA;
- plasma velocity: 20-50 km/s;
- plasma electron density: $10^{17} - 10^{18}$ cm^{-3}, and
- plasma temperature: 2-4 eV.

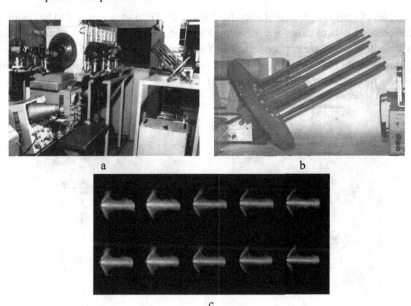

Figure 4. General views of the experimental setup (a), discharge device (b) and compression plasma flow (c)

All the described experimental installations are equipped with diagnostic systems which enable measurements of different parameters with temporal and spatial resolution. The following methods were developed: high-speed photo-registration techniques, interferometric (in particular, with visualization of viewing field), shadowgraphic and spectroscopic methods as well as probe methods for electrical and magnetic field measurement, methods of photoelectric detection of plasma radiation, calorimetric methods for energy measurements of plasma flow, and optical methods for measurement of pressure in the plasma flow [7, 10, 16, 20].

3. Application of Quasi-stationary Plasma Accelerators for Solid Surface Modifications

We investigated the opportunities of using of compression plasma flow generated mention above quasi-stationary plasma accelerators for surface structural-morphological modifications of materials having wide application in engineering (carbon steels) and in microelectronics (silicon).

3.1. MODIFICATION OF CARBON STEELS

Samples of carbon steel (initial microhardness is 2.5 GPa) were exposed to high-energy compression plasma flows generated by P50-type QHPA. The action of compression plasma flow on sample surface results to form modified layer (Fig. 5) whose thickness (from ~ 50 up to 300 microns) depends on energy density, duration of compression plasma flow action, and surface potential [1, 2].

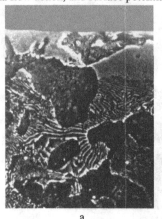 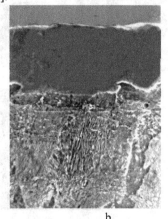

a   b

Figure 5. Surface structure before (a) and after (b) the action of a high-energy plasma flow. Magnification 500X

According to results of investigations, there are three zones in modified layer. These zones correspond to latent-acicular martensite ($\sim$ 11.4 GPa), martensite and bainite ($\sim$ 9.3 GPa), and bainite ferrite ($\sim$ 4.1 GPa), respectively. Following modified layer, is transition zone whose microhardness falls from 3.5 GPa to 2.6 GPa [20].

Primary factors ensuring changes in properties of surface exposed to the action of high-energy plasma flows, are alternate high-speed processes of heating and cooling of treated material layer. On thermal treatment accompanied by surface melting, the resulting structure is formed at the stage of the molten metal cooling.

It was demonstrated that the use of pulse-periodic regime of action causes the thickness of hardened surface layer to increase i.e., the effect of cumulative growth of hardened layer thickness was observed [1].

3.2. FORMATION OF SUBMICRON HIGHLY ORIENTED CYLINDRICAL STRUCTURES ON SILICON SURFACE

As a result of our studies, for the first time the highly oriented submicron-sized cylindrical structures have been formed at silicon surface under the action of compression plasma flow generated by MPC-type plasma accelerator [3, 4, 21].

The single-crystalline silicon samples (10 mm x 10 mm x 0.28 mm) of (111) and (100) crystallographic orientations were mounted at an axis of the system normally to a compression flow at distances from the tip of MPC discharge device ranging between 6 and 16 cm. The surface microrelief and the slices of single-crystal silicon samples were photographed using high-resolution scanning electron microscopy on a Hitachi S806 microscope.

Action of compression plasma flow on silicon surface results in the formation of near surface plasma layer. According to conducted calorimetric measurements, values of energy absorbed by silicon surface (depending on sample location and MPC initial voltage) range from 5 to 25 J per pulse, which corresponds (in our experimental conditions) to an increase in power density of a plasma flow from $0.5 \cdot 10^5$ to $3 \cdot 10^5$ W/cm^2.

As follows from the analysis of experimental data, the action of the compression plasma flow on the sample results in the melting and subsequent modification of silicon material down to the depth of 6 μm. Cylindrical fragments 50-100 μm in length and 0.3-1 μm in diameter are formed on silicon surface (Fig. 6).

In order to check the cylindrical structures originate from Si but not from SiO$_2$, the sample after exposure to plasma flow was treated by HF acid. It has been known in microelectronics, that the HF acid dissolves SiO$_2$ and does not affect Si substrate. After HF-acid treatment the sample was reexamined with SEM and no changes were detected. This proves the resulting structures are derived from pure silicon.

252

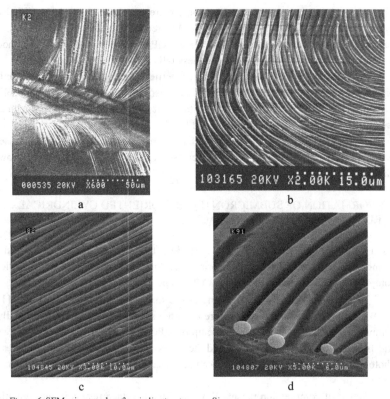

Figure 6. SEM micrographs of periodic structures on Si

In order to clarify whether cylindrical structures have tubular form, silicon surface was exposed to compression plasma flow with excess energy. Under such conditions cylindrical structures are destroyed showing their interior arrangement (Fig. 7). As clearly demonstrates this SEM micrograph, the cylindrical structures are hollow.

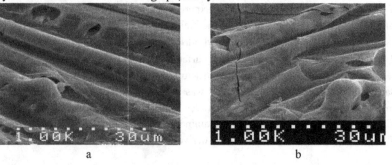

Figure 7. Micrographs of "destroyed" tubes

Compression plasma flow action on silicon sample in the presence of the steady external magnetic field B = 0.1 mT results in a decrease in diameter of surface cylindrical structures and enhance their surface density (Fig. 8) [21].

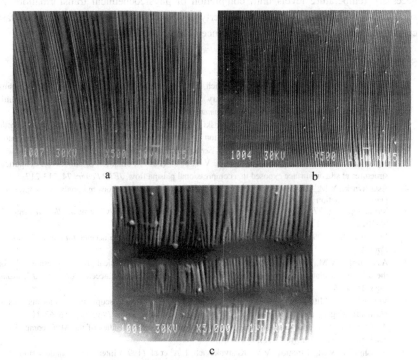

Figure 8. Influence of magnetic field on silicon surface structures: a – B=0; b,c – B=0.1 T

The formation of observed structures is primarily caused by energetic action of compression flow on the surface, leading to fast heating and melting of the surface layer, development of thermoelastic stresses, and plasma spreading over the surface under the action of both the pressure of compression flow and the gradient of plasma parameters in near surface plasma layer. Finally, the structural phase transformations of silicon surface state may be associated with fast crystallization of molten layer against the backgrounds of various instabilities that develop in the presence of the induced magnetic field.

4. Conclusions

The performed experiments showed a high efficiency of material surface modifications under the action of compression plasma flows. The main factors ensuring the efficient modification of various materials are the rapid heating of surface (up to temperatures exceeding melting points of the most of refractory materials) under thermalization of kinetic energy of compression plasma flow during its deceleration, the maintaining of necessary temperature levels until completion of physicochemical transformations in surface layer without disturbing underlying bulk material structure, and fast crystallization of melted layer in the presence of induced electromagnetic field.

5. References

1. Ananin, S.I., Astashynski, V.M., Kostyukevich, E.A. et al. (1999) The action of compression plasma flow generated by two-stage quasi-stationary high-current plasma accelerator (QHPA) on multi-profile surfaces, *Proc. 24 Int. Conf. on Phenomena in Ionized Gases* 1, Warsaw, p.103.

2. Ananin, S.I., Astashynski, V.M., Astashynski, V.V. et al. (2000) Modification of multi-profile surfaces by compression plasma flows action of quasistationary high-current plasma accelerator, *Problems of Atomic Science and Technology, Series: Plasma Physics* 6, 152-157

3. Uglov, V.V., Anishchik, V.M., Astashynski, V.V. et al. (2001) Formation of submicron cylindrical structures at silicon surface exposed to a compressional plasma flow, *JETP Letters* 74, 213-217.

4. Astashynski, V.M., Kuraica, M.M., Puric, J. (2002) Modification of various materials by compression plasma flow action , *SFIN, Series A: Conferences* A15 , 317-322.

5. Morozov, A.I.(1975) On processes in magnetoplasma compressor (MPC), *Sov. J. Plasma Phys.* 1, 95-101.

6. Morozov, A.I. (1990) Principles of coaxial (quasi-) steady-state plasma accelerators, *Sov. J. Plasma Phys.* 16, 69-77.

7. Astashinskii, V.M., Man'kovskii, A.A., Min'ko, L.Ya. et al. (1992) Physical processes responsible for the different operating regimes of quasistationary high-current plasma accelerators, *Sov. J. Plasma Phys.* 18, 47-53.

8. Volkov, Ya.F., Mitina, N.I., Solyakov, D.G. et al. (1994) Optic-spectroscopy methods for analysis of plasma flows generated by quasi-stationary plasma accelerators, *Plasma Phys. Rep.* 16, 67-74.

9. Astashinskii, V.M. and Kostyukevich, E.A. (1981) Interferometric studies of the MPC compression zone, *Sov. J. Plasma Phys.* 7, 282-289.

10. Astashinskii, V.M., Efremov, V.V., Kostyukevich, E.A. et al. (1991) Interference-shadow studies of the processes in a magnetoplasma compressor, *Sov. J. Plasma Phys.* 17, 545-553.

11. Astashinskii, V. M., Bakanovich, G. I., Kuz'mitskii, A. M. et al. (1992) Choice of operating conditions and plasma parameters of a magnetoplasma compressor, *J. Engineering Physics and Thermophysics* 62, 281-285.

12. Doichinovich, I.P., Gemishich, M.P., Obradovich, B.M., Kuraitsa, M.M., Astashinskii, V.M., Purich, Ya. (2001) Investigations of plasma parameters in magneto plasma compressor, *J. Appl. Spectr.* 68, 529-633.

13. Dojcinovic, I.P., Astashynski, V.M., Kuraica M.M. and. Puric, J. (2002) Silicon nanostructures created by compression plasma flow, *Contributed papers of Applied Physics in Serbia*, Belgrade, 217- 222.

14. Ananin, S.I., Astashinskii, V.M., Bakanovich, G.I. et al. (1990) Study of the formation of plasma streams in a quasistationary high-current plasma accelerator, *Sov. J. Plasma Phys.* 16, 102-108.

15. Ananin, S.I., Astashinskii, V.M., Kostyukevich E.A., et al. (1998) Interferometric Studies of the processes occurring in a quasi-steady high-current plasma accelerator, *Plasma Physics Reports* 24, 936-941.

16. Astashynski, V.M., Min'ko, L.Y. (1999) Physical processes in quasistationary plasma accelerators with ion current transfer, in N. Konjevic, M. Cuk and S. Djurovic (eds.), *The Physics of ionized Gases*, University of Belgrade, pp. 285-294.

17. Astashynski, V.M. (1999) Formation of compression plasma flow (plasma focus) in two-stage quasi-stationary high-current plasma accelerator, *Proc. Int. Conf. on Phenomena in Ionized Gases* **2**, Warsaw, 35.

18. Kuraica, M.M., Astashynski, V.M., Dojčinović, I. and Purić, J. (2002) New generation of quasi-stationary plasma accelerators of compact geometry. *Contributed papers 21th Intern. Symp. Phys. Ionized Gases*, Nis, pp. 510-513.

19. Astashinskii, V.M. (2000) Formation of compression plasma fluxes of a given composition in dense gases, *J. Appl. Spectr.* **67**, 312-319.

20. Anishchik, V.M., Uglov, V.V., Astashynski, V.V. et al. (2003) Compressive plasma flows interaction with steel surface: structure and sechanical properties of modified layer, *Vacuum* **70**, 269-274.

21. Astashynski, V.M., Ananin, S.I., Askerko, V.V. et al. (2002) Structural modification of silicon surface by compression plasma flows action, *Proc. 16 Europ. Conf. Atomic & Molecular Phys. Ionized Gases* **1**, Grenoble, P.149.